AF749469

Discrete Control Systems

Yoshifumi Okuyama

Discrete Control Systems

Springer

Yoshifumi Okuyama
Humanitech Laboratory Co. Ltd.
Tokushima, Japan

Additional material to this book can be downloaded from http://extras.springer.com

ISBN 978-1-4471-5666-6 ISBN 978-1-4471-5667-3 (eBook)
DOI 10.1007/978-1-4471-5667-3
Springer London Heidelberg New York Dordrecht

Library of Congress Control Number: 2013957065

© Springer-Verlag London 2014
This work is subject to copyright. All rights are reserved by the Publisher, whether the whole or part of the material is concerned, specifically the rights of translation, reprinting, reuse of illustrations, recitation, broadcasting, reproduction on microfilms or in any other physical way, and transmission or information storage and retrieval, electronic adaptation, computer software, or by similar or dissimilar methodology now known or hereafter developed. Exempted from this legal reservation are brief excerpts in connection with reviews or scholarly analysis or material supplied specifically for the purpose of being entered and executed on a computer system, for exclusive use by the purchaser of the work. Duplication of this publication or parts thereof is permitted only under the provisions of the Copyright Law of the Publisher's location, in its current version, and permission for use must always be obtained from Springer. Permissions for use may be obtained through RightsLink at the Copyright Clearance Center. Violations are liable to prosecution under the respective Copyright Law.
The use of general descriptive names, registered names, trademarks, service marks, etc. in this publication does not imply, even in the absence of a specific statement, that such names are exempt from the relevant protective laws and regulations and therefore free for general use.
While the advice and information in this book are believed to be true and accurate at the date of publication, neither the authors nor the editors nor the publisher can accept any legal responsibility for any errors or omissions that may be made. The publisher makes no warranty, express or implied, with respect to the material contained herein.

Printed on acid-free paper

Springer is part of Springer Science+Business Media (www.springer.com)

Dedicated to my wife, Yoko

Preface

Almost all control systems are realized using discretized (discrete-time and discrete-value, i.e., digital) signals. However, analysis and design methods for discretized/quantized control systems are not always established. The analytical treatment of linear discrete-time (sampled-data) control systems was developed in the 1950s and the 1960s, and is covered in several classical textbooks. Nevertheless, since the characteristics of control systems with discrete-value (discretized) signals become nonlinear, the analysis and design of these discrete control systems has not been elucidated. The aim of this book is to establish a basis for the analysis and design of discretized/quantized control systems for continuous physical systems.

Chapter 1 surveys the mathematical descriptions of discrete-time and also "discrete-value" systems. In Chap. 2, beginning with the necessary mathematical foundations and system model descriptions, the analysis in function spaces for these discretized (nonlinear) control systems is developed. Chapter 3 analyzes the robust stability of discretized nonlinear feedback systems in the frequency domain based on the input-output stability concept. In order to keep a practical perspective on the uncertain physical systems, most of the methods are carried out in the frequency domain. As part of the design procedure, modified Nyquist-Hall and Nichols diagrams are presented. In Chap. 4, first, a discretized version of traditional proportional-integral-derivative (PID) control schemes is reconsidered. Next, schemes for a model reference feedback that corresponds to a discrete observer feedback are proposed. It is shown that the model reference feedback approximately becomes a PID control scheme.

Although single-loop feedback systems form the core of the text, in Chap. 5, some considerations are given to multiple loops and nonlinearities. Furthermore, Chap. 6 discusses the robust control performance and stability of discrete interval systems (with multiple uncertainties) from the viewpoint of the characteristic roots area based on Sturm's theorem. Finally, in Chap. 7, the relationship between feedback control and discrete event systems is outlined. The nonlinear phenomena associated with practically important event-driven systems are elucidated, and the dynamics and stability of finite state and discrete event systems are defined.

The author's thoughts on recent control theory are as follows.

(1) The state-space representation is useful for the analysis and simulation of control systems in the time domain. However, the method is not always appropriate for the design of control systems.
(2) To keep a practical perspective on uncertain physical systems, modeling/identification and control should be carried out in a limited frequency range.
(3) There will be a large difference in time scale for feedback control systems (transient responses) and adaptive control loops. Therefore, adaptive and learning control processes should be discussed separately.
(4) This book treats discrete signals; therefore, differential and integral techniques are not used in principle.

I would like to thank Mr. Oliver Jackson and Ms. Charlotte Cross at Springer UK for giving me the opportunity to write this book and for helping me to complete it.

Tokushima, Japan
October 2013

Yoshifumi Okuyama

Contents

Chapter 1
Mathematical Descriptions and Models

1.1 Introduction

Almost all control systems are realized using discretized (discrete-time and discrete-value, i.e., digital) signals. However, analysis and design methods for discretized/quantized control systems are not always established. An analytical treatment of linear discrete-time (sampled-data) control systems was developed in the 1950s and the 1960s, and is covered in several classical textbooks [4, 10]. Nevertheless, since the characteristics of control systems with discrete-value (discretized) signals become nonlinear, the analysis and design of these discrete control systems has not been elucidated [3]. In this chapter, the mathematical descriptions of discrete-time and also "discrete-value" systems are surveyed [5, 6].

1.2 Input–Output Representation

The control systems to be considered here are composed of some (dynamic) elements or subsystems. Each element or subsystem (and whole system) is regarded as a relationship between the cause and the effect (i.e., the input and the output).

1.2.1 The System

First, it is assumed that the system can be expressed by a "relation" between one input and one output. In general, the relation is given by

$$y = S(u), \tag{1.1}$$

where S is an operator (or a transformation) which defines the system. This may be represented in block diagram form, as shown in Fig. 1.1. Here, u is the input, and y

Y. Okuyama, *Discrete Control Systems*,
DOI 10.1007/978-1-4471-5667-3_1, © Springer-Verlag London 2014

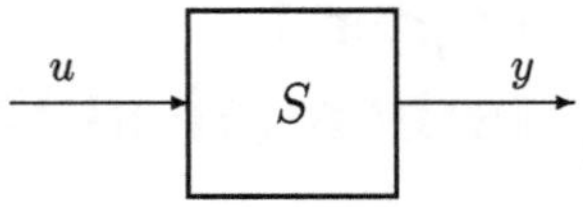

Fig. 1.1 The discrete system

is the output. In other words, u is an independent variable, and y is a dependent variable. In general, u is considered a time sequence of real numbers $u_0, u_1, \cdots, u_k, \cdots$, and y is a time sequence of real numbers y_0, y_1, $\cdots$, y_k, $\cdots$. Of course, these may be considered sequences in space.

1.2.2 Linear and Nonlinear

The transformation S is linear if it follows the principle of superposition, i.e.,

$$y = S(au + bv) = aS(u) + bS(v), \tag{1.2}$$

where a and b are constant parameters. In this case, y is a linear function of u and v; otherwise, it is a nonlinear function. Here, the former transformation S is referred to as a linear operator; otherwise, S is referred to as a nonlinear operator.

1.2.3 Static and Dynamic Systems, Causality

If a system is given by

$$y_k = S(u_k), \qquad k = 0, 1, 2, \cdots, \tag{1.3}$$

the system is called static. On the other hand, if a system is affected by time sequence $u_0, u_1, u_2, \cdots, u_N$ as

$$y_k = S(u_0, u_1, \cdots, u_N), \qquad 0 \le k \le N, \tag{1.4}$$

the system is considered dynamic. When system (1.4) is written as

$$y_k = S(u_0, u_1, \cdots, u_k) \tag{1.5}$$

the system may be called causal.

1.2.4 Time Invariance

The system S is time-invariant if its response is independent of the moment of application of the input u. If i and k represent arbitrary time instances, then

$$y_{i-k} = S(u_{i-k}), \quad \forall i, k \ge 0,$$

$$\text{where}\quad y_{i-k} = 0, \quad \text{for}\ i < k.$$

Otherwise, the system is time-variant.

1.2.5 Discrete and Continuous Time

When a variable is in the form of a time sequence, e.g., $u_0, u_1, u_2, \cdots, u_k, \cdots$, it can be assigned to $u(0), u(1), u(2), \cdots, u(k), \cdots$ as a function of discrete time $0, 1, 2, \cdots, k, \cdots$.

In general, the input and the output of the system S can be written as

$$u_k = u(k), \quad v_k = v(k),$$
$$k \in \mathbb{Z}_+ := \{0, 1, 2, 3, \cdots\}.$$

Functions of continuous time, $u(t)$ and $v(t)$, can be given as

$$u = u(t), \quad v = v(t),$$
$$t \in \mathbb{R}_+ := [0,\ \infty).$$

1.2.6 Discrete and Continuous Values

In the above expression, values of $u(k)$, $u(t)$ may be either discrete or continuous. However, in this section, the set of real numbers (an uncountably infinite set) $\mathbb{R}$ will be identified with a countably infinite set $\mathbb{Z}$,

$$u(k), y(k) \in \mathbb{Z} \subset \mathbb{R},$$
$$\mathbb{Z} := (-\infty, +\infty),$$

in spite of the strict mathematical definition. The values of $u(k)$ and $y(k)$ may belong to finite sets $\boldsymbol{U}, \boldsymbol{Y} \subset \mathbb{Z}$ due to some limiter or saturation in the system $\boldsymbol{S}$. The behavior of finite value and state systems will be discussed in the next subsection and in Chap. 7.

1.2.7 Concept of the State

In order to express the internal structure strictly, the concept of "state" will be introduced. In general, a system $\boldsymbol{S}$ can be written as follows:[1]

[1] In the following, boldface will be used in principle for vector (multivariable) representation and also for set representation.

$$\begin{cases} \boldsymbol{x}(k+1) = \boldsymbol{f}(\boldsymbol{x}(k), u(k)) \\ y(k) = \boldsymbol{g}(\boldsymbol{x}(k+1)) \end{cases} \tag{1.6}$$

$$\boldsymbol{x}(k) \in \mathbb{Z}^n, \quad u(k), y(k) \in \mathbb{Z}$$

$$\boldsymbol{f} : \mathbb{Z}^n \times \mathbb{Z} \to \mathbb{Z}^n, \quad \boldsymbol{g} : \mathbb{Z}^n \to \mathbb{Z}.$$

The system equations (1.6) can be rewritten as follows:

$$\begin{cases} \boldsymbol{x}(k+1) = \boldsymbol{f}(\boldsymbol{x}(k), u(k)) \\ y(k) = \boldsymbol{g}(\boldsymbol{x}(k+1))) = \boldsymbol{g}^*(\boldsymbol{x}(k), u(k)) \end{cases} \tag{1.7}$$

$$g^* : \mathbb{Z}^n \times \mathbb{Z} \to \mathbb{Z}.$$

When $\mathbb{Z}$ is considered as finite sets $\boldsymbol{U}$, $\boldsymbol{X}$, and $\boldsymbol{Y}$, the system (1.6) can be written as:

$$\begin{cases} \boldsymbol{x}(k+1) = \boldsymbol{f}(\boldsymbol{x}(k), u(k)) \\ y(k) = \boldsymbol{g}(\boldsymbol{x}(k+1)) \end{cases} \tag{1.8}$$

$$\boldsymbol{x}(k) \in \boldsymbol{X}, \quad u(k) \in \boldsymbol{U}, \quad y(k) \in \boldsymbol{Y}$$

$$\boldsymbol{f} : \boldsymbol{X} \times \boldsymbol{U} \to \boldsymbol{X}, \quad \boldsymbol{g} : \boldsymbol{X} \to \boldsymbol{Y}.$$

In the automata theory, Eq. (1.6) is called the *Moore* machine expression. On the other hand, Eq. (1.7) is called the *Mealy* machine expression with respect to $\boldsymbol{g}^*(\cdot,\cdot)$. However, in this chapter, only "time-driven"-type systems are considered. In these expressions, $k \in \mathbb{N}$ is an independent variable that corresponds to an elapsed time t. With respect to discrete event ("event-driven"-type) systems, the definition and the behavior of finite state systems will be described in Chap. 7.

1.3 Linear Discrete Equations

As in differential equations, the discrete equation (system model) for linear dynamic systems is described in the relationship between a dependent variable (sometimes called "output" y) and a forcing function (sometimes called "input" u) with respect to an independent variable (discrete time series k). In this section, the method of solving such discrete equations is presented. It helps to formulate, synthesize, and analyze discrete-time linear systems similarly to linear differential equations.

1.3.1 Forward Discrete-Time Equation

The discrete equation based on a forward-type expression is given as follows:

$$a_0 y(k+n) + a_1 y(k+n-1) + \cdots + a_{n-1} y(k+1) + a_n y(k) = u(k). \tag{1.9}$$

Here, independent variable k may or may not be time elapsed. It may be in the form of discrete-space (e.g., integer grid) coordinates. However, throughout this book, it is assumed that the independent variable is a discrete-time variable, as shown in the sampling process (Fig. 1.7). In general, (1.9) can be written as

$$y(k+n)+\cdots+a_{n-1}y(k-1)+a_n y(k)=b_0 u(k+n)+\cdots+b_{n-1}u(k-1)+b_n u(k), \tag{1.10}$$

where n is the order of the equation.[2]

In these equations, dependent variables $y(k)$ and $u(k)$ may or may not be of continuous value, i.e.,

$$y(k), u(k) \in \mathbb{R}, \quad \text{or} \quad y(k) = y^\dagger(k), u(k) = u^\dagger(k) \in \mathbb{Z}.$$

However, in this section, the dependent variables are considered to be continuous, unless otherwise specified (or to belong to an uncountably infinite set).

Example 1.1 In the case of a first-order system, the equation can be written as[3]

$$x(k+1)+a_1 x(k) = b_0 u(k+1) + b_1 u(k).$$

When it is homogeneous, the following simple expression is given:

$$x(k+1)+a_1 x(k) = 0.$$

Thus, the first-order homogeneous equation can be written as

$$x(k+1) = Ax(k), \quad A = -a_1, \quad k = 0, 1, 2, \cdots,$$

where $x(k+1)$ is the next stage of the above discrete equation. When $k=0$, $x(0)$ is the initial condition, and $x(1)$ is the next stage. If $x(k) \in \mathbb{R}$ and $A \in \mathbb{R}$, $x(k+1) \in \mathbb{R}$ is satisfied.

In general, by using a vector-matrix form, the homogeneous discrete equation can be expressed as

$$\boldsymbol{x}(k+1) = \boldsymbol{A}\boldsymbol{x}(k), \quad \boldsymbol{x} \in \mathbb{R}^n, \quad \boldsymbol{A} \in \mathbb{R}^{n\times n}. \tag{1.11}$$

If there exists an exogenous (forced) input, the vector-matrix form can be written as

$$\boldsymbol{x}(k+1) = \boldsymbol{A}\boldsymbol{x}(k) + \boldsymbol{B}u(k), \quad u \in \mathbb{R}, \quad \boldsymbol{B} \in \mathbb{R}^n. \tag{1.12}$$

---

[2]Without loss of generality, the leading coefficient is assumed to be 1 (i.e., $a_0 = 1$). This assumption can easily lead to a vector-matrix form and a "monic" polynomial of z-transform variables.

[3]Symbol x was used for y in order to introduce a state-variable representation.

A set of first-order equations can be represented by using the vector-matrix form (1.11) as follows:

$$\begin{bmatrix} x_1(k+1) \\ x_2(k+1) \\ \vdots \\ x_n(k+1) \end{bmatrix} = \begin{bmatrix} a_{11} & a_{12} & \dots & a_{1n} \\ a_{21} & a_{22} & \dots & a_{2n} \\ \vdots & \vdots & \ddots & \vdots \\ a_{n1} & a_{n2} & \dots & a_{nn} \end{bmatrix} \begin{bmatrix} x_1(k) \\ x_2(k) \\ \vdots \\ x_n(k) \end{bmatrix} + \begin{bmatrix} b_1 \\ b_2 \\ \vdots \\ b_n \end{bmatrix} u(k). \tag{1.13}$$

Since an n-th-order discrete-time equation (1.10) can be written as a set of first-order equations, e.g.,

$$\begin{cases} x_1(k+1) = x_2(k) - a_1 x_1(k) + (b_1 - a_1 b_0)u(k) \\ x_2(k+1) = x_3(k) - a_2 x_1(k) + (b_2 - a_2 b_0)u(k) \\ \quad\vdots \\ x_n(k+1) = -a_n x_1(k) + (b_n - a_n b_0)u(k), \end{cases} \tag{1.14}$$

it can be given in the following vector-matrix form:

$$\begin{bmatrix} x_1(k+1) \\ x_2(k+1) \\ \vdots \\ x_{n-1}(k+1) \\ x_n(k+1) \end{bmatrix} = \begin{bmatrix} -a_1 & 1 & 0 & \dots & 0 \\ -a_2 & 0 & 1 & \dots & 0 \\ \vdots & \vdots & \vdots & \ddots & \vdots \\ -a_{n-1} & 0 & 0 & \dots & 1 \\ -a_n & 0 & 0 & \dots & 0 \end{bmatrix} \begin{bmatrix} x_1(k) \\ x_2(k) \\ \vdots \\ x_{n-1}(k) \\ x_n(k) \end{bmatrix} + \begin{bmatrix} b_1 - a_1 b_0 \\ b_2 - a_2 b_0 \\ \vdots \\ b_{n-1} - a_{n-1} b_0 \\ b_n - a_n b_0 \end{bmatrix} u(k) \tag{1.15}$$

and

$$y(k) = x_1(k) + b_0 u(k). \tag{1.16}$$

Equations (1.15) and (1.16) correspond to a linear constant (time-invariant) expression for the general system equation (1.7). The above expression is also called an "observable canonical form." With respect to vector-matrix form (i.e., state-space representation), there are many possible expressions. As an example, the following expression, which is also called a "controllable canonical form," can be shown:

$$\begin{bmatrix} x_1(k+1) \\ x_2(k+1) \\ \vdots \\ x_{n-1}(k+1) \\ x_n(k+1) \end{bmatrix} = \begin{bmatrix} 0 & 1 & 0 & \dots & 0 \\ 0 & 0 & 1 & \dots & 0 \\ \vdots & \vdots & \vdots & \ddots & \vdots \\ 0 & 0 & 0 & \dots & 1 \\ -a_n & -a_{n-1} & -a_{n-2} & \dots & -a_1 \end{bmatrix} \begin{bmatrix} x_1(k) \\ x_2(k) \\ \vdots \\ x_{n-1}(k) \\ x_n(k) \end{bmatrix} + \begin{bmatrix} 0 \\ 0 \\ \vdots \\ 0 \\ 1 \end{bmatrix} u(k) \tag{1.17}$$

and

$$y(k) = \begin{bmatrix} b_n - a_n & b_{n-1} - a_{n-1} & \cdots & b_1 - a_1 \end{bmatrix} \begin{bmatrix} x_1(k) \\ x_2(k) \\ \vdots \\ x_n(k) \end{bmatrix} + u(k). \tag{1.18}$$

1.3.2 Backward Discrete-Time Equation

When considering time shifting $k+n \to k$ the discrete-time equation given by (1.10) is rewritten as follows:

$$y(k)+\cdots+a_{n-1}y(k-n+1)+a_n y(k-n)=b_0u(k)+\cdots+b_{n-1}u(k-n+1)+b_nu(k-n). \tag{1.19}$$

For a first-order system, it can be written as

$$x(k)+a_1x(k-1)=b_0u(k)+b_1u(k-1).$$

Therefore, the vector-matrix discrete-time equation which corresponds to (1.11) will be expressed as

$$\boldsymbol{x}(k)=\boldsymbol{A}\boldsymbol{x}(k-1), \quad \boldsymbol{x}\in\mathbb{R}^n, \quad \boldsymbol{A}\in\mathbb{R}^{n\times n}. \tag{1.20}$$

Of course, if there exists an exogenous input, the vector-matrix discrete equation which corresponds to (1.12) will be written as follows:

$$\boldsymbol{x}(k)=\boldsymbol{A}\boldsymbol{x}(k-1)+\boldsymbol{B}u(k), \quad u\in\mathbb{R}, \quad \boldsymbol{B}\in\mathbb{R}^n. \tag{1.21}$$

Since future data cannot be used, these backward expressions (1.19) and (1.21) are actually used on computer simulation models instead of the forward expressions (1.10) and (1.12).

1.3.3 Difference Equation

Another approach for formulating a discrete equation is to analyze the behavior of the differences between two successive values of the dependent variable. The first forward difference,

$$\Delta y(k)=y(k+1)-y(k),$$

transforms the discrete equation to the following difference equation:

$$d_0\Delta^n y(k)+d_1\Delta^{n-1}y(k)+\cdots+d_{n-1}\Delta y(k)+d_n y(k)=u(k). \tag{1.22}$$

Here, higher differences are defined similarly to the first difference, i.e.,

$$\Delta^n y(k)=\Delta(\Delta^{n-1}y(k)). \tag{1.23}$$

For example, the second forward difference is given by

$$\Delta^2 y(k)=\Delta(\Delta y(k))=\Delta y(k+1)-\Delta y(k).$$

Thus, it can be defined as

$$\Delta^2 y(k) = y(k+2) - 2y(k+1) + y(k).$$

Properties

(i) The first forward difference of the product of two discrete functions is given by

$$\Delta(f(k)g(k)) = f(k+1)\Delta g(k) + g(k)\Delta f(k). \tag{1.24}$$

Proof As is obvious from the definition of the first forward difference,

$$\begin{aligned}\Delta(f(k)g(k)) &= f(k+1)g(k+1) - f(k)g(k)\\ &= f(k+1)(g(k+1) - g(k)) + g(k)(f(k+1) - f(k)),\end{aligned}$$

(1.24) is proved.

(ii) The relationship between the discrete equation (1.9) and difference equation (1.22) is given as follows:

$$\begin{cases} a_p = \sum\limits_{q=0}^{p} (-1)^{p-q} C_{p-q}^{n-q} d_q \\ d_p = \sum\limits_{q=0}^{p} C_{p-q}^{n-q} a_q \end{cases} \tag{1.25}$$

Equation (1.25) relates the coefficients of the difference equation to the coefficients of the discrete equation and vice versa.

Backward Difference Equation The first backward difference is defined as

$$\varDelta y(k) = y(k) - y(k-1).$$

An n-th-order backward difference equation which corresponds to forward difference equation (1.22) is given as follows:

$$d_0 \varDelta^n y(k) + d_1 \varDelta^{n-1} y(k) + \cdots + d_{n-1} \varDelta y(k) + d_n y(k) = u(k). \tag{1.26}$$

Higher differences are defined similarly to the first difference, i.e.,

$$\varDelta^n y(k) = \varDelta(\varDelta^{n-1} y(k)). \tag{1.27}$$

The second backward difference is given by

$$\varDelta^2 y(k) = \varDelta(\varDelta y(k)) = \varDelta y(k) - \varDelta y(k-1).$$

Thus, it can be defined as

$$\varDelta^2 y(k) = y(k) - 2y(k-1) + y(k-2).$$

Properties

(i) The first backward difference of the product of two discrete functions is given by

$$\Delta(f(k)g(k)) = f(k)\Delta g(k) + g(k-1)\Delta f(k). \tag{1.28}$$

Proof From the definition of the first backward difference,

$$\begin{aligned}\Delta(f(k)g(k)) &= f(k)g(k) - f(k-1)g(k-1)\\ &= f(k)(g(k)-g(k-1)) + g(k-1)(f(k)-f(k-1)),\end{aligned}$$

(1.28) is proved.

(ii) The relationship between the discrete equation (1.19) and difference equation (1.26) is given as follows:

$$\begin{cases} a_p = \sum_{q=0}^{p} (-1)^{p-q} C_{p-q}^{n-q} d_q \\ d_p = \sum_{q=0}^{p} C_{p-q}^{n-q} a_q \end{cases} \tag{1.29}$$

Equation (1.29) relates the coefficients of the difference equation to the coefficients of the discrete equation and vice versa.

1.3.4 Solution of Discrete and Difference Equations

The classical methods for the solution of discrete equations are similar to the methods for differential equations. Here, homogeneous equations are considered.

Solution of Homogeneous Equation Consider a general homogeneous equation,

$$a_0 y(k+n) + a_1 y(k+n-1) + \cdots + a_n y(k) = 0.$$

If we assume $y(k) = \lambda^k$ as a solution, then

$$y(k+n-p) = \lambda^{k+n-p}.$$

For all $0 \le p \le n$, the following is obtained:

$$(a_0\lambda^n + a_1\lambda^{n-1} + \cdots + a_n)\lambda^k = 0.$$

The characteristic equation can be defined as follows:

$$a_0\lambda^n + a_1\lambda^{n-1} + \cdots + a_n = 0.$$

The general solution is a linear combination of solutions, based on the root of the characteristic equation.

Table 1.1 The numerical behavior for real λ

λ	Values of $y(k)$ for $k = 0, 1, 2, \cdots$
$\lambda > 1$	Increasing
$\lambda = 1$	Constant
$0 < \lambda < 1$	Decreasing
$-1 < \lambda < 0$	Decreasing, alternating sign
$\lambda = -1$	Alternating value
$\lambda < -1$	Increasing, alternating sign

Example 1.2 Consider the following example:

$$y(k+2) - 1.5y(k+1) + 0.5y(k) = 0$$

The characteristic equation is obtained as

$$\lambda^2 - 1.5\lambda + 0.5 = 0, \quad \lambda = 1.0, \ 0.5.$$

Then, the solution is given by

$$y(k) = C_1 + C_2 \cdot (0.5)^k,$$

where C_1 and C_2 are arbitrary (real) constants.

In general, the values of λ describe the natural behavior of the solution $y(k)$.[4] The numerical behavior of $y(k)$ is summarized in Table 1.1 when one of the roots is real.

In the case of complex or imaginary roots, the solution is given as

$$y(k) = C_1\lambda + C_2\bar{\lambda},$$

where

$$\lambda = a + jb, \quad \bar{\lambda} = a - jb, \quad j = \sqrt{-1},$$

where a and b are real numbers.

Similarly to continuous systems, the behavior of $y(k)$ is described by a combination of two sinusoids (damped or undamped), and for the case of a conjugate imaginary pair it is a pure oscillation. Multiple real roots generate behavior which consists of the term $k\lambda^k$. The graphical description of the natural behavior of the homogeneous equation is given in the following sections on the z-plane.

A practical method for solving the discrete or difference equations is to use the z-transform approach [2, 6]. This is the subject of the following sections.

---

[4] λ^k plays the same role in discrete equations as $e^{\lambda t}$ in linear, time-invariant differential equations.

1.4 The z-Transform and Transfer Characteristics

The z-transform is a valuable approach for formulating, analyzing, and solving problems in the time-invariant, linear, discrete-time domain.

1.4.1 Definition of the z-Transform

The z-transform is defined as

$$\mathcal{Z}[f(kh)] := F(z) = \sum_{k=0}^{\infty} f(kh)z^{-k}, \tag{1.30}$$

where it is assumed that $f(kh)$ is defined only in $k \geq 0$.

The right side of (1.30) exists if and only if the following infinite series converges:

$$F(z) = f(0) + f(h)z^{-1} + f(2h)z^{-2} + \cdots. \tag{1.31}$$

That is, the summation of series (1.31) should be expressed in a closed form.

Example 1.3 Consider the following exponential function:

$$f(kh) = \mathrm{e}^{-akh}, \quad \text{for } k \geq 0,$$

$$F(z) = \sum_{k=0}^{\infty} \mathrm{e}^{-akh} z^{-k} = 1 + \mathrm{e}^{-ah} z^{-1} + \mathrm{e}^{-2ah} z^{-2} + \cdots. \tag{1.32}$$

If $|z| > \mathrm{e}^{-ah}$ is satisfied, the following closed form is obtained:

$$F(z) = \frac{1}{1 - \mathrm{e}^{-akh} z^{-1}} = \frac{z}{z - \mathrm{e}^{-akh}}. \tag{1.33}$$

As a special case, when $a = 0$, the above can be written as

$$f(kh) = 1, \quad \text{for } k \geq 0$$

$$F(z) = \sum_{k=0}^{\infty} z^{-k} = 1 + z^{-1} + z^{-2} + \cdots.$$

Thus, if $|z| > 1$ is satisfied, the following is obtained:

$$F(z) = \frac{1}{1 - z^{-1}} = \frac{z}{z - 1}. \tag{1.34}$$

1.4.2 Properties of the z-Transform

Some properties of the z-transform are given here:

(i) Superposition.

$$\mathcal{Z}[af(kh)+bg(kh)] = a\mathcal{Z}[f(kh)]+b\mathcal{Z}[g(kh)] = aF(z)+bG(z).$$

(ii) Real backward translation (shifting theorem).

$$\mathcal{Z}[f(kh-\ell h)] = z^{-\ell}\mathcal{Z}[f(kh)] = z^{-\ell}F(z), \quad \ell > 0.$$

(iii) Real forward translation (shifting theorem).

$$\mathcal{Z}[f(kh+\ell h)] = z^{\ell}F(z) - \sum_{k=0}^{\ell-1} f(kh)z^{\ell-k}, \quad \ell \geq 1.$$

(iv) Complex translation.

$$\mathcal{Z}[\mathrm{e}^{akh}f(kh)] = F(\mathrm{e}^{-ah}z)$$

(v) Initial value theorem.

$$f(0) = \lim_{z\to\infty} F(z).$$

(vi) Final value theorem.

$$\lim_{k\to\infty} f(kh) = \lim_{z\to 1}(1-z^{-1})F(z).$$

Proof From the linearity of z-transform (1.30), (i) and (ii) are easily proved. For (iii) through (vi), the following proof is given:

(iii) As is obvious from (1.30),

$$\mathcal{Z}[f(kh+k\ell)] = \sum_{k=0}^{\infty} f(kh+k\ell)z^{-k} = z^{\ell}\sum_{k=0}^{\infty} f((k+\ell)h)z^{-(k+\ell)}$$

$$= z^{\ell}\sum_{\kappa=\ell}^{\infty} f(\kappa h)z^{-\kappa} = z^{\ell}\left\{F(z) - \sum_{\kappa=0}^{\ell-1} f(\kappa h)z^{-\kappa}\right\}, \quad \kappa = k+\ell.$$

(iv) From (1.30), the following is obtained:

$$\mathcal{Z}[\mathrm{e}^{akh}f(kh)] = \sum_{k=0}^{\infty} f(kh)(\mathrm{e}^{-ah}z)^{-k} = F(\mathrm{e}^{-ah}z).$$

(v) From (1.31), the following is easily obtained:

$$\lim_{z\to\infty} F(z) = f(0).$$

(vi) In regard to the right side of the equation, the following can be obtained:

$$(1 - z^{-1})F(z) = f(0) + (f(h) - f(0))z^{-1} + ((f(2h) - f(h))z^{-2}$$
$$+ \cdots + (f(kh) - f((k-1)h))z^{-k} + \cdots .$$

Thus

$$\lim_{z \to 1}(1 - z^{-1})F(z) = \lim_{k \to \infty} f(kh).$$

1.4.3 Solving Discrete-Time Equations

Using the definition of the z-transform and its properties, an approach for solving a discrete equation can be formulated. Consider the following equation:[5]

$$y(k+n) + \cdots + a_{n-1}y(k+1) + a_n y(k) = b_0 u(k+n) + \cdots + a_{n-1}z + a_n u(k). \tag{1.35}$$

If each term on the left side of this equation is transformed, the following can be obtained based on the shifting theorem:

$$\begin{cases} \mathcal{Z}\{y(k+n)\} = z^n \hat{y}(z) - (y(0)z^{n-1} + \cdots + y(n-1)z^{n-1}) \\ \quad \cdots \\ \mathcal{Z}\{a_{n-1}y(k+1)\} = a_{n-1}(z\hat{y}(z) - y(0)) \\ \mathcal{Z}\{a_n y(k)\} = a_n \hat{y}(z). \end{cases}$$

The transformed function of each term on the right side of (1.35) will be obtained in the same way. For simplicity, the initial conditions are assumed to be zero (i.e., $y(0) = y(1) = \cdots = y(n-1) = 0$ and also $u(0) = u(1) = \cdots = u(n-1) = 0$). Then, the following transformed equation can be given:

$$(z^n + a_1 z^{n-1} + \cdots + a_{n-1}z + a_n)\hat{y}(z) = (b_0 z^n + b1 z^{n-1} + \cdots + a_{n-1}z + b_n)\hat{u}(z). \tag{1.36}$$

Equation (1.36) may be written as

$$\hat{y}(z) = \frac{b_0 z^n + b_1 z^{n-1} + \cdots + b_{n-1}z + b_n}{z^n + a_1 z^{n-1} + \cdots + a_{n-1}z + a_n} \cdot \hat{u}(z),$$

where $z^n + a_1 z^{n-1} + \cdots + a_{n-1}z + a_n$ is the characteristic polynomial. The roots of the characteristic polynomial correspond to the roots λ's of the homogeneous solution.

---

[5]Hereafter, sampling period h is omitted, and to indicate z-transformed functions a hat symbol, e.g., $\hat{y}(z)$ is used instead of a capital letter $Y(z)$, because a capital letter is usually used for a matrix (or set) expression.

Example 1.4 Consider the following discrete-time equation:

$$y(k+2) - y(k+1) + 0.5y(k) = u(k+1) + u(k). \tag{1.37}$$

The z-transformed equation is given as

$$(z^2 - z + 0.5)\hat{y}(z) - zy(0) - y(1) + y(0) = (z+1)\hat{u}(k) - u(0). \tag{1.38}$$

If the input (forced term) is assumed to be a step function, then

$$(z^2 - z + 0.5)\hat{y}(z) - zy(0) - y(1) + y(0) = \frac{z(z+1)}{z-1} - u(0).$$

When the initial conditions are zero (i.e., $y(0) = y(1) = 0$ and $u(0) = 0$), the following transformed equation is obtained:

$$\hat{y}(z) = \frac{z(z+1)}{(z-1)(z^2 - z + 0.5)}. \tag{1.39}$$

The method used to find the explicit solution $y(k)$ will be presented in Sect. 1.4.5 (Example 1.6).

Incidentally, when considering time shifting $k+2 \to k$ $(k = 2, 3, \cdots)$ as shown in (1.19), discrete-time equation (1.37) can also be written as

$$y(k) - y(k-1) + 0.5y(k-2) = u(k-1) + u(k-2), \quad k = 2, 3, \cdots. \tag{1.40}$$

Since data in a computer are obtained with a time delay, the discrete-time equation might have to be written as shown in (1.40). Using the backward shifting theorem, the transformed equation is given as

$$(1 - z^{-1} + 0.5z^{-2})\hat{y}(z) = (z^{-1} + z^{-2})\hat{u}(z). \tag{1.41}$$

Thus,

$$\hat{y}(z) = \frac{z^{-1} + z^{-2}}{1 - z^{-1} + 0.5z^{-2}}\hat{u}(z). \tag{1.42}$$

When a unit step function $\dfrac{z}{z-1} = \dfrac{1}{1-z^{-1}}$ is added as an input, the following is obtained:

$$\hat{y}(z) = \frac{(1+z^{-1})z^{-1}}{(1-z^{-1})(1-z^{-1}+0.5z^{-2})}. \tag{1.43}$$

Obviously, (1.42) is the same as (1.39).

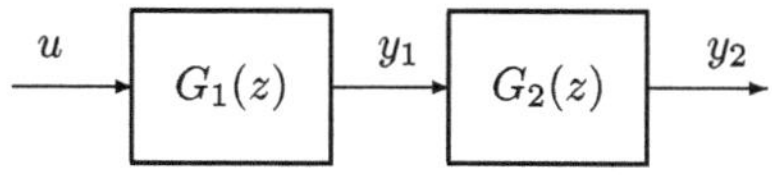

Fig. 1.2 Connected systems

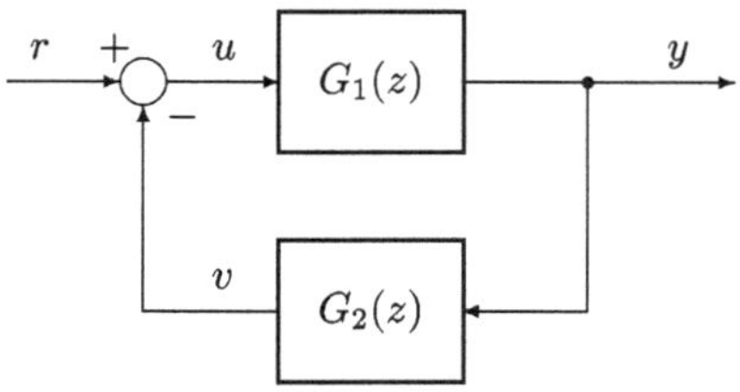

Fig. 1.3 Feedback system

1.4.4 Transfer Function of Connected Systems

The concept of transfer function may be developed independently by using the discrete equation formulation, e.g., Eq. (1.10),

$$y(k+n)+\cdots+a_{n-1}y(k-1)+a_n y(k)=b_0u(k+n)+\cdots+b_{n-1}u(k-1)+b_n u(k).$$

In this case, the z-transformation yields

$$(z^n+a_1z^{n-1}+\cdots+a_n)\hat{y}(z)=(b_0z^n+b_1z^{n-1}+\cdots+b_n)\hat{u}(z).$$

Here, all the initial conditions are assumed to be zero. Thus, the transfer function is defined as follows:

$$G(z)=\frac{\hat{y}(z)}{\hat{u}(z)}=\frac{b_0z^n+b_1z^{n-1}+\cdots+b_n}{z^n+a_1z^{n-1}+\cdots+a_n}.$$

If several discrete-time systems are connected to each other, the output of one of the systems serves as the input to the others. As was described in continuous-time systems, the transfer function of cascade connected systems (Fig. 1.2) is given by

$$\hat{y}_2(z)=G_2(z)\hat{y}_1(z)=G_2(z)G_1(z)\hat{u}(z).$$

For a feedback connection (Fig. 1.3),

$$\hat{y}(z)=G_1(z)\hat{u}(z)$$
$$\hat{v}(z)=G_2(z)\hat{y}(z),\quad \hat{u}(z)=\hat{r}(z)-\hat{v}(z).$$

Thus, the transfer function of the feedback system is given by

$$\frac{\hat{y}(z)}{\hat{r}(z)}=\frac{G_1(z)}{1+G_1(z)G_2(z)}.$$

In general, the usual block diagram algebra can be applied.

1.4.5 The Inverse Transformation

The inverse transformation is used to obtain the explicit behavior $f(kh)$ ($k = 1, 2, \cdots$) in the time domain from the z-transform $F(z)$. The inverse transformation formula is given as follows:

$$f(kh) = \mathcal{Z}^{-1}[F(z)] = \frac{1}{2\pi j}\oint_{\mathbb{C}} F(z)z^{k-1}\mathrm{d}z, \tag{1.44}$$

where $\mathbb{C}$ is a Jordan curve in the complex plane. The solution of inverse transformation (1.44) can be obtained from Cauchy's residue theorem, i.e.,

$$\frac{1}{2\pi j}\oint_{\mathbb{C}} F(z)\mathrm{d}z = \sum_{i=1}^{n} \mathrm{Res}(\mathcal{F};\, p_i), \tag{1.45}$$

where $\mathrm{Res}(\mathcal{F};\, p_i)$ indicates a residue of function $\mathcal{F}$ with respect to the pole, p_i. Although the definition of inverse transformation is given by (1.45), the direct contour integration is scarcely used. Usually, the following partial fraction or power series expansions are applied.

Example 1.5 Consider a simple example,

$$F(z) = \frac{z+0.5}{(z-1)(z-0.5)} = \frac{z+0.5}{z^2 - 1.5z + 0.5}. \tag{1.46}$$

(i) First, the direct formula (1.45) is applied:

$$\begin{aligned} f(k) &= \mathrm{Res}(\mathcal{F};\, 1) + \mathrm{Res}(\mathcal{F};\, 0.5) \\ &= \lim_{z\to 1}(z-1)F(z)z^{k-1} + \lim_{z\to 0.5}(z-0.5)F(z)z^{k-1} \\ &= 3 - 2(0.5)^{k-1}, \quad k = 1, 2, \cdots. \end{aligned} \tag{1.47}$$

(ii) Using partial fraction expansion, z-transform (1.46) can be written as

$$\begin{aligned} F(z) &= \frac{3}{z-1} + \frac{-2}{z-0.5} = \frac{3z^{-1}}{1-z^{-1}} - \frac{2z^{-1}}{1-0.5z^{-1}} \\ &= 3z^{-1}(1 + z^{-1} + z^{-2} + \cdots) - 2z^{-1}(1 + 0.5z^{-1} + 0.25z^{-2} + \cdots) \end{aligned} \tag{1.48}$$

with respect to $|z| > 1$. From the coefficients of power series (1.48), the result, (1.47), is obtained.

(iii) In general, if z-transform function $F(z)$ is represented by the following rational function:[6]

$$F(z) = \frac{b_0 z^n + b_1 z^{n-1} + \cdots + b_n}{z^n + a_1 z^{n-1} + \cdots + a_n}, \tag{1.49}$$

a power series can be obtained by using the division algorithm,

$$F(z) = c_0 + c_1 z^{-1} + c_2 z^{-2} + \cdots. \tag{1.50}$$

Here, the coefficients of power series (1.50) are determined as follows:

$$\begin{aligned} c_0 &= b_0 \\ c_1 &= b_1 - a_1 c_0 \\ c_2 &= b_2 - a_2 c_0 - a_1 c_1 \\ c_3 &= b_3 - a_3 c_0 - a_2 c_1 - a_1 c_2 \\ &\cdots \\ c_k &= b_k - \sum_{i=0}^{k-1} a_{k-i} c_i \\ &\cdots, \quad (a_k, b_k = 0, \text{ for } k > n). \end{aligned} \tag{1.51}$$

Thus, the time sequence $f(k)$ is given as

$$f(k) = c_k.$$

In the case of (1.46), the following coefficients are determined:

$$\begin{aligned} c_0 &= b_0 = 0 \\ c_1 &= b_1 = 1 \\ c_2 &= b_2 - a_1 c_1 = 2 \\ c_3 &= -a_2 c_1 - a_1 c_2 = 2.5 \\ c_4 &= -a_2 c_2 - a_1 c_3 = 2.75 \\ &\cdots. \end{aligned}$$

That is, the power series of (1.46) is given as

$$F(z) = z^{-1} + 2z^{-2} + 2.5z^{-3} + 2.75z^{-4} + \cdots, \tag{1.52}$$

and the time sequence, $f(k)$ (vs. $k = 0, 1, 2, \cdots$), is depicted as shown in Fig. 1.4.

---

[6] Also in this case, for simplicity, a leading coefficient of the denominator polynomial of $a_0 = 1$ is assumed.

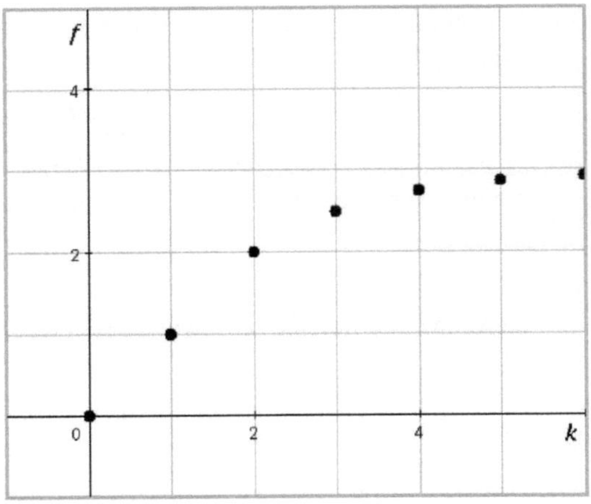

Fig. 1.4 Time sequence of the solution for Example 1.5

Example 1.6 Consider transformed function (1.39) shown in Example 1.4. Since it can also be written as

$$F(z) = \frac{z^2 + z)}{z^3 - 2z^2 + 1.5z - 0.5}, \tag{1.53}$$

the following sequence is obtained based on (1.51):

$$\begin{aligned}
c_0 &= b_0 = 0 \\
c_1 &= b_1 = 1 \\
c_2 &= b_2 - a_1c_1 = 3 \\
c_3 &= -a_2c_1 - a_1c_2 = 4.5 \\
c_4 &= -a_3c_1 - a_2c_2 - a_1c_3 = 5 \\
c_5 &= -a_3c_2 - a_2c_3 - a_1c_4 = 4.75 \\
c_6 &= -a_3c_3 - a_2c_4 - a_1c_5 = 4.25 \\
&\cdots .
\end{aligned}$$

Thus,

$$F(z) = z^{-1} + 3z^{-2} + 4.5z^{-3} + 5z^{-4} + 4.75z^{-5} + 4.25z^{-6} + \cdots . \tag{1.54}$$

The time sequence, $f(k)$ (vs. $k = 0, 1, 2, \cdots$), is as shown in Fig. 1.5. The same result is obtained on a computer simulation based on the vector-matrix form (e.g., (1.15) and (1.16)). The state-space representation is given as

$$\begin{cases}
\begin{bmatrix} x_1(k+1) \\ x_2(k+1) \end{bmatrix} = \begin{bmatrix} -a_1 & 1 \\ -a_2 & 0 \end{bmatrix} \begin{bmatrix} x_1(k) \\ x_2(k) \end{bmatrix} + \begin{bmatrix} b_1 \\ b_2 \end{bmatrix} u(k), \\
y(k) = x_1(k), \quad \text{where } a_1 = -1, \ a_2 = 0.5.
\end{cases} \tag{1.55}$$

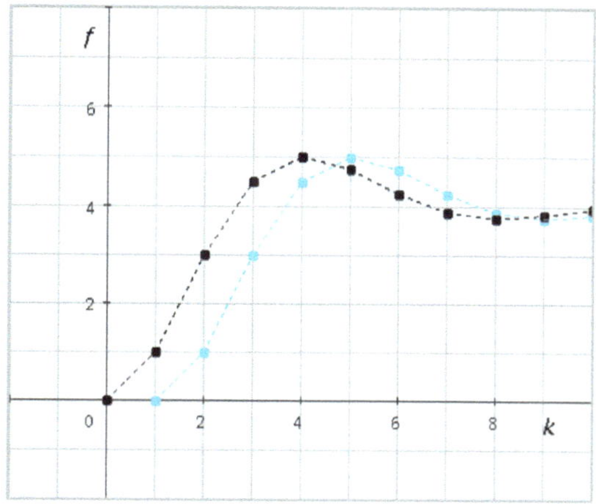

Fig. 1.5 Time sequence of the solution for Example 1.6

Thus, if $u(k) = 1$ for $k \geq 0$ is applied, sequence $y(k)$ can easily be calculated. (Note that $a_0 = 1$ and $b_0 = 0$ in this example.) A block diagram representation for a computer simulation is as shown in Fig. 1.6. Note that the response is delayed by one step as shown in Fig. 1.5 if $y(k) = x_1(k+1)$ is applied to the computer program for (1.55).

1.5 Sampling/Holding and Discrete-Time Signals

In this section, the sampling and holding process in the time domain for continuous signals is described. The sampled signals are analyzed by using Laplace and z-transformation.

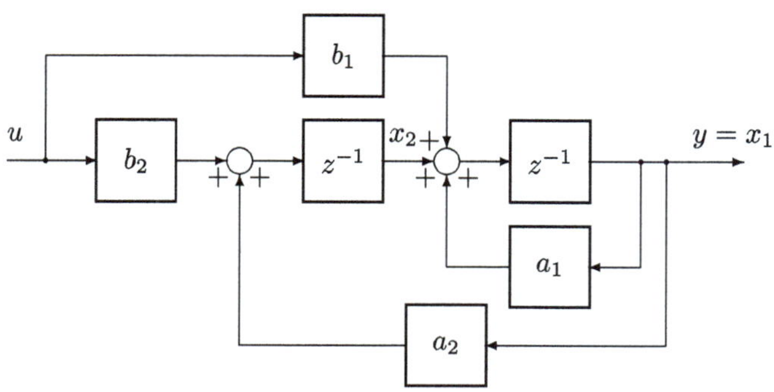

Fig. 1.6 Block diagram for Example 1.6, where $a_1 = 1, a_2 = 0.5$, and $b_1 = b_2 = 1$

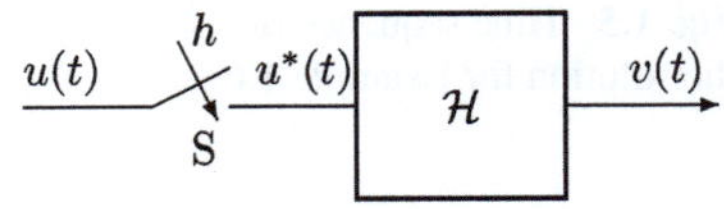

Fig. 1.7 Sampling (time discretization) and holding process

1.5.1 The Sampling and Holding Process

Consider the sampling and holding process represented by Fig. 1.7. The sampling process consists of sampling a continuous-time signal $v(t)$ every h seconds (with small time width) as shown in Fig. 1.8(a), (b). The sampled signal $u^*(t)$ may be held on a constant value in the small interval as shown in Fig. 1.9(a). In practice, the sampled signal will be stored in an electronic circuit until the next sample occurs (i.e., ZOH: zero-order hold).[7] Therefore, the output signal $v(t)$ becomes a stepwise function of continuous time t as shown in Fig. 1.9(b).[8] In this figure, an example of the output with a relatively large resolution is represented. By using expressions of countably infinite set $\mathbb{Z}$, the output can therefore be written as

$$v(t) = v(kh) : \mathbb{Z}_+ \to \mathbb{Z}, \tag{1.56}$$
$$\mathbb{Z}_+ := [0, +\infty),$$

where $h \in \mathbb{Z}_+$ is the sampling period. The analysis of time (and also space) discretized signals (i.e., discrete-time and discrete-value signals) will be discussed in Sect. 1.6.

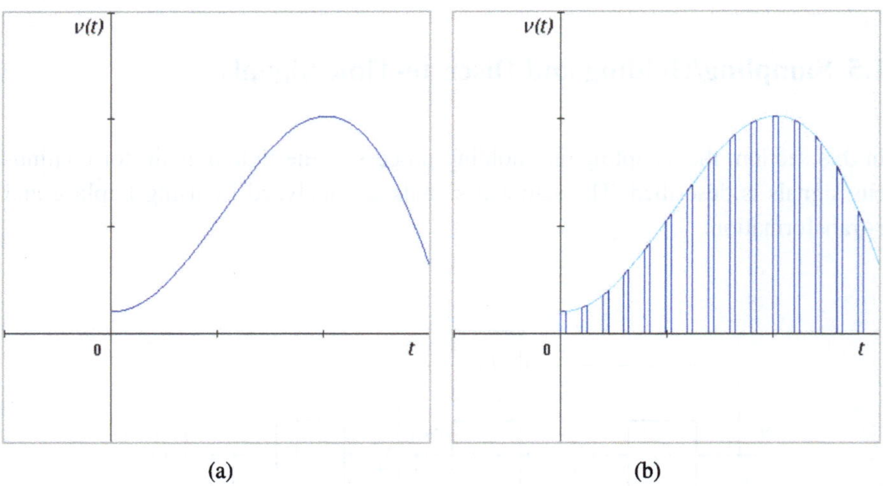

Fig. 1.8 Sampling with small time width

[7]In this book, the symbol $\mathcal{H}$ is used.

[8]The sampling period may be time-varying. Usually, the sampling process is an electrical signal in an analog-to-digital (A/D) converter, but optical, mechanical, and other forms of signals are possible.

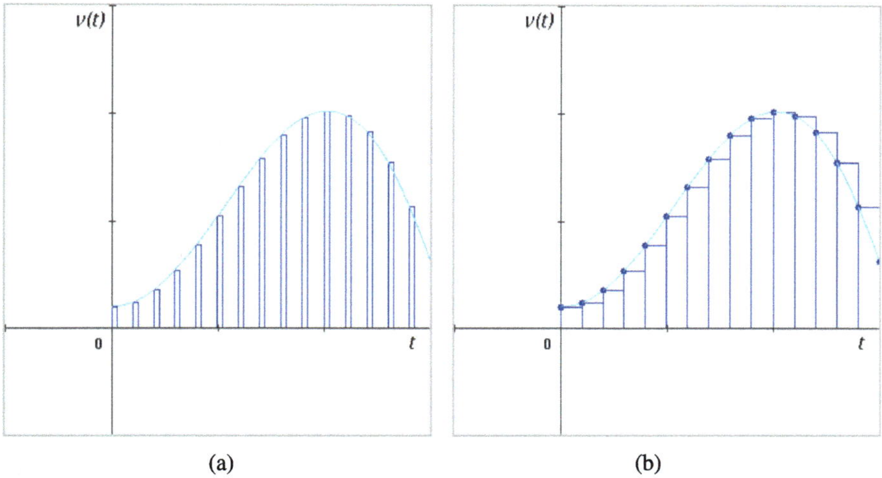

Fig. 1.9 Sampling and holding functions

Fig. 1.10 Continuous plant and sampling/holding function

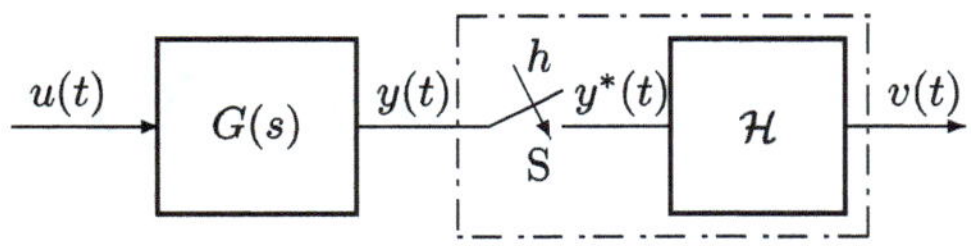

1.5.2 Transfer Function of Sampling/Holding Process

The output of the sampling and holding process, as shown in Fig. 1.10, is given by

$$y_p^*(t) = \sum_{k=0}^{\infty} y(kh)\Delta_p(t-kh).$$

Here, Δ_p is a rectangular pulse, as shown in Fig. 1.11. The Laplace transform of the pulse is written as

$$\mathcal{L}[\Delta_p(t)] = \frac{1-\mathrm{e}^{-ps}}{s}.$$

Therefore,

$$\hat{y}_p^*(s) = \sum_{k=0}^{\infty} y(kh)\mathcal{L}[\Delta_p(t-kh)] = \sum_{k=0}^{\infty} y(kh)\left(\frac{1-\mathrm{e}^{-ps}}{s}\right)\mathrm{e}^{-khs}. \tag{1.57}$$

When $p \to 0$, the following approximation holds:

$$1-\mathrm{e}^{-ps} = 1-\left(1-ps+\frac{(ps)^2}{2!}-\cdots\right) \approx ps.$$

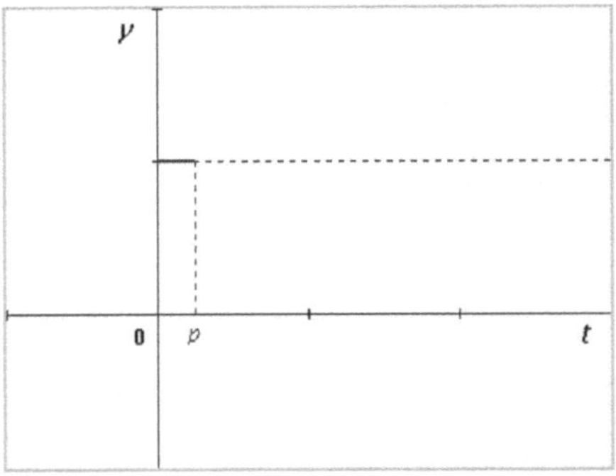

Fig. 1.11 Rectangular pulse function

Thus,

$$\hat{y}_p^*(s) \approx p \sum_{k=0}^{\infty} y(kh) \mathrm{e}^{-khs}. \tag{1.58}$$

The time function can be written as

$$y_p^*(t) \approx p \sum_{k=0}^{\infty} y(kh) \delta(t - kh), \tag{1.59}$$

where $\delta(t)$ is the unit impulse function. The right side of (1.59) is an impulse train that has magnitude $p \cdot y(kh)$ at $t = kh$.

Consider an ideal sampler which operates at each sampling instant h and has a width of $p = 0$. When considering $y_{op}^*(t) = y_p^*(t)/p$, the Laplace transform is written as follows:

$$\hat{y}_{op}^*(s) = \frac{1}{p} \hat{y}_p^*(s) = \sum_{k=0}^{\infty} y(kh) \left(\frac{1 - \mathrm{e}^{-ps}}{ps} \right) \cdot \mathrm{e}^{-khs}. \tag{1.60}$$

If the limitation of $p \to 0$ is considered,

$$\hat{y}^*(s) = \lim_{p \to 0} \hat{y}_{op}^*(s) = \sum_{k=0}^{\infty} y(kh) \mathrm{e}^{-khs}. \tag{1.61}$$

The output of an ideal sampler (a hypothetical sampling signal) is, therefore, written as

$$y^*(t) = \lim_{p \to 0} y_{op}^*(t) = \sum_{k=0}^{\infty} y(kh) \delta(k - kh) = y(t) \sum_{k=0}^{\infty} \delta(t - kh). \tag{1.62}$$

The output of the (zero-order) hold circuit is

$$v(t) = y(kh), \quad kh \leq t \leq (k + 1)h.$$

The transfer function of the continuous plant and holding circuit is given as

$$G^*(s) = \frac{\hat{v}(s)}{\hat{u}(s)} = \frac{1 - \mathrm{e}^{-hs}}{s} \cdot G(s). \tag{1.63}$$

In the z-transform expression, the following relation can be obtained:

$$G(z) = \tilde{\mathcal{Z}}\left\{\frac{1 - \mathrm{e}^{-hs}}{s} \cdot G(s)\right\} = (1 - z^{-1})\tilde{\mathcal{Z}}\left\{\frac{G(s)}{s}\right\}. \tag{1.64}$$

Here, the symbol $\tilde{\mathcal{Z}}\{\cdot\}$ means the z-transform of the sampled sequence for the inverse Laplace transform.

If the denominator of the transfer function $G(s)$ is factored as

$$G(s) = \frac{N(s)}{(s - p_1)(s - p_2)\cdots(s - p_n)}, \tag{1.65}$$

the transformed function can be written in the following partial fraction form:

$$G_1(s) := \frac{G(s)}{s} = \frac{K_0}{s} + \frac{K_1}{s - p_1} + \cdots + \frac{K_n}{s - p_n}. \tag{1.66}$$

It is assumed that p_i $(i = 1, 2, \cdots)$ are simple poles (in other words, p_i are all different real or complex constants) and are not equal to zero. Here, $N(s)$ is a numerator polynomial. Obviously,

$$K_0 = sG_1(s)|_{s=0} \quad \text{and} \quad K_i = (s - p_i)G_1(s)|_{s=p_i}, \quad i = 1, 2, \cdots, n. \tag{1.67}$$

From (1.66) and (1.32), the z-transform expression is given as follows:

$$G_1(z) := \tilde{\mathcal{Z}}\{G_1(s)\} = \frac{K_0}{1 - z^{-1}} + \frac{K_1}{1 - \mathrm{e}^{p_1 h} z^{-1}} + \cdots \frac{K_n}{1 - \mathrm{e}^{p_n h}} z^{-1}. \tag{1.68}$$

If reducing to a common denominator, (1.68) becomes

$$G_1(z) = \frac{K_0[(1 - \mathrm{e}^{p_1 h} z^{-1}) \cdots (1 - \mathrm{e}^{p_n h} z^{-1})] + \cdots + K_n[(1 - z^{-1}) \cdots (1 - \mathrm{e}^{p_{n-1} h} z^{-1})]}{(1 - z^{-1})(1 - \mathrm{e}^{p_1 h} z^{-1}) \cdots (1 - \mathrm{e}^{p_n h} z^{-1})}.$$

Thus, (1.64) is expressed as

$$\begin{aligned} G(z) &= (1 - z^{-1})G_1(z) \\ &= \frac{K_0[(1 - z^{-1}\mathrm{e}^{p_1 h}) \cdots (1 - \mathrm{e}^{p_n h} z^{-1})] + \cdots + K_n[(1 - z^{-1}) \cdots (1 - \mathrm{e}^{p_{n-1} h} z^{-1})]}{(1 - \mathrm{e}^{p_1 h} z^{-1}) \cdots (1 - \mathrm{e}^{p_n h} z^{-1})}. \end{aligned}$$

On the other hand, if the transfer function $G(s)$ has a pole at the origin, i.e.,

$$G(s) = \frac{N(s)}{s(s - p_1)(s - p_2)\cdots(s - p_n)}, \tag{1.69}$$

Table 1.2 z-Transform table

Continuous time ($t \geq 0$)	Discrete time ($k \geq 0$)	Laplace transform ($\Re s > 0$)	z-Transform ($\lvert z\rvert > 1$)
1	1	$\frac{1}{s}$	$\frac{z}{z-1}$
t	kh	$\frac{1}{s^2}$	$\frac{hz}{(z-1)^2}$
e^{pt}	khe^{pkh}	$\frac{1}{s-p}$	$\frac{z}{z-e^{ph}}$
te^{pt}	e^{pkh}	$\frac{1}{(s-p)^2}$	$\frac{he^{ph}z}{(z-e^{ph})^2}$
$\sin \omega t$	$\sin \omega kh$	$\frac{\omega}{s^2+\omega^2}$	$\frac{(\sin \omega h)z}{z^2-2(\cos \omega h)z+1}$
$\cos \omega t$	$\cos \omega kh$	$\frac{s}{s^2+\omega^2}$	$\frac{z^2-(\cos \omega h)z}{z^2-2(\cos \omega h)z+1}$

the partial fraction expansion should be given as

$$G_1(s) = \frac{K_0}{s^2} + \frac{K_{01}}{s} + \frac{K_1}{s-p_1} + \cdots + \frac{K_n}{s-p_n}, \tag{1.70}$$

where

$$K_0 = sG_1(s)|_{s=0} \quad \text{and} \quad K_i = (s-p_i)G_1(s)|_{s=p_i}, \quad i = 1, 2, \cdots, n$$

and

$$K_{01} = \left.\frac{\mathrm{d}s^2 G_1(s)}{\mathrm{d}s}\right|_{s=0}. \tag{1.71}$$

From (1.70) and Example 1.3, the following z-transformed function is obtained:

$$\begin{aligned} G_1(z) &= \frac{K_0 h z^{-1}}{(1-z^{-1})^2} + \frac{K_{01}}{1-z^{-1}} + \frac{K_1}{1-e^{p_1 h}z^{-1}} + \cdots + \frac{K_n}{1-e^{p_n h}z^{-1}} \\ &= \frac{K_0 h z}{(z-1)^2} + \frac{K_{01}}{z-1} + \frac{K_1 z}{z-e^{p_1 h}} + \cdots + \frac{K_n z}{z-e^{p_n h}}. \end{aligned} \tag{1.72}$$

The relationship between time sequences, Laplace transforms, and z-transforms is given in Table 1.2 for the reader's reference.

In any case, the z-transform of $G(s)$ with a zero-order holding can be written as follows:[9]

[9]These operations can be provided in C-language functions.

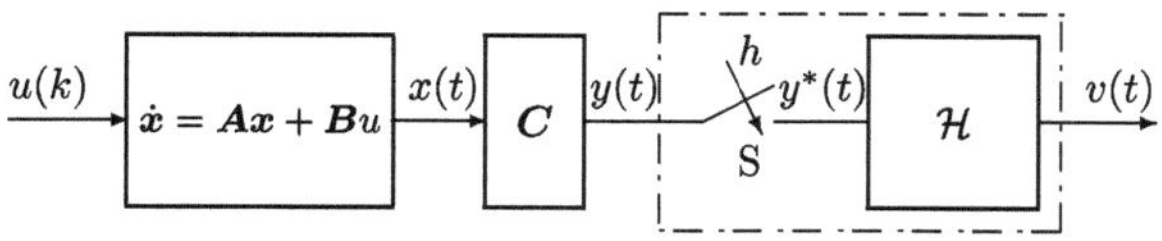

Fig. 1.12 State-space representation and sampling/holding function

$$G(z) = (1 - z^{-1})G_1(z) = \frac{b_0 + b_1 z^{-1} + b_2 z^{-2} \cdots + b_n z^{-n}}{1 + a_1 z^{-1} + a_2 z^{-2} \cdots + a_n z^{-n}}$$
$$= \frac{b_0 z^n + b_1 z^{n-1} + \cdots + b_{n-1} z + b_n}{z^n + a_1 z^{n-1} + \cdots + a_{n-1} z + a_n}. \tag{1.73}$$

From the last expression, the difference equation (1.10) can be considered. Furthermore, the vector-matrix form (1.12) (i.e., the state-space representation) and, for example, (1.15) can be given.

1.5.3 Discretization for State-Space Representation

In this subsection, the direct transformation to a state-space system is represented. Consider a discretized system as shown in Fig. 1.12. Here, $\boldsymbol{x}$ is an n-dimensional state vector (i.e., $\boldsymbol{x} \in \mathbb{R}^n$). The continuous-time linear system can be written in the following state representation:[10]

$$\begin{cases} \dfrac{\mathrm{d}\boldsymbol{x}(t)}{\mathrm{d}t} = \boldsymbol{A}\boldsymbol{x}(t) + \boldsymbol{B}u(t) \\ y(t) = \boldsymbol{C}\boldsymbol{x}(t) \end{cases}. \tag{1.74}$$

Its discretized version is obtained as

$$\boldsymbol{x}(k+1) = \boldsymbol{\Phi}(h)\boldsymbol{x}(k) + \int_{kh}^{(k+1)h} \boldsymbol{\Phi}[(k+1)h - \tau]\boldsymbol{B}u(\tau)\mathrm{d}\tau, \tag{1.75}$$

where the transition matrix, $\boldsymbol{\Phi}(\tau)$, is given by

$$\boldsymbol{\Phi}(\tau) := \mathrm{e}^{\boldsymbol{A}\tau} = 1 + \boldsymbol{A}\tau + \frac{\boldsymbol{A}\tau^2}{2!} + \cdots. \tag{1.76}$$

If the system is time-invariant, (1.75) is simply written as

$$\boldsymbol{x}(k+1) = \boldsymbol{\Phi}(h)\boldsymbol{x}(k) + \int_0^h \boldsymbol{\Phi}(\tau)\boldsymbol{B}u(\tau)\mathrm{d}\tau. \tag{1.77}$$

[10] In order to clarify the relationship between continuous and discrete systems, a traditional state-space expression is considered here.

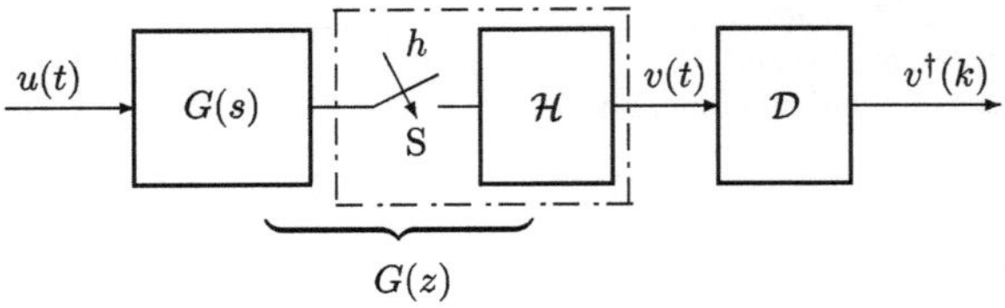

Fig. 1.13 Continuous plant, sampling/holding, and discretization

For $u(\tau) = \text{const.}$ during period h, the discrete-time version of (1.74) can be expressed as

$$\begin{cases} \boldsymbol{x}(k+1) = \boldsymbol{\Phi}(h)\boldsymbol{x}(k) + \boldsymbol{\Gamma}(h)u(k), \quad \boldsymbol{\Gamma}(h) = \int_0^h \boldsymbol{\Phi}(\tau)\boldsymbol{B}\mathrm{d}\tau \\ y(k) = \boldsymbol{C}\boldsymbol{x}(k+1). \end{cases} \tag{1.78}$$

It can be seen that (1.78) corresponds to a linear constant version of (1.6).

Thus, the z-transform of the system is given by

$$\begin{cases} z\hat{\boldsymbol{x}}(z) = \boldsymbol{\Phi}\hat{\boldsymbol{x}}(z) + \boldsymbol{\Gamma}\hat{u}(z) \\ \hat{y}(z) = \boldsymbol{C}z\hat{\boldsymbol{x}}(z). \end{cases} \tag{1.79}$$

Rearranging the above, the following expression can be obtained:

$$\hat{y}(z) = \boldsymbol{C}[I - \boldsymbol{\Phi}z^{-1}]^{-1}\boldsymbol{\Gamma}\hat{u}(z). \tag{1.80}$$

1.6 Space Discretization of Continuous Signals

So far the sampling and holding process of continuous signals in the time domain (i.e., the method of processing discrete-time signals) has been described. In this section, the space discretization of continuous signals (i.e., the method of processing discrete-value signals) is also discussed.

1.6.1 Sampling/Holding and Discretization Process

As was described in Sect. 1.5, the sampling and holding process consists of sampling a continuous signal $u(t)$ every h (e.g., seconds) and holding its value constant. Usually this is executed in A/D and digital-to-analog (D/A) converters. Figure 1.13 shows the sampling/holding and space discretization process of an A/D (D/A) converter. An example of output $u(k)$ is drawn as shown in Fig. 1.14. Here, the sampling period and the resolution value are chosen as $h = \gamma = 1$. Figure 1.14(a) shows the sampling/holding signals with and without (space) discretization, respectively. The discretized output is clearly depicted in Fig. 1.14(b).

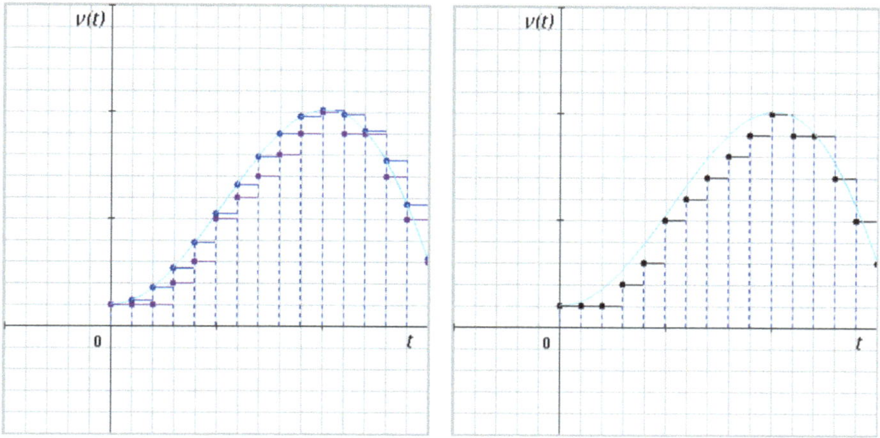

(a) (b)

Fig. 1.14 Output of sampling/holding and discretization

1.6.2 Discrete-Value Signals

It has been assumed that the values of both continuous-time function $v(t)$ and discrete-time function $v(k)$ are continuous. However, those values in computerized or automated digital systems are not continuous in practice. In this section, the method of processing these discrete-time and discrete-value systems is described. In a discrete system, a signal $v(k)$ that is discretized in a broad sense can be written as

$$v(k) \in \{\cdots, -2\gamma, -\gamma, 0, \ \gamma, \ 2\gamma, \cdots\},$$

where γ is the resolution of the signal.[11] In the following, such a discretized signal will be denoted as $v^{\dagger}(k)$, and without loss of generality it may be written as

$$v^{\dagger}(k) \in \mathbb{Z} := \{\cdots, -2, -1, 0, 1, 2, \cdots\},$$

with the assumption of $\gamma = 1$.

1.6.3 Connection of Sampling/Holding Process

Figure 1.15 shows the cascade connection of sampling and holding processes. Unfortunately, since the characteristic of discretizations $\mathcal{D}_\ell$ ($\ell = 1, 2$) is nonlinear, the

[11]For example, in TV scanning or image processing, the resolution is given as $\gamma = R/N$, where R and N are the operating range and the pixel number, respectively.

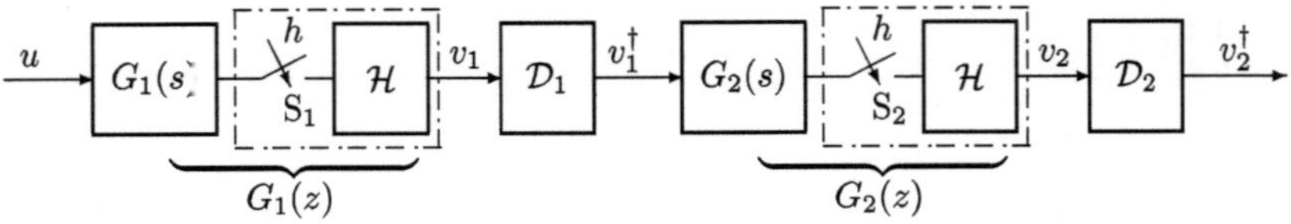

Fig. 1.15 Sampling/holding processes and cascade connection

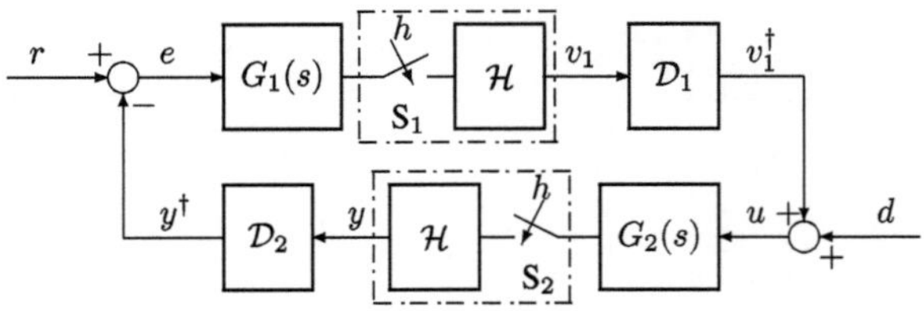

Fig. 1.16 Sampling/holding processes and feedback connection

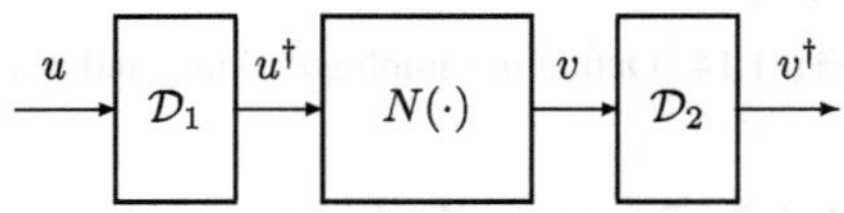

Fig. 1.17 The space discretizing process

transfer function of the connected system cannot be obtained using the method described in Sect. 1.4.4. Moreover, a feedback connection which corresponds to an infinitely connected system is always found in control systems. Figure 1.16 shows the feedback connection of sampling and holding processes [1, 9]. These problems are discussed in Chap. 2 and later in the book.

1.6.4 Space Discretizing Process

A discretization process as shown in Fig. 1.17 is considered here. In this process, $\mathcal{D}_1$ and $\mathcal{D}_2$ are discretizing elements on the input and output sides of a continuous static and memoryless (frequency-independent) nonlinear characteristic $N(\cdot)$. The output of the discretizing elements is written as

$$u^\dagger(k) \in \{\cdots, -2\gamma_1, -\gamma_1, 0, \ \gamma_1, \ 2\gamma_1, \cdots\},$$
$$v^\dagger(k) \in \{\cdots, -2\gamma_2, -\gamma_2, 0, \ \gamma_2, \ 2\gamma_2, \cdots\}.$$

In the following chapter, the resolutions will be assumed to be $\gamma_1 = \gamma_2 = 1.0$, without loss of generality. Thus, variables $u^\dagger(k)$ and $v^\dagger(k)$ may be written on integer grid coordinates as follows:

$$u^\dagger(k), \ u^\dagger(k) \in \mathbb{Z}$$
$$\mathbb{Z} := \{\cdots, -3, -2, -1, 0, 1, 2, 3, \cdots\}.$$

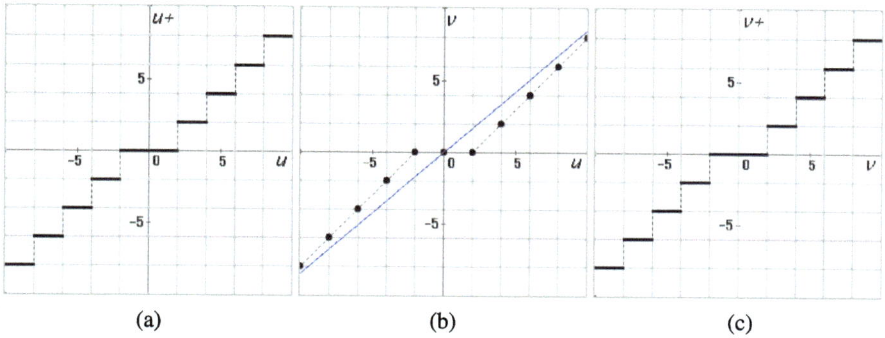

Fig. 1.18 Discretizing process for a linear continuous characteristic

Table 1.3 Discretized signals

u	$u^\dagger$	v	$v^\dagger$
(−12.0, −10.0]	−10.0	−8.5	−8.0
(−10.0, −8.0]	−8.0	−6.8	−6.0
(−8.0, −6.0]	−6.0	−5.1	−4.0
(−6.0, −4.0]	−4.0	−3.4	−2.0
(−4.0, −2.0]	−2.0	−1.7	0.0
(−2.0, 0.0]	0.0	0.0	0.0
[0.0, 2.0)	0.0	0.0	0.0
[2.0, 4.0)	2.0	1.7	0.0
[4.0, 6.0)	4.0	3.4	2.0
[6.0, 8.0)	6.0	5.1	4.0
[8.0, 10.0)	8.0	6.8	6.0
[10.0, 12.0)	10.0	8.5	8.0

Example 1.7 Figure 1.18 shows a numerical example of the space discretizing process for a case where continuous characteristic $N(\cdot)$ is linear. The input/output characteristic of input-side discretization $\mathcal{D}_1$ for resolution $\gamma_1 = 2.0$ is as shown in Fig. 1.18(a). In this example, for output-side discretization $\mathcal{D}_2$ the resolution is chosen the same as for the input-side discretization, as shown in Fig. 1.18(c). Figure 1.18(b) shows the discretized (point-to-point) characteristic when the linear continuous characteristic is $v = 0.85u$.

As is shown in the figure, the discretization performed in this process is based on a round-down procedure. Table 1.3 shows each variable of the discretizing process.

Example 1.8 Figure 1.19 shows a numerical example of the space discretizing process for a case where continuous characteristic $N(\cdot)$ is nonlinear. In this example, a sigmoid function is applied for the continuous characteristic. Using a programming

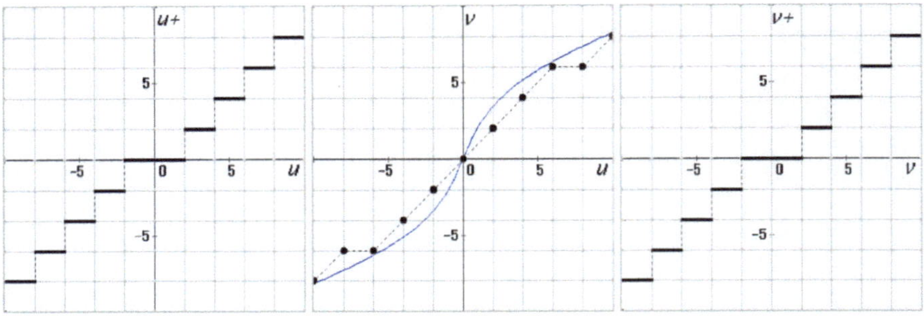

(a) (b) (c)

Fig. 1.19 Discretized process for a nonlinear characteristic

Table 1.4 Discretized signals for nonlinear characteristic

u	$u^\dagger$	v	$v^\dagger$
$(-12.0, -10.0]$	-10.0	-8.22	-8.0
$(-10.0, -8.0]$	-8.0	-7.31	-6.0
$(-8.0, -6.0]$	-6.0	-6.30	-6.0
$(-6.0, -4.0]$	-4.0	-5.13	-4.0
$(-4.0, -2.0]$	-2.0	-3.43	-2.0
$(-2.0, 0.0]$	0.0	0.0	0.0
$[0.0, 2.0)$	0.0	0.0	0.0
$[2.0, 4.0)$	2.0	3.43	2.0
$[4.0, 6.0)$	4.0	5.13	4.0
$[6.0, 8.0)$	6.0	6.30	6.0
$[8.0, 10.0)$	8.0	7.31	6.0
$[10.0, 12.0)$	10.0	8.22	8.0

language such as C, the following expression can be given:

$$\begin{aligned} u^\dagger &= \gamma * (\text{double})(\text{int})(u/\gamma), \\ v &= 0.3 * u^\dagger + 3.0 * \text{atan}(0.6 * u^\dagger), \\ v^\dagger &= \gamma * (\text{double})(\text{int})(v/\gamma). \end{aligned} \tag{1.81}$$

The input-output side discretization elements, $\mathcal{D}_1$, $\mathcal{D}_2$, are the same as in Example 1.7. Figure 1.19(b) shows the discretized (point-to-point) characteristic when the continuous characteristic is the above sigmoid function. Table 1.4 shows each variable of the discretizing process.

1.6.5 Binary Arithmetic with a Finite Word Length

The discussion in this subsection concentrates on the methods most used in microprocessor digital control [6]. It will be assumed that a word length of $C+1$ bits is chosen to represent a number, C bits for the numerical value and one bit for the sign. Therefore, 2^C different numbers may be represented with a C-bit word for the positive axis and the same for the negative axis. 2^{-C} is the least significant bit of the binary number and represents the limit of the resolution.

In fixed point representation, the binary point is fixed, e.g., for $C=4$ the following is obtained:

$$\begin{aligned} 11.01 &= 1\times 2^1 + 1\times 2^0 + 0\times 2^{-1} + 1\times 2^{-2} \\ &= 2+1+0+0.25 \\ &= 3.25. \end{aligned}$$

Depending on the way in which negative numbers are represented, there are three different forms of fixed point arithmetic.

(i) The sign-magnitude representation, in which the reading bit represents the sign, 0 for positive values and 1 for negative values, e.g.,

$$-3.25 \simeq 011.01$$
$$+3.25 \simeq 111.01$$

In sign-magnitude form the number 0 has two representations, 000.00 and 100.00.

(ii) The 2's-complement representation, in which the positive numbers are identical to the sign-magnitude representation. The negative of a positive number is obtained by complementing all bits, and adding 1 in the least significant bit, e.g.,

$$\begin{aligned} -(111.01) &= (000.10) + (000.01) \\ &= 000.11. \end{aligned}$$

Using positive numbers, the following is obtained:

$$\begin{aligned} -(000.11) &= (111.00) + (000.01) \\ &= 111.01. \end{aligned}$$

(iii) The 1's-complement representation, where positive numbers are represented as in sign magnitude and 2's-complement, and the negative of a positive number is obtained by complementing all the bits of the positive number, e.g.,

$$-(111.01) = 000.10.$$

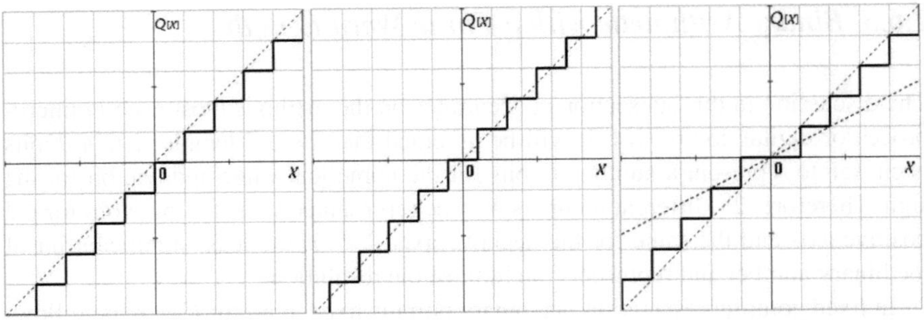

(a). (b) (c)

Fig. 1.20 Truncation and rounding

The position of the binary point in fixed point arithmetic in microprocessors is just to the right of the first bit. Proper scaling is needed to represent all quantities in the range -1.0 to $+1.0 - 2^{-C}$.

1.6.6 Truncation and Rounding of Binary Numbers

It is assumed that there is no overflow during A/D conversion and during arithmetic operations. However, there is a limit on the resolution, because the width of the resolution is the value of the least significant bit (2^{-C}).

Truncation In truncation all bits less than the least significant bit are discarded. The relationship between the untruncated value x and the truncated number $Q(x)$ is depicted in Fig. 1.20(a). For 2's-complement representation (fixed point) the truncation error ϵ_T, defined as

$$\epsilon_T = Q_T[x] - x,$$

is

$$0 \geq \epsilon_T > -2^{-C}.$$

For 1's complement and sign-magnitude representations the truncation error is

$$0 \leq \epsilon_T < 2^{-C} \quad \text{for} \quad x < 0$$
$$0 \geq \epsilon_T > -2^{-C} \quad \text{for} \quad x > 0.$$

Rounding Rounding of a binary number to C bits is accomplished by choosing the number in the C-bit closest to the unrounded quantity, e.g., 0.01101 rounded to a 4-bit number is 0.011. A choice must be made for rounding of numbers; e.g.,

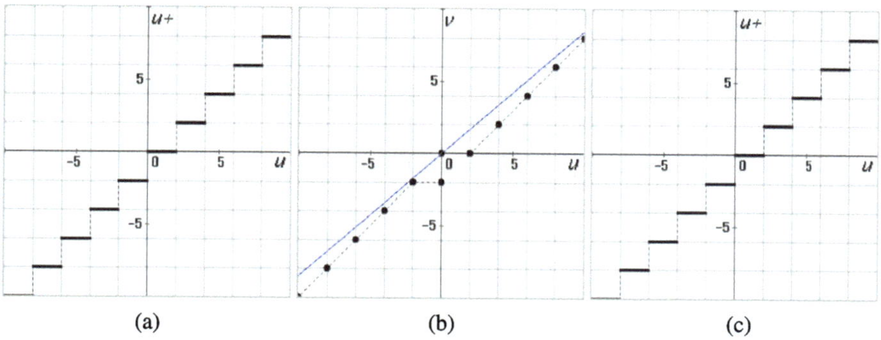

Fig. 1.21 Truncation process for a linear characteristic

Table 1.5 Discretized signals

u	$u^\dagger$	v	$v^\dagger$
(−12.0, −10.0]	−10.0	−10.5	−10.0
(−10.0, −8.0]	−8.0	−8.8	−8.0
(−8.0, −6.0]	−6.0	−7.1	−6.0
(−6.0, −4.0]	−4.0	−5.4	−4.0
(−4.0, −2.0]	−2.0	−3.7	−2.0
(−2.0, 0.0]	0.0	−2.0	2.0
[0.0, 2.0)	0.0	0.0	0.0
[2.0, 4.0)	2.0	1.7	0.0
[4.0, 6.0)	4.0	3.4	2.0
[6.0, 8.0)	6.0	5.1	4.0
[8.0, 10.0)	8.0	6.8	6.0
[10.0, 12.0)	10.0	8.5	8.0

0.01010 rounded to a 4-bit number is either 0.010 or 0.011. For fixed point arithmetic, the error made by rounding is the same for all three types of number representations (sign-magnitude, 1's-complement, and 2's-complement). The error is

$$\epsilon_R = Q_R[x] - x$$
$$\frac{-2^{-C}}{2} \le \epsilon_R \le \frac{2^{-C}}{2}.$$

The behavior of $Q_R[x]$ compared to x is depicted in Fig. 1.20(b). Figure 1.20(c) shows a stepwise characteristic which corresponds to the discretizing process shown in Fig. 1.14 and C programming (1.81). Figure 1.21 shows an example of the truncation process for a linear characteristic based on the discretization as shown in Fig. 1.20(a). Table 1.5 shows each variable of the discretizing process for the truncation.

1.7 Exercises

(1) Find the homogeneous solutions to each of the following difference equations:

(i) $y(k+2) - 3y(k+1) + 2y(k) = u(k)$,
(ii) $y(k+2) - 2y(k+1) + y(k) = u(k)$.

(2) Prove that

$$\mathcal{Z}[kf(kh)] = -z\frac{\mathrm{d}F(z)}{\mathrm{d}z} \quad \text{for } |z| > 1.$$

(3) Determine the z-transform of the discrete ramp function

$$f(k) = \begin{cases} k & \text{for } k \geq 0, \\ 0 & \text{for } k < 0. \end{cases}$$

(4) Show that an n-th-order discrete-time equation (1.99) can also be written in the following vector-matrix form:

$$\begin{bmatrix} x_1(k+1) \\ x_2(k+1) \\ \vdots \\ x_n(k+1) \end{bmatrix} = \begin{bmatrix} 0 & 1 & \dots & 0 \\ \vdots & \vdots & \ddots & 0 \\ 0 & 0 & \dots & 1 \\ -a_n & -a_{n-1} & \dots & a_1 \end{bmatrix} \begin{bmatrix} x_1(k) \\ x_2(k) \\ \vdots \\ x_n(k) \end{bmatrix} + \begin{bmatrix} 0 \\ 0 \\ \vdots \\ 1 \end{bmatrix} u(k)$$

and

$$y(k) = \begin{bmatrix} b_n - a_n & b_{n-1} - a_{n-1} & \dots & b_1 - a_1 \end{bmatrix} \begin{bmatrix} x_1(k) \\ x_2(k) \\ \vdots \\ x_n(k) \end{bmatrix} + u(k).$$

(5) Determine the discrete-time version (1.78) of the following state-space representation:

$$\begin{cases} \dfrac{\mathrm{d}}{\mathrm{d}t}\begin{bmatrix} x_1 \\ x_2 \end{bmatrix} = \begin{bmatrix} 0 & 1 \\ -2 & -3 \end{bmatrix} \begin{bmatrix} x_1 \\ x_2 \end{bmatrix} + \begin{bmatrix} 0 \\ 1 \end{bmatrix} u(t) \\ y(t) = \begin{bmatrix} 1 & 0 \end{bmatrix} \begin{bmatrix} x_1 \\ x_2 \end{bmatrix}. \end{cases}$$

Assume $h = 1$, and $u =$ const. through $0 \leq t \leq h$.

(6) Determine the z-transform of the following time function as a closed form:

(i) $f(t) = t \cdot \mathrm{e}^{-at}$
(ii) $f(t) = \sin \omega t$

(7) Determine $f(k)$ for

$$F(z) = \frac{z+1}{z^3 - 2z^2 + 1.5z - 0.5},$$

and confirm that it corresponds to the delayed response as shown in light blue in Fig. 1.4.

Appendix A: Simultaneous Linear Equations and Matrix Expressions

Consider the following linear system of m equations in n unknowns:

$$\begin{cases} a_{11}x_1 + a_{12}x_2 + \cdots + a_{1n}x_n = y_1 \\ a_{21}x_1 + a_{22}x_2 + \cdots + a_{2n}x_n = y_2 \\ \quad \cdots \\ a_{m1}x_1 + a_{m2}x_2 + \cdots + a_{mn}x_n = y_m \end{cases} . \tag{1.82}$$

As is well known, (1.82) can be simply written as follows:

$$\boldsymbol{A}\boldsymbol{x} = \boldsymbol{y}. \tag{1.83}$$

Here,

$$\boldsymbol{A} = \begin{bmatrix} a_{11} & a_{12} & \cdots & a_{1n} \\ a_{21} & a_{22} & \cdots & a_{2n} \\ \vdots & \vdots & \ddots & \vdots \\ a_{m1} & a_{m2} & \cdots & a_{mn} \end{bmatrix}, \quad \boldsymbol{x} = \begin{bmatrix} x_1 \\ x_2 \\ \vdots \\ x_n \end{bmatrix}, \quad \boldsymbol{y} = \begin{bmatrix} y_1 \\ y_2 \\ \vdots \\ y_m \end{bmatrix}.$$

There are appropriate reasons why the vector-matrix expression (1.83) is considered from the original simultaneous equations (1.82). It can be seen from (1.82) that the following operations do not influence the solution of simultaneous equations [12]:

(1) interchanging two equations,
(2) multiplying each term in one equation by a nonzero constant,
(3) adding a constant multiple of one equation to another.

In the matrix expression $\boldsymbol{A}$, these are called *elementary row operations*, and they become the following operations:

(a) interchanging two rows,
(b) multiplying each entry in one row by a nonzero constant,
(c) adding a constant multiple of one row to another row.

Note that the above operations are related not only to entries of $\boldsymbol{A}$ but also to variables x_i and y_j $(i, j = 1, 2, \cdots, n)$.

Consider a square matrix (i.e., $m = n$). If operation (c) is applied to the left side matrix $\boldsymbol{A}$ in (1.83), the following matrices can be obtained:[12]

$$\rightarrow \boldsymbol{A}_1 = \begin{bmatrix} a_{11} & a_{12} & \dots & a_{1n} \\ 0 & a_{22} & \dots & a_{2n} \\ \vdots & \vdots & \ddots & \vdots \\ 0 & 0 & \dots & a_{nn} \end{bmatrix}, \quad \rightarrow \boldsymbol{A}_2 = \begin{bmatrix} a_{11} & 0 & \dots & 0 \\ 0 & a_{22} & \dots & 0 \\ \vdots & \vdots & \ddots & \vdots \\ 0 & 0 & \dots & a_{nn} \end{bmatrix}.$$

Here, $\boldsymbol{A}_1$ is called an upper[13] *triangular matrix*, and $\boldsymbol{A}_2$ is a *diagonal matrix*. Moreover, the following diagonal matrix (having all diagonal elements equal to 1) is called a *unity matrix* (or an *identity matrix*):

$$\boldsymbol{I} = \begin{bmatrix} 1 & 0 & \dots & 0 \\ 0 & 1 & \dots & 0 \\ \vdots & \vdots & \ddots & \vdots \\ 0 & 0 & \dots & 1 \end{bmatrix}.$$

(It is also written as $\boldsymbol{I}_n$ or $\boldsymbol{E}$.) If the solution $\boldsymbol{x}$ is obtained by using the above process, it can be written as

$$\boldsymbol{x} = \boldsymbol{B}\boldsymbol{y} = \boldsymbol{A}^{-1}\boldsymbol{y}, \tag{1.84}$$

where $\boldsymbol{B} = \boldsymbol{A}^{-1}$ is called an *inverse matrix*.

Determinant With respect to a square matrix $\boldsymbol{A}$, the *determinant* is defined as follows:

$$\det \mathrm{A} = |\boldsymbol{A}| = \begin{vmatrix} a_{11} & a_{12} & \dots & a_{1n} \\ a_{21} & a_{22} & \dots & a_{2n} \\ \vdots & \vdots & \ddots & \vdots \\ a_{n1} & a_{n2} & \dots & a_{nn} \end{vmatrix}, \tag{1.85}$$

which is a generalization of the area of a parallelogram in the two-dimensional plane, i.e.,

$$\begin{vmatrix} a_{11} & a_{12} \\ a_{21} & a_{22} \end{vmatrix} = a_{11}a_{22} - a_{12}a_{21}.$$

In general for $n \geq 3$, the determinant of a square matrix $\boldsymbol{A}$ is defined for the $(n-1) \times (n-1)$ matrix $\boldsymbol{A}_{ij}$ that is obtained by deleting the i-th row and j-th column from $\boldsymbol{A}$:

$$\det \boldsymbol{A} = \sum_{j=1}^{n} a_{ij}(-1)^{i+j} \det \boldsymbol{A}_{ij}. \tag{1.86}$$

[12] This operation is referred to as the sweep-out method.

[13] If all the entries above the main diagonal are zero, it is called a lower triangular matrix.

Since the calculation of (1.86) becomes a permutation problem, it is difficult to obtain the value of the determinant. Then, by applying the elementary row operation (c), the matrix A should be led to an upper triangular matrix. Note that the value of the determinant is invariant for this operation. As a result, it is obtained in regard to the diagonal elements of the triangular matrix as follows:

$$\det \boldsymbol{A} = a_{11}a_{22}\cdots a_{nn}. \tag{1.87}$$

Here, we define the matrix of cofactors, i.e.,

$$\mathrm{adj}\boldsymbol{A} = [(-1)^{i+j} \det \boldsymbol{A}_{ji}]. \tag{1.88}$$

Thus, it can be shown that

$$\boldsymbol{A} \cdot \mathrm{adj}\boldsymbol{A} = \det \boldsymbol{A} \cdot \boldsymbol{I} = \mathrm{adj}\boldsymbol{A} \cdot \boldsymbol{A}.$$

For $\det \boldsymbol{A}$, the inverse matrix in (1.84) is given by

$$\boldsymbol{A}^{-1} = \frac{\mathrm{adj}\boldsymbol{A}}{\det \boldsymbol{A}}. \tag{1.89}$$

When $\boldsymbol{A}^{-1}$ is obtainable as shown in (1.89), the matrix $\boldsymbol{A}$ is referred to as a *regular* or *nonsingular* matrix.

Eigenvalue/Eigenvector When the direction of a vector $\boldsymbol{Ax}$ is the same as that of $\boldsymbol{x}$, i.e.,

$$\boldsymbol{Ax} = \lambda \boldsymbol{x}, \tag{1.90}$$

the vector $\boldsymbol{x}$ is called an *eigenvector*, and in that case λ is called an *eigenvalue*. The idea in (1.90) is also applied to complex vectors (and matrices) in general. Equation (1.90) can be written as

$$(\lambda \boldsymbol{I} - \boldsymbol{A})\boldsymbol{x} = \boldsymbol{0}. \tag{1.91}$$

In this expression, (1.91) is considered to be simultaneous homogeneous equations. The trivial solution of (1.91) is $\boldsymbol{x} = \boldsymbol{0}$. Obviously, the condition having nontrivial solutions of (1.91) is given as follows:

$$\det(\lambda \boldsymbol{I} - \boldsymbol{A}) = 0. \tag{1.92}$$

Equation (1.92) is called a *characteristic equation*, and λ is called a *characteristic root*.

Quadratic Form With respect to a square matrix $\boldsymbol{A}$, the following function, $\mathbb{R}^n \to \mathbb{R}$, can be defined:

$$\begin{aligned} \mathcal{Q}(\boldsymbol{x}) = \boldsymbol{x}^T\boldsymbol{Ax} &= \begin{bmatrix} x_1 & x_2 & \dots & x_n \end{bmatrix} \begin{bmatrix} a_{11} & a_{12} & \dots & a_{1n} \\ a_{21} & a_{22} & \dots & a_{2n} \\ \vdots & \vdots & \ddots & \vdots \\ a_{n1} & a_{n2} & \dots & a_{nn} \end{bmatrix} \begin{bmatrix} x_1 \\ x_2 \\ \vdots \\ x_n \end{bmatrix} \\ &= \sum_{i=1}^{n}\sum_{j=1}^{n} a_{ij}x_ix_j \end{aligned} \tag{1.93}$$

which corresponds to the equation of an ellipse and a hyperbola. The value of the function is scalar, and is considered the inner product of $\boldsymbol{Ax}$ and $\boldsymbol{x}$. The above mathematical form is called a *quadratic form*.

If considering x_{ij}, the coefficients a_{ij} and a_{ji} can be replaced with $a'_{ij} = a'_{ji} = (a_{ij} + a_{ji})/2$. Therefore, the matrix $\boldsymbol{A}$ in (1.93) can be considered a symmetry matrix. Since such an expression is clear in mathematics, it is often applied to control theory, e.g., as a candidate of a Lyapunov function. However, the mathematical form in (1.93) is not always suitable for the expression of real complex systems.

Appendix B: Function Space, H_p, L_p, and ℓ_p Spaces

A linear space X is called a normed linear space (or simply a *normed space*) [11], if for every $x \in X$, there is associated a real number $\|x\|$, the norm of the vector x, such that

$$\|x\| \geq 0 \text{ and } \|x\| = 0 \text{ if and only if } x = 0, \tag{1.94}$$

$$\|x + y\| \leq \|x\| + \|y\| \ \text{: triangle inequality}, \tag{1.95}$$

$$\|\alpha x\| = |\alpha| \cdot \|x\|. \tag{1.96}$$

The topology of a normed space X is defined by the distance (or *metric*)[14]

$$\rho(x, y) = \|x - y\|. \tag{1.97}$$

Here, $\rho(x, y)$ satisfies the following axiom of distance:

$$\rho(x, y) \geq 0 \text{ and } \rho(x, y) = 0 \text{ if and only if } x = y, \tag{1.98}$$

$$\rho(x, y) \leq \rho(x, z) + \rho(z, y) \ \text{: triangle inequality}, \tag{1.99}$$

$$\rho(x, y) = \rho(y, x). \tag{1.100}$$

Obviously, $\rho(x, y) = \|x - y\| = \|y - x\| = \rho(y, x)$ and

$$\rho(x, y) = \|x - y\| = \|x - z + z - y\| \leq \|x - z\| + \|z - y\| = \rho(x, z) + \rho(z, y).$$

The convergence $\lim_{n\to\infty} \rho(x_n, x) = 0$ in a normed space X is denoted by $x_n \to x$, and sequence $\{x_n\}$ is said to converge to x. Thus $\lim_{n\to\infty} \|x_n\| = \|x\|$, if $x_n \to x$.

(1) Let X be a real or complex vector space (i.e., $\boldsymbol{x} = (x_1, x_2, \cdots, x_n)^T$ with $x_i \in \mathbb{R}$ or $\mathbb{C}$). The following norms are defined:

---

[14] See also Appendix A in Chap. 7.

$$\|\boldsymbol{x}\|_1 := \sum_{i=1}^{n} |x_i| \tag{1.101}$$

$$\|\boldsymbol{x}\|_p := \left(\sum_{i=1}^{n} |x_i|^p\right)^{1/p}, 1 \le p < \infty \tag{1.102}$$

$$\|\boldsymbol{x}\|_\infty := \max_i |x_i|. \tag{1.103}$$

Here, $\|\boldsymbol{x}\|_2$ is called the *Euclidean norm* of $\boldsymbol{x}$. Thus, the normed space corresponds to an n-dimensional Euclidean space.

(2) Let X be the space of sequences of real numbers ($x = (x(1), x(2), \cdots, x(N))$). The following norms can be defined:

$$\|x(k)\|_1 := \sum_{k=1}^{N} |x(k)| \tag{1.104}$$

$$\|x(k)\|_p := \left(\sum_{k=1}^{N} |x(k)|^p\right)^{1/p}, 1 \le p < \infty \tag{1.105}$$

$$\|x(k)\|_\infty := \sup_{1 \le k \le N} |x(k)|. \tag{1.106}$$

The corresponding normed spaces for $N \to \infty$ are called ℓ_1, ℓ_p, and ℓ_∞, respectively. In this book, the following norm is especially considered for discrete signals ($x = x(0), x(1), x(2), \cdots$) with $x : \mathbb{Z}_+ \to \mathbb{R}$:

$$\|x(k)\|_2 = \left(\sum_{k=0}^{\infty} |x(k)|^2\right)^{1/2} < \infty. \tag{1.107}$$

The normed space corresponds to an ℓ_2 space.

(3) Let X be the space of continuous signals with $x : \mathbb{R}_+ \to \mathbb{R}$. The following norms are defined:

$$\|x(t)\|_1 := \int_0^\infty |x(t)|\mathrm{d}t \tag{1.108}$$

$$\|x(t)\|_p := \left(\int_0^\infty |x(t)|^p \mathrm{d}t\right)^{1/p}, 1 \le p < \infty \tag{1.109}$$

$$\|x(t)\|_\infty := \text{ess} \sup_{t \in \mathbb{R}} |x(t)|. \tag{1.110}$$

The corresponding normed spaces are called L_1, L_p, and L_∞, respectively.

(4) Let X be the space of transformed variables with $\hat{x} : \mathbb{C} \to \mathbb{C}$. The following norms are defined:

$$\|\hat{x}(\zeta)\|_1 := \sup_{r \in [0,1)} \|\hat{x}_r(\zeta)\|_1 = \sup_{r \in [0,1)} \frac{1}{2\pi} \int_{-\pi}^{\pi} |\hat{x}(\zeta)|\mathrm{d}\theta \tag{1.111}$$

$$\|\hat{x}(\zeta)\|_p := \sup_{r\in[0,1)} \|\hat{x}_r(\zeta)\|_p = \sup_{r\in[0,1)} \left(\frac{1}{2\pi}\int_{-\pi}^{\pi} |\hat{x}(\zeta)|^p \mathrm{d}\theta\right)^{1/p} \quad (1.112)$$

$$\|\hat{x}(\zeta)\|_\infty := \sup_{r\in[0,1)} |\hat{x}(\zeta)|, \quad \text{where } \zeta = r\cdot \mathrm{e}^{j\theta}. \quad (1.113)$$

The corresponding spaces are called the Hardy spaces, H_1, H_p, and H_∞, respectively [7]. In particular, H_∞ has recently been used in control theory. As for continuous systems, the following domain is usually considered:

$$\zeta = \mathrm{e}^{-Ts}, \quad T > 0.$$

The right half-plane of $s = \sigma + j\omega$ (i.e., $\sigma > 0$) corresponds to the open unit disk,

$$\mathbb{D} = \{\zeta \in \mathbb{C} : |\zeta| < 1\}.$$

The boundary line of domain $\mathbb{D}$ (i.e., the imaginary axis $s = j\omega$) corresponds to the unit circle,

$$\mathbb{L} = \{\zeta \in \mathbb{C} : |\zeta| = 1\}.$$

In this book, since discrete-time signals are treated in the system analysis, the following norm on the boundary line is considered for $\zeta = z^{-1} = \mathrm{e}^{-hs}$ (h : sampling period):

$$\|\hat{x}(\zeta)\|_2 := \sup_{r\in[0,1)} \left(\frac{1}{2\pi}\int_{-\pi}^{\pi} |\hat{x}(\zeta)|^2 \mathrm{d}\theta\right)^{1/2}.$$

Obviously, $r = \mathrm{e}^{-h\sigma}$ and $\theta = \omega h$. Thus, the L_2 norm becomes

$$\|\hat{x}(z)\|_2 = \left(\frac{h}{2\pi}\int_{-\pi}^{\pi} |\hat{x}(\mathrm{e}^{-j\omega h})|^2 \mathrm{d}\omega\right)^{1/2}.$$

Appendix C: Inverse z-Transform

The z-transform and the inverse z-transform are defined by the following pair of equations:

$$F(z) = \sum_{k=0}^{\infty} f(kh) z^{-k}, \quad (1.114)$$

$$f(kh) = \frac{1}{2\pi j}\oint_{\mathbb{C}} F(z) z^{k-1} \mathrm{d}z, \quad (1.115)$$

where $\mathbb{C}$ is a Jordan curve in the complex plane. Here, as a reference, the definition of the Laplace transform and its inverse transform are given below:

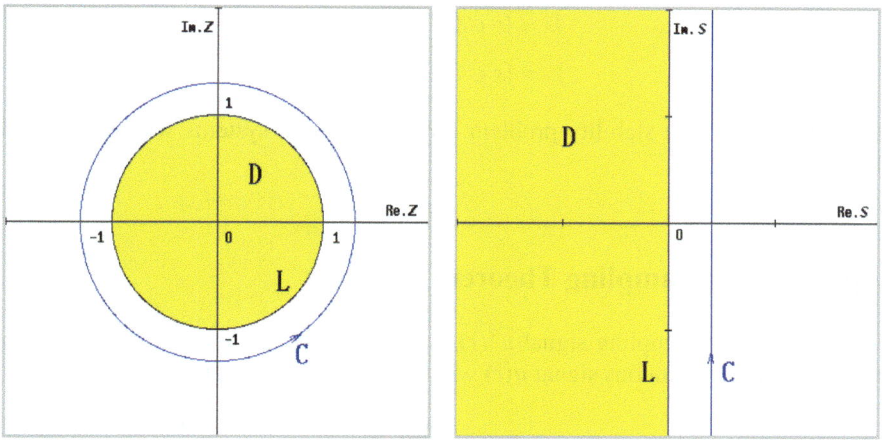

Fig. 1.22 Relationship between z-plane and s-plane, and contours

$$F(s) = \int_0^{\infty} f(t)\, \mathrm{e}^{-st} \mathrm{d}t, \tag{1.116}$$

$$f(t) = \frac{1}{2\pi j} \int_{c-j\infty}^{c+j\infty} F(s)\, \mathrm{e}^{st} \mathrm{d}s, \qquad s = \sigma + j\omega. \tag{1.117}$$

If $f(t)$ is a sample/hold (stepwise) function $\tilde{f}(kh)$, then $\tilde{F}(s) = \mathcal{L}\{\tilde{f}(kh)\}$ becomes

$$\tilde{F}(s) = h \sum_{k=0}^{\infty} \tilde{f}(kh) z^{-k}.$$

On the other hand, from (1.117) the following expression can be obtained:

$$\tilde{f}(kh) = \frac{1}{2\pi hj} \oint_{\mathbb{C}} \tilde{F}(s) z^{k-1} \mathrm{d}z,$$

because $\mathrm{d}z = h\mathrm{e}^{hs}\mathrm{d}s$. Thus, the relationship between the z-transform of $\tilde{f}(kh)$ and the inverse one hold with respect to arbitrary h.

Relation Between z-Plane and s-Plane As was defined in Sect. 1.4, the relationship between the z-transform variable and Laplace transform variable is given as $z = \mathrm{e}^{hs}$. This is shown graphically in Fig. 1.22. In these complex planes, two domains and contours,

$$\mathbb{D} = \{z \in \mathbb{C} : |z| < 1\},$$
$$\mathbb{L} = \{z \in \mathbb{C} : |z| = 1\},$$

and

$$\mathbb{D} = \{s \in \mathbb{C} : \Re(s) < 0\},$$
$$\mathbb{L} = \{s \in \mathbb{C} : \Re(s) = 0\},$$

are defined [7]. The stability problem for discrete-time systems will be clarified from these diagrams.

Appendix D: Sampling Theorem

The hypothetical sampling signal $u^*(t)$ is expressed by an impulse function train that modulates continuous signal $u(t)$ with a carrier wave signal as follows [8]:

$$u^*(t) = u(t) \sum_{k=-\infty}^{\infty} \delta(t - kh). \tag{1.118}$$

Since $\sum_{k=-\infty}^{\infty} \delta(t - kh)$ is a periodic function, the following expression can be given by using the complex Fourier series:

$$\sum_{k=-\infty}^{\infty} \delta(t - kh) = \frac{1}{h} \sum_{k=-\infty}^{\infty} \mathrm{e}^{jk\omega_s t}, \tag{1.119}$$

where $\omega_s = 2\pi / h$ is the sampling angular frequency. Therefore, the Laplace transform of $u^*(t)$ is given by using the shifting theorem in the s-plane as follows:

$$\hat{u}^*(s) = \frac{1}{h} \sum_{k=-\infty}^{\infty} \hat{u}(s + jk\omega_s). \tag{1.120}$$

That is, the Laplace transform of hypothetical sampling signal $\hat{u}(s)$ is the Laplace transform of a modulated signal $\hat{u}(s)$ that is shifted to the imaginary axis by ω_s (plus or minus). When the frequency spectrum of $|\hat{u}(j\omega)|$ is given as shown in Fig. 1.23, the frequency spectrum of the sampler output is repeated as shown in Fig. 1.24(a). Therefore, if cut-off frequency ω_c[15] is given by

$$\omega_c < \frac{\omega_s}{2}, \tag{1.121}$$

the modulated signal $u(t)$ is transmitted on the sampling process without losing its information, and the original signal can be recovered perfectly. However, if

$$\omega_c > \frac{\omega_s}{2}, \tag{1.122}$$

[15] A cut-off frequency means that the frequency spectrum is negligible there.

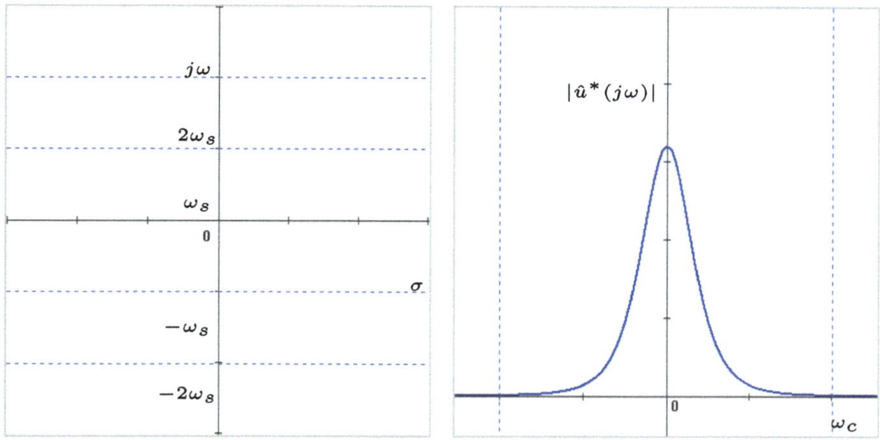

(a) (b)

Fig. 1.23 Frequency shifting and spectrum

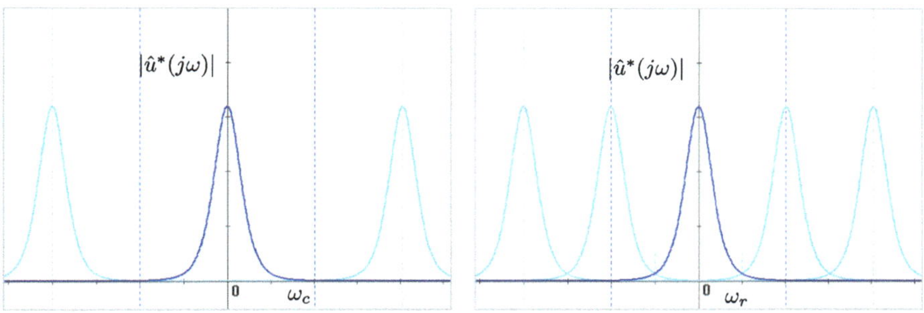

(a) (b)

Fig. 1.24 Frequency responses of the sampling

the frequency spectrum is duplicated (in other words "folded") as shown in Fig. 1.24(b), and the modulated signal cannot be transmitted without losing its information. Thus, it is difficult to recover the original signal. The sampling (angular) frequency ω_s that can recover a continuous signal must be at least twice the highest component frequency, i.e.,

$$\frac{\omega_s}{2} > \omega_c. \tag{1.123}$$

This is called *Shannon's sampling theorem.*

Systems with Transmission Delay In regard to control systems with transmission delay, the above concept should be revised as follows. Assume that the continuous signal is written as

$$u_d(t) = u(t - L),$$

where L is a time delay. When the signal is in discrete time, it can be written as

$$u_d(kh) = u((k - d_\ell)h),$$

where $d_\ell = L/h$ ($d_\ell \in \mathbb{Z}_+$). The Laplace transform expression of $u_d^*(t)$ that corresponds to (1.120) is given as

$$\hat{u}_d^*(s) = \hat{u}^*(s)\mathrm{e}^{-Ls} = \frac{1}{h}\sum_{k=-\infty}^{\infty} \hat{u}(s + jk\omega_s)\mathrm{e}^{-L(s+jk\omega_s)}. \tag{1.124}$$

References

1. Bartoszewics A (ed) (2011) Robust control, theory and application, pp 243–260. Chap 11
2. Cadzow JA (1973) Discrete-time systems. Prentice-Hall, New York
3. Desoer CA, Vidyasagar M (1975) Feedback systems—input-output properties. Academic Press, New York, republished by SIAM, 2009
4. Jury EI (1964) Theory and application of the z-transform method. Wiley, New York, revised by R.E. Krieger, 1973
5. Kalman RE, Falb PL, Arbib MA (1969) Topics in mathematical system theory. McGraw-Hill, New York
6. Katz P (1981) Digital control using microprocessors. Prentice-Hall, New York
7. Mashreghi J (2009) Representation theorem in Hardy spaces. Cambridge University Press, Cambridge
8. Ogata K (2006) Discrete-time control systems, 2nd edn. Prentice-Hall, New York
9. Okuyama Y (2006) Robust stability analysis for discretized nonlinear control systems in a global sense. In: Proc of the 2006 American control conference, Minneapolis, MN, USA, pp 2321–2326
10. Ragazzini SP, Franklin GF (1958) Sampled-data control systems. McGraw-Hill, New York
11. Rudin W (1987) Real and complex analysis, 3rd edn. McGraw-Hill, New York
12. Wright DJ (1999) Introduction to linear algebra. McGraw-Hill, New York

Chapter 2
Discretized Feedback Systems

2.1 Introduction

As we described in the previous chapter, since discretized/quantized feedback systems become nonlinear, the analysis and design of those types of systems has not been elucidated. The first attempt to clarify these problems was described in a paper of Kalman [3]. However, few results have been obtained for the stability analysis of nonlinear discrete-time feedback systems [2, 9]. In this chapter, the analysis in an ℓ_2 space for such a discrete-time and discrete-value system is discussed.

2.2 Discretized Control Systems

A discretized nonlinear control system can be represented by a sampled-data control system with two samplers, S_1, S_2, and a continuous nonlinear characteristic, $N(\cdot)$, as shown in Fig. 2.1. Here, $\mathcal{D}_1$, $\mathcal{D}_2$, and $\mathcal{H}$ denote the discretization and zero-order hold elements, which are usually performed in A/D (D/A) conversion, and $G(s)$ is the transfer function (the Laplace-transformed one) of a linear (continuous-time) controlled system. It is assumed that the two samplers with sampling period h operate synchronously. The feedback structure corresponds to the sampling/holding system shown in Fig. 1.16, when $G_1(s)$ is considered to be a static nonlinear characteristic. The sampled-data control system can be equivalently transformed into a discretized control system, as shown in Fig. 2.2. Here, $G(z)$ is the z-transform of $G(s)$ together with a zero-order hold, and $\mathcal{D}_1$ and $\mathcal{D}_2$ are the discretizing units (static quantizers) on the input and output sides of the nonlinear element, respectively.

In Fig. 2.2, each symbol e, u, v, ... indicates the sequence $e(k)$, $u(k)$, $v(k)$, ..., $(k = 0, 1, 2, \cdots)$ in discrete time, but in continuous values. On the other hand, each symbol $e^\dagger$, $u^\dagger$, $v^\dagger$, ... indicates a discrete value that can be assigned to an integer number, e.g.,

$$e^\dagger \in \{\cdots, -3\gamma_1, -2\gamma_1, -\gamma_1, 0, \gamma_1, 2\gamma_1, 3\gamma_1, \cdots\},$$

Y. Okuyama, *Discrete Control Systems*,
DOI 10.1007/978-1-4471-5667-3_2, © Springer-Verlag London 2014

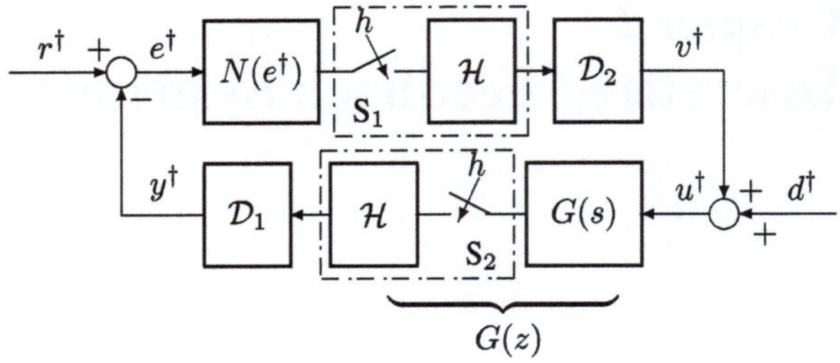

Fig. 2.1 Nonlinear sampled-data feedback system

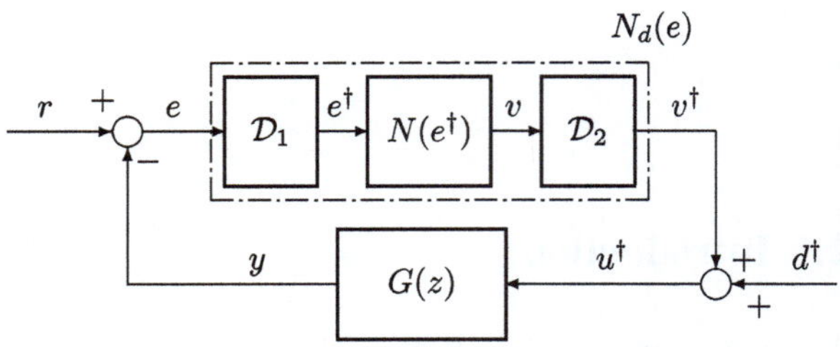

Fig. 2.2 Discretized nonlinear control system

$$v^{\dagger} \in \{\cdots, -3\gamma_2, -2\gamma_2, -\gamma_2, 0, \ \gamma_2, \ 2\gamma_2, \ 3\gamma_2, \cdots\},$$

where γ_1 and γ_2 are the resolution values of each variable. In the above expressions, it is assumed that the input and output signals of the nonlinearity have the same resolution in the discretization (i.e., $\gamma = \gamma_1 = \gamma_2 > 0$) [1, 5, 6]. Here, $e^{\dagger}$, $u^{\dagger}$, and $v^{\dagger}$ also represent the time sequences $e^{\dagger}(k)$, $u^{\dagger}(k)$, and $v^{\dagger}(k)$.

The relationship between e and $v^{\dagger} = N_d(e)$ in the figure becomes a stepwise nonlinear characteristic on integer grid coordinates, as shown in Fig. 2.3(a). In this chapter, a round-down discretization, which is usually executed on a computer, is

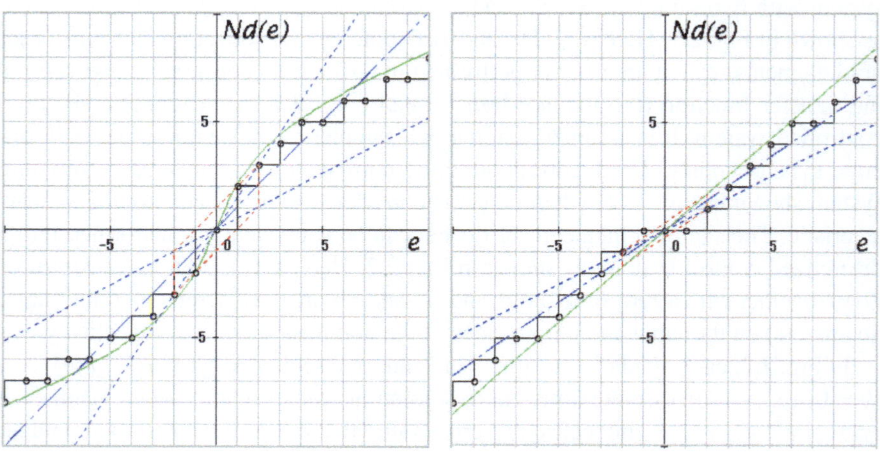

(a) (b)

Fig. 2.3 Discretization for nonlinear and linear characteristics

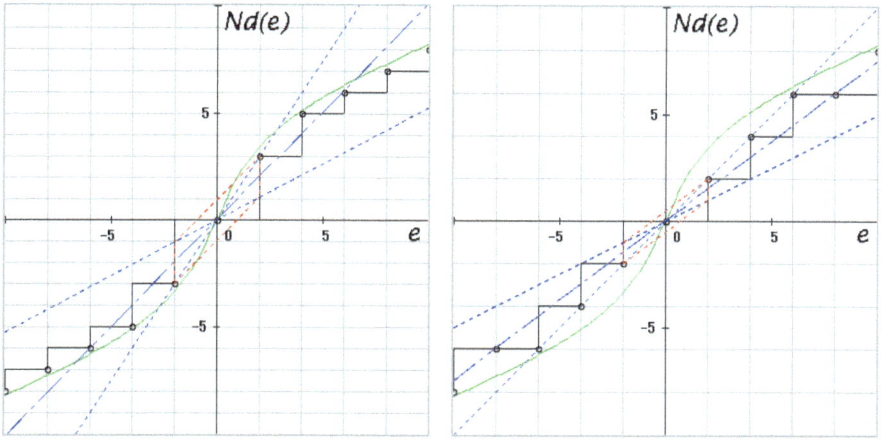

(a) (b)

Fig. 2.4 Effect of resolution values

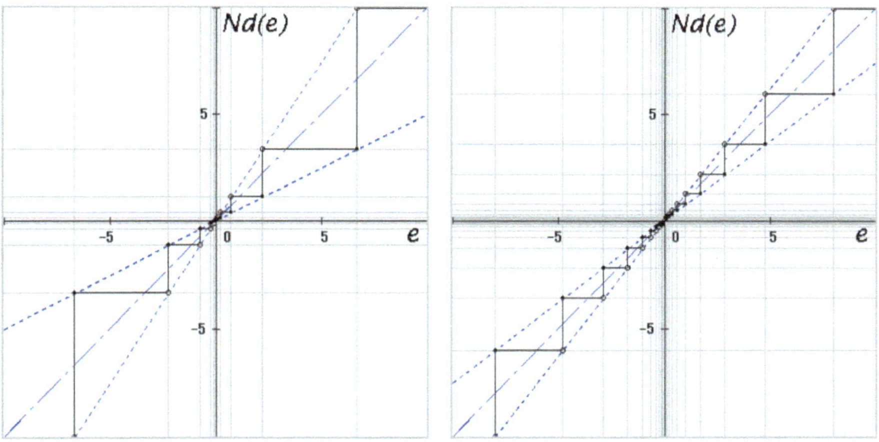

(a) (b)

Fig. 2.5 Logarithmic quantizers

applied. Therefore, the relationship between e and $v^{\dagger}$ is indicated by small circles (i.e., a point-to-point transition) on the stepwise nonlinear characteristic. Even if the continuous characteristic $N(\cdot)$ is linear, the discretized characteristic $v^{\dagger}$ becomes nonlinear on integer grid coordinates, as shown in Fig. 2.3(b). In order to compare the discretization, Figs. 2.4(a) and (b) show the effect of resolution values. In these figures, two cases are depicted: (a) input resolution $\gamma = 2$ and output resolution $\gamma = 1$, (b) input and output resolutions $\gamma = 2$.

Some authors have investigated a logarithmic quantizer in relation to the robust stability. Figs. 2.5(a) and (b) show two cases of the logarithmic quantizer. However, the applications of these discretizations are scarcely known in practice.

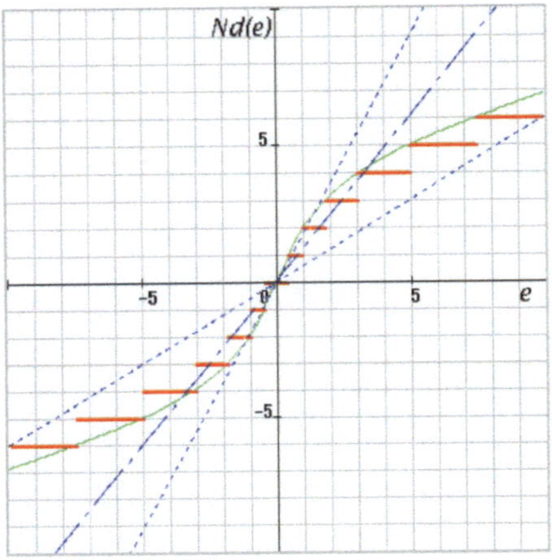

(a)

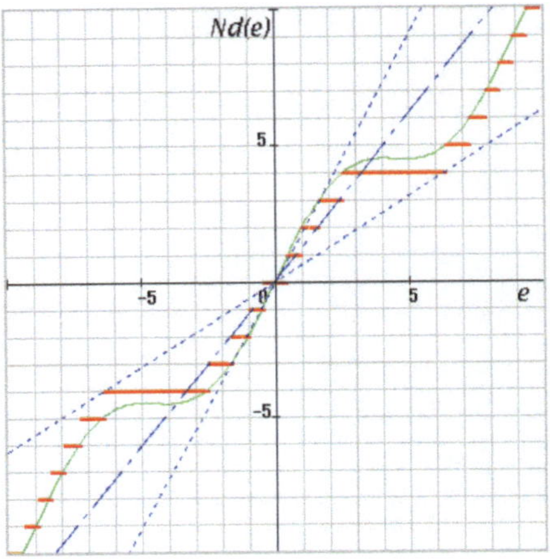

(b)

Fig. 2.6 Nonlinear characteristics and output-side discretizations

Hereafter, without loss of generality, it will be assumed that $\gamma = 1$. That is, the variables $e^\dagger, u^\dagger, \cdots$ are defined by integers as follows:

$$e^\dagger, u^\dagger \in \mathbb{Z}, \quad \mathbb{Z} := \{\cdots -3, -2, -1, 0, 1, 2, 3, \cdots\}. \tag{2.1}$$

On the other hand, the time variable t is given as $t \in \{0, h, 2h, 3h, \cdots\}$ for the sampling period h. When assuming $h = 1.0$, the following expression can be defined:

$$t \in \mathbb{Z}_+, \quad \mathbb{Z}_+ := \{0, 1, 2, 3, \cdots\}. \tag{2.2}$$

Therefore, each signal $e^\dagger(t), u^\dagger(t), \cdots$ traces on a grid pattern that is composed of integers in the time and (controller variables) space.

2.3 Discretization and Nonlinear Sector

2.3.1 Three Types of Discretization

Output-Side Discretization When a signal is discretized only on the output side of the nonlinear characteristic, the relationship between e and $v^\dagger$ becomes a stepwise nonlinear characteristic with step height 1, as shown in Figs. 2.6(a) and (b). Figure 2.6(a) is the output-side discretization for a saturation-type nonlinear characteristic (arctangent sigmoid function). On the other hand, Fig. 2.6(b) is the output-side discretization for a sinusoidal nonlinear characteristic. Without loss of generality, it is assumed that the nonlinear characteristics have origin symmetry and exist in the first and third quadrants.

Input-Side Discretization When a signal is discretized only on the input side of the nonlinear characteristic, the relationship between e and $v^\dagger$ becomes a stepwise

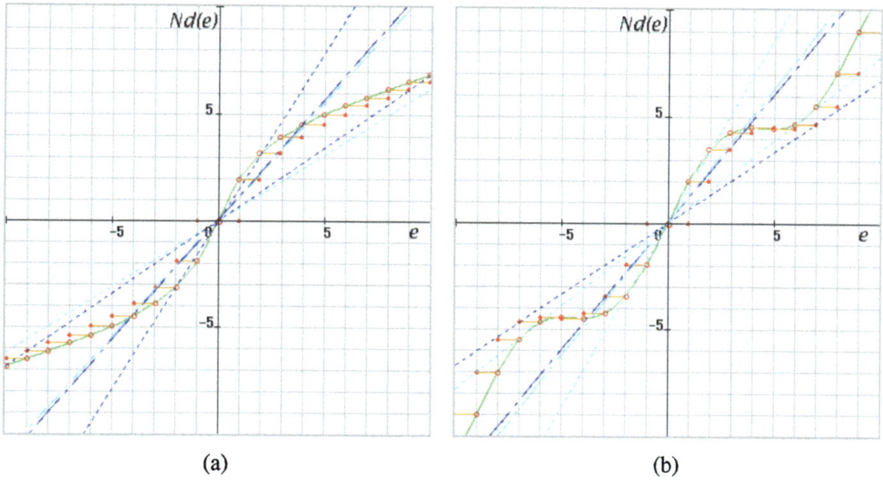

Fig. 2.7 Nonlinear characteristics and input-side discretizations

nonlinear characteristic with step width 1, as shown in Figs. 2.7(a) and (b). Figure 2.7(a) is the input-side discretization for a saturation-type nonlinear characteristic (arctangent sigmoid function), whereas Fig. 2.7(b) is the input-side discretization for a sinusoidal nonlinear characteristic.

Input and Output Sides Discretization When a signal is discretized on the input and output sides of the nonlinear characteristic, the relationship between e and $v^\dagger$ becomes a stepwise nonlinear characteristic with step hight and width 1 (i.e., broken line on integer coordinates) as shown in Figs. 2.8(a) and (b). Figure 2.8(a) is the input and output side discretization for a saturation-type nonlinear characteristic (arctangent sigmoid function). On the other hand, Fig. 2.8(b) is the input and output side discretization for a sinusoidal nonlinear characteristic.

2.3.2 *Nominal Gains and Sector Parameters*

In general, the discretized nonlinear characteristic

$$v^\dagger = N_d(e) = Ke + g(e), \qquad 0 < K < \infty, \tag{2.3}$$

can be partitioned into the following two sections:

$$|g(e)| \le \overline{g} < \infty, \tag{2.4}$$

for $|e| < \varepsilon$, and

$$|g(e)| \le \beta |e|, \qquad 0 \le \beta \le K, \tag{2.5}$$

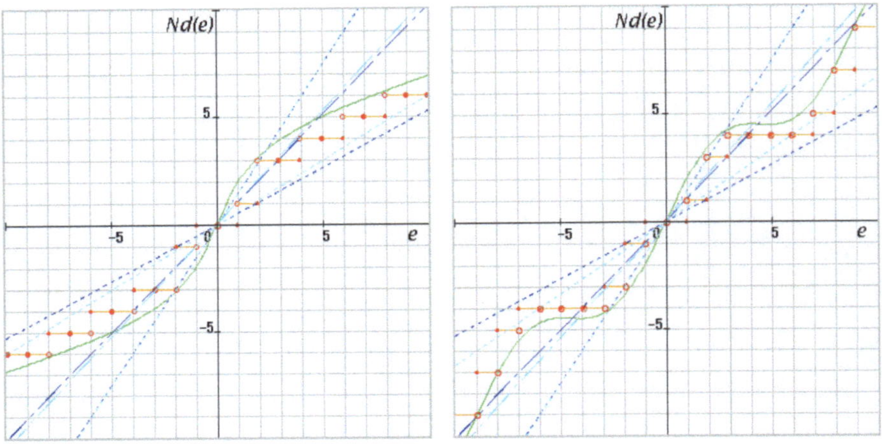

Fig. 2.8 Nonlinear characteristics and input and output side discretizations

for $|e| \geq \varepsilon$. When considering relative nonlinear characteristics, the partitioned expression is given as follows:

$$\begin{aligned} v^{\dagger} &= N_d(e) = K(e + n(e)), \\ |n(e)| &\leq \alpha |e|. \end{aligned} \tag{2.6}$$

Clearly, the sector parameter is considered to be $\beta = K\alpha$.

Equation (2.4) represents a bounded nonlinear characteristic that exists in a finite region. On the other hand, Eq. (2.5) represents a sectorial nonlinearity for which the equivalent linear gain exists in a limited range. It can also be expressed as follows:

$$0 \leq g(e)e \leq \beta e^2. \tag{2.7}$$

When dealing with the robust stability in a global sense, it is sufficient to consider the nonlinear term (2.5) for $|e| \geq \varepsilon$ because the nonlinear term (2.4) can be treated as a disturbance signal. (In the stability problem, a fluctuation or an offset of error is assumed to be allowable in $|e| < \varepsilon$.) Figures 2.9(a) and (b) show the discretization characteristics and the nonlinear parts $g(e)$ of two examples. In these examples, the thresholds are chosen as $\varepsilon = 2$.

In partitioning (2.3), nominal gain K and sectorial nonlinearity $g(e)$ can be chosen appropriately. For example, if K is chosen in integer numbers, $w^{\dagger} = g(e)$ also becomes an integer number. Figures 2.10(a) and (b) show examples of the partitioning of nonlinear characteristics when $K = 1$ (i.e., an integer number). In these cases, sector parameter β should be given by the larger value of $K_{\max} - K$ and $K - K_{\min}$. In general, the nominal gain and the sector parameter should be determined as follows:

$$K = \frac{K_{\max} + K_{\min}}{2} \tag{2.8}$$

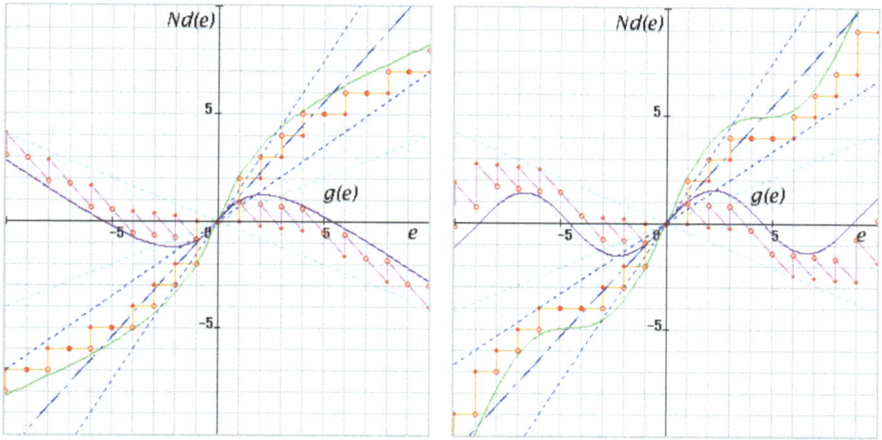

(a) (b)

Fig. 2.9 Nonlinear characteristics and discretized outputs

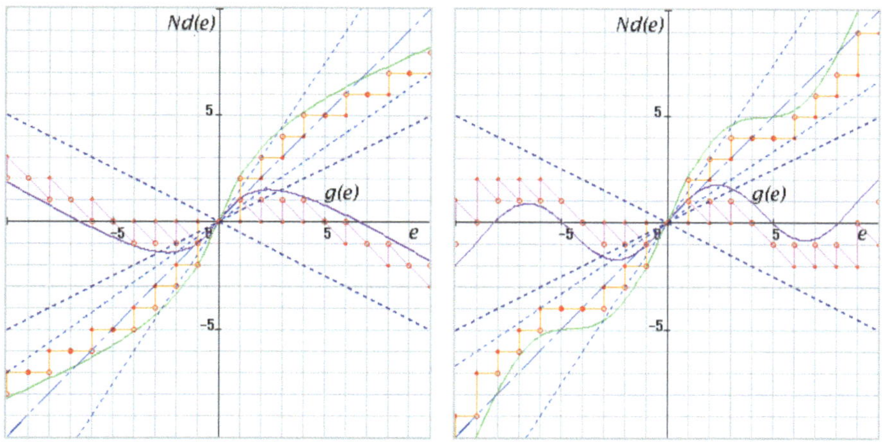

(a) (b)

Fig. 2.10 Nonlinear characteristics and discretized outputs

$$\beta = \frac{K_{\max} - K_{\min}}{2} = K_{\max} - K. \tag{2.9}$$

By partitioning nonlinear characteristic $N_d(\cdot)$, a single-loop control system can be drawn as shown in Fig. 2.11. It can also be drawn in regard to the discretized input as shown in Fig. 2.12. In the figure, discretized input $e^\dagger$ is assumed to be determined as follows:

$$e = e^\dagger, \quad \text{when } \ e^\dagger \le e < e^\dagger + \gamma.$$

In this case, the nominal gain and the sector parameter will be given as follows:

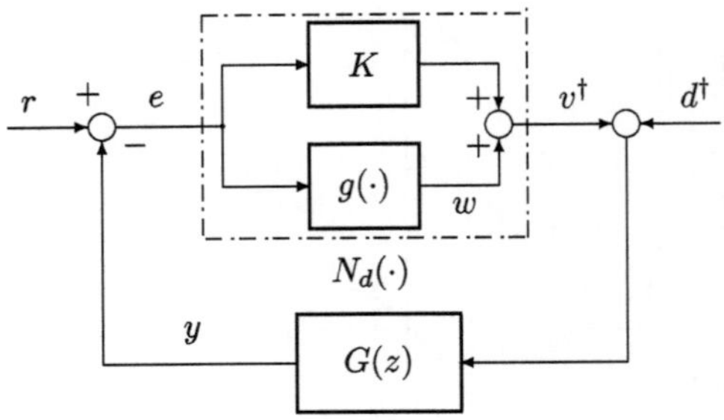

Fig. 2.11 Discretized nonlinear control system

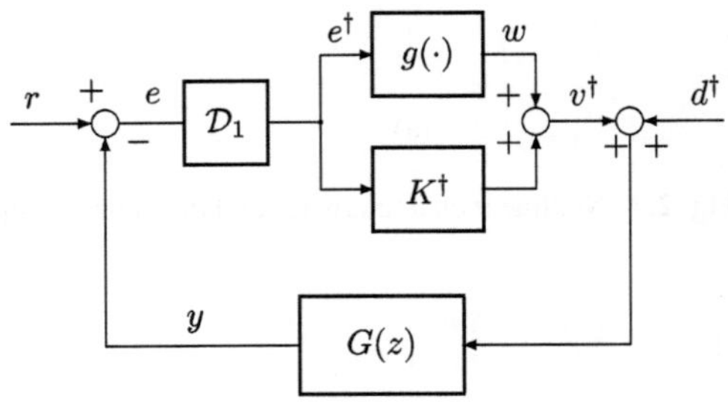

Fig. 2.12 Discrete-input nonlinear control system

$$K_n^\dagger = \frac{K_{\max}^\dagger + K_{\min}^\dagger}{2}, \tag{2.10}$$

$$\beta^\dagger = \frac{K_{\max}^\dagger - K_{\min}^\dagger}{2}, \tag{2.11}$$

where the following relations hold:[1]

$$K_{\max}^\dagger = K_{\max}, \quad K_{\min}^\dagger \le K_{\min}.$$

Figure 2.13(a) shows the difference between $K_{\min}$ and $K_{\min}^\dagger$ for the sinusoidal nonlinearity shown in Figs. 2.9(b) and 2.10(b). In this figure, nominal gains K and $K^\dagger$ are also drawn in chain lines. The control system shown in Fig. 2.12 can also be drawn as shown in Fig. 2.14. From Fig. 2.14, the relationship between $e(k)$ and $e^\dagger(k)$ and the equivalent exogenous input $\epsilon(k)$ are drawn in Fig. 2.13(b). As is clear from the figure, $\epsilon(k) = e^\dagger(k) - e(k)$ may be considered a bounded disturbance signal.

[1] β will be used instead of $\beta^\dagger$ in the following discussion.

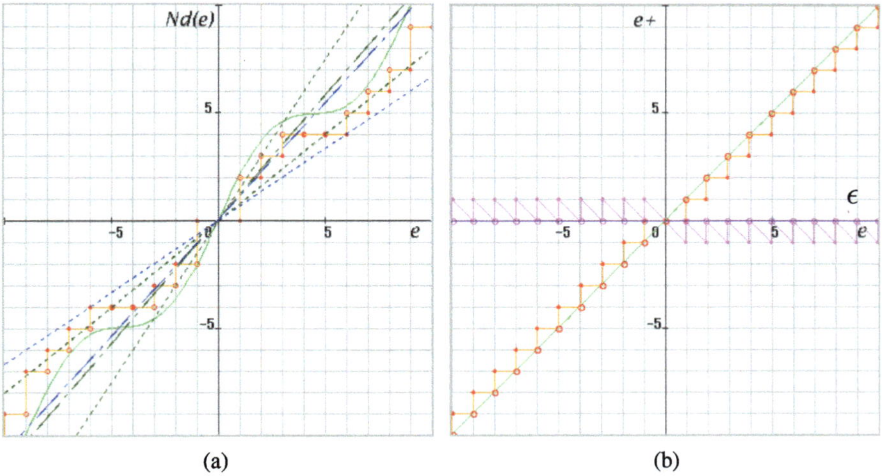

Fig. 2.13 Different sectors and equivalent input

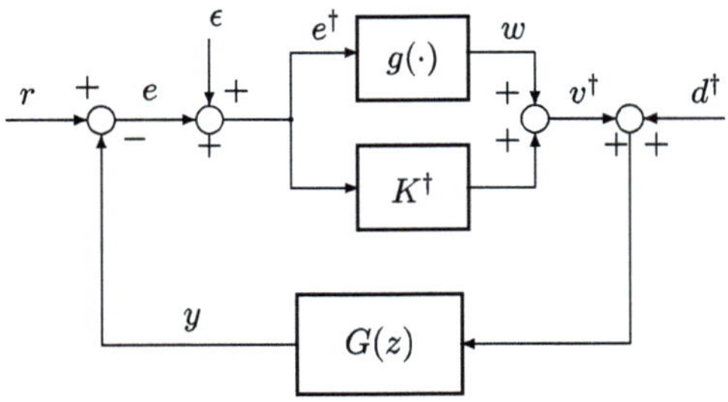

Fig. 2.14 Equivalent discrete-input control system

2.4 Equivalent Transformation

Based on the above considerations, the following new sequences $\overline{e}^*(k)$ and $\overline{w}^*(k)$ are defined:

$$\overline{e}^*(k) = \overline{e}(k) + q \cdot \frac{\Delta e(k)}{h}, \tag{2.12}$$

$$\overline{w}^*(k) = \overline{w}(k) - \beta q \cdot \frac{\Delta e(k)}{h}, \tag{2.13}$$

where q is a non-negative number, $\overline{e}(k)$ and $\overline{w}(k)$ are neutral points of sequences $e(k)$ and $w(k)$,

$$\overline{e}(k) = \frac{e(k) + e(k-1)}{2}, \tag{2.14}$$

$$\overline{w}(k) = \frac{w(k) + w(k-1)}{2}, \tag{2.15}$$

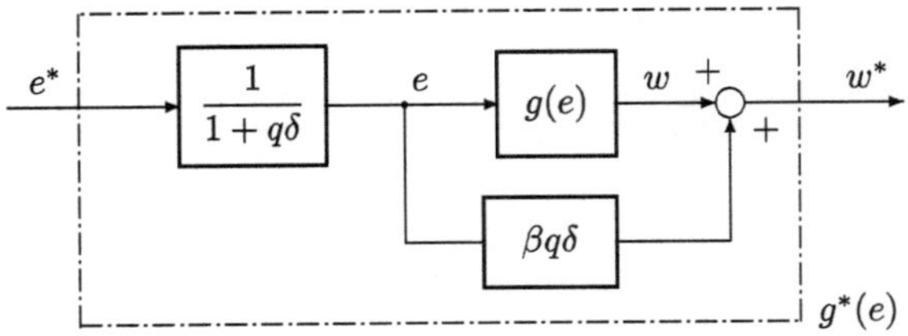

Fig. 2.15 Transformation of nonlinear element $g(e)$

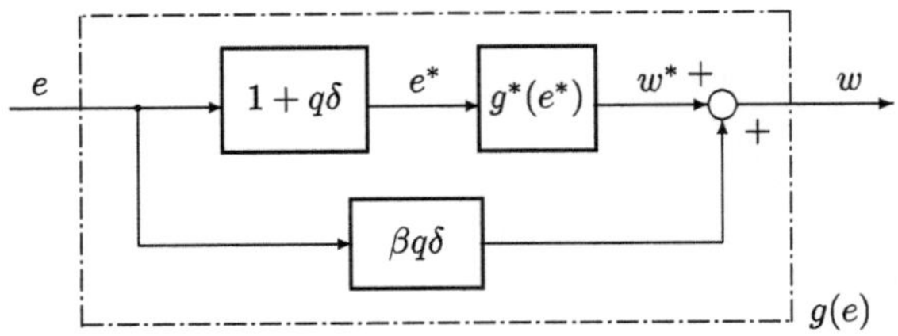

Fig. 2.16 Equivalent nonlinear subsystem

and $\Delta e(k)$ is the backward difference of sequence $e(k)$, that is,

$$\Delta e(k) := e(k) - e(k-1). \tag{2.16}$$

When using delay operator z^{-1}, Eqs. (2.14), (2.15), and (2.16) may be given as follows:

$$\overline{e}(k) = \frac{(1+z^{-1})}{2} e(k), \tag{2.17}$$

$$\overline{w}(k) = \frac{(1+z^{-1})}{2} w(k), \tag{2.18}$$

and then

$$\Delta e(k) = (1-z^{-1})\, e(k). \tag{2.19}$$

By using z-transform expressions, Eqs. (2.12) and (2.13) can be written as follows:

$$\frac{(1+z^{-1})}{2}\hat{e}^*(z) = \frac{(1+z^{-1})}{2}\hat{e}(z) + q\cdot\frac{(1-z^{-1})}{h}\hat{e}(z), \tag{2.20}$$

$$\frac{(1+z^{-1})}{2}\hat{w}^*(z) = \frac{(1+z^{-1})}{2}\hat{w}(z) - \beta q\cdot\frac{(1-z^{-1})}{h}\hat{e}(z). \tag{2.21}$$

The relationship between Eqs. (2.20) and (2.21) is shown by the block diagram in Fig. 2.15, and by the equivalent subsystem in Fig. 2.16. In these figures, operator δ is defined by a bilinear transformation as follows:

$$\delta(z) := \frac{2}{h}\cdot\frac{1-z^{-1}}{1+z^{-1}}. \tag{2.22}$$

Therefore, if the subsystem shown in Fig. 2.16 is used instead of $g(e)$, the whole control system is drawn as shown in Fig. 2.17. The control system represented by

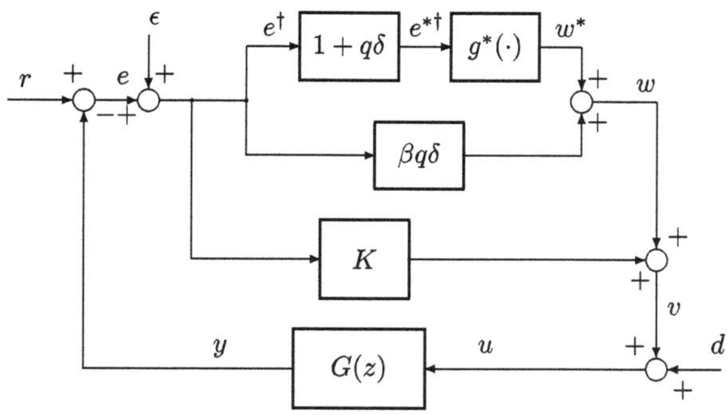

Fig. 2.17 Discrete nonlinear control system

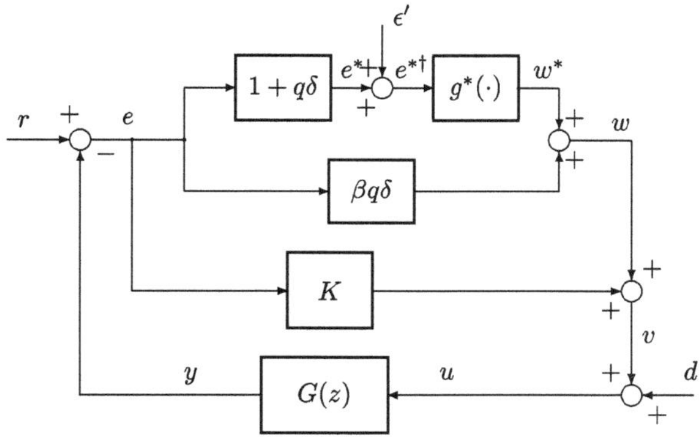

Fig. 2.18 Discrete nonlinear control system 2

Fig. 2.17 is equivalently transformed into Fig. 2.18. In this figure, since

$$\begin{cases} \hat{e}^{\dagger}(z) = \hat{e}(z) + \hat{\epsilon}(z) \\ \hat{e}^{*\dagger}(z) = (1+q\delta(z))\hat{e}^{\dagger}(z), \end{cases}$$

the equivalent exogenous input ϵ' can be given by

$$\hat{\epsilon}'(z) = (1+q\delta(z))\hat{\epsilon}(z). \tag{2.23}$$

From these figures,

$$\hat{e}(z) = \hat{r}(z) - G(z)[\hat{w}^*(z) + (K+\beta q\delta(z))\hat{e}(z) + \hat{d}'(z)].$$

Furthermore,

$$[1+(K+\beta q\delta(z)]G(z)\hat{e}(z) = G(z)\hat{w}^*(z) + \hat{r}(z) + G(z)\hat{d}'(z).$$

Thus,

$$\hat{e}(z) = \frac{G(z)\hat{w}^*(z) + \hat{r}(z) + G(z)\hat{d}'(z)}{1+(K+\beta q\delta(z))G(z)}.$$

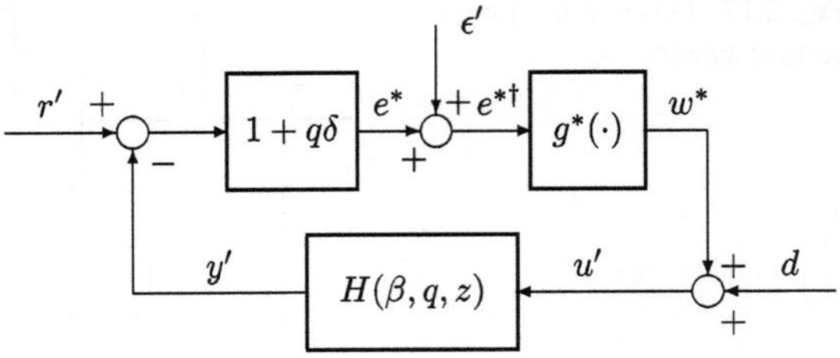

Fig. 2.19 Equivalent nonlinear control system

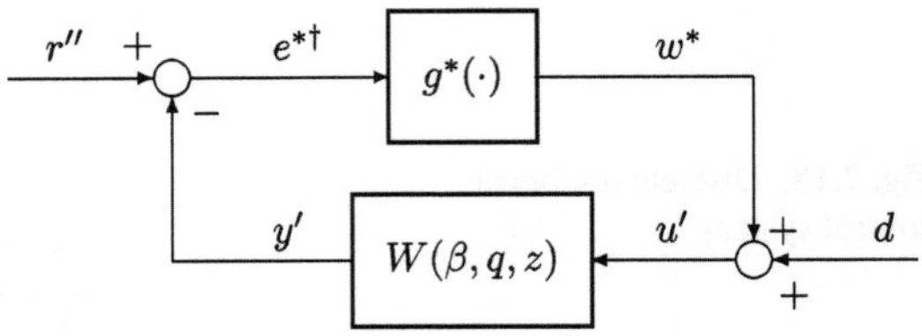

Fig. 2.20 Small-gain feedback system

From these equations, we can obtain

$$\hat{e}^*(z) = \frac{(1+q\delta(z))G(z)}{1+(K+\beta q\delta(z))G(z)}\hat{w}^*(z) + \frac{(1+q\delta(z))}{1+(K+\beta q\delta(z))G(z)}\hat{r}(z) + \frac{(1+q\delta(z))G(z)}{1+(K+\beta q\delta(z))G(z)}\hat{d}(z). \tag{2.24}$$

Therefore,

$$e^{*\dagger}(z) = \frac{(1+q\delta(z))G(z)}{1+(K+\beta q\delta(z))G(z)}\hat{w}^*(z) + \hat{\epsilon}'(z) + \frac{1+q\delta(z)}{1+(K+\beta q\delta(z))G(z)}\hat{r}(z) + \frac{(1+q\delta(z))G(z)}{1+(K+\beta q\delta(z))G(z)}\hat{d}(z) \tag{2.25}$$

Consequently, we have the block diagram of the discretized control system shown in Fig. 2.19, where

$$H(\beta,q,z) = \frac{G(z)}{1+(K+\beta q\delta(z))G(z)}. \tag{2.26}$$

Thus, the loop transfer function from w^* to e^* can be given by

$$W(\beta,q,z) = \frac{(1+q\delta(z))G(z)}{1+(K+\beta q\delta(z))G(z)}, \tag{2.27}$$

as shown in Fig. 2.20. Here, r' is given by

$$\hat{r}'(z) = \frac{1}{1+(K+\beta q\delta(z))G(z)}\hat{r}(z).$$

Furthermore, the reference input r'' in Fig. 2.20 is equivalently expressed as

$$\hat{r}''(z) = (1 + q\delta(z))\hat{r}'(z) + \hat{\epsilon}(z).$$

2.5 Norm Inequalities

We now provide an assumption with respect to the behavior of control systems.

Assumption The absolute value of the backward difference of sequence $e(k)$ does not exceed γ, i.e.,

$$|\Delta e(k)| = |e(k) - e(k-1)| \leq \gamma. \tag{2.28}$$

If condition (2.28) is satisfied, $\Delta e(k)$ becomes exactly $\pm\gamma$ or 0 because of the discretization $\mathcal{D}_1$. That is, the absolute value of the backward difference can be given as

$$|\Delta e(k)| = |e(k) - e(k-1)| = \gamma \ \text{ or } \ 0.$$

The assumption stated above will be satisfied in some examples given in the following chapters. These examples will include figures illustrating the phase trace of the backward difference Δe.

In this subsection, some lemmas with respect to an ℓ_2 norm of the sequences are presented. Here, we define a new nonlinear function

$$f(e) := g(e) + \beta\, e. \tag{2.29}$$

When considering the discretized output of the nonlinear characteristic, $w^\dagger = v^\dagger - Ke^\dagger$, the following expression can be given:

$$f(e^\dagger(k)) = w^\dagger(k) + \beta e^\dagger(k). \tag{2.30}$$

In expression (2.30), we note that $w^\dagger \notin \mathbb{Z}$ in general. From inequality (2.5), it can be seen that the function (2.30) belongs to the first and third quadrants. Figures 2.21(a) and (b) show the discretized outputs $v^\dagger = N_d(e)$ and $f(e)$ for the examples given in Fig. 2.9(a) and (b) when the discretized error ϵ was not considered and the point-to-point transition was executed.

Considering the equivalent linear characteristic, the following inequality can be defined:

$$0 \leq \psi(k) := \frac{f(e(k))}{e(k)} \leq 2\beta. \tag{2.31}$$

When this type of nonlinearity $\psi(k)$ is used, inequality (2.5) can be written as

$$w^\dagger(k) = g(e(k)) = (\psi(k) - \beta)e(k). \tag{2.32}$$

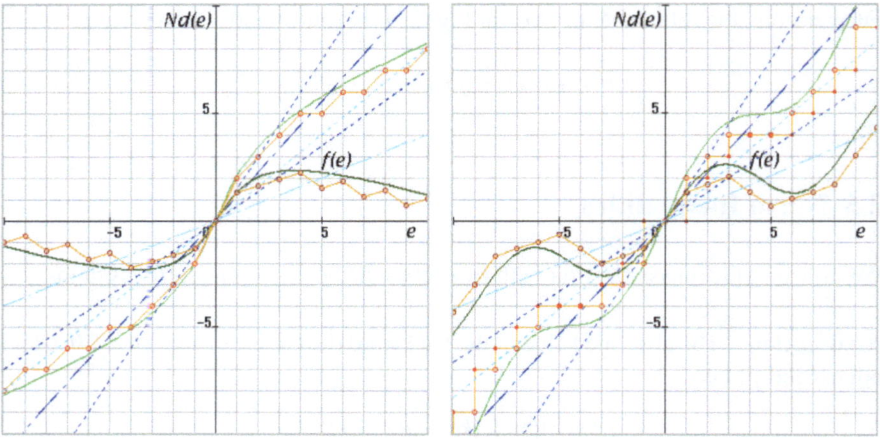

(a) (b)

Fig. 2.21 Nonlinear characteristics and discretized outputs

For the neutral points of $e(k)$ and $w^{\dagger}(k)$, the following expression is given from (2.30):

$$\frac{1}{2}(f(e(k)) + f(e(k-1))) = \overline{w}^{\dagger}(k) + \beta\overline{e}^{\dagger}(k). \tag{2.33}$$

Moreover, Eq. (2.32) is rewritten as $\overline{w}^{\dagger}(k) = (\psi(k) - \beta)\overline{e}^{\dagger}(k)$. Since $|\overline{e}^{\dagger}(k)| \le |\overline{e}(k)|$, the following inequality is satisfied when a round-down discretization is executed:

$$|\overline{w}^{\dagger}(k)| \le \beta|\overline{e}^{\dagger}(k)| \le \beta|\overline{e}(k)|. \tag{2.34}$$

Based on this premise, the following norm conditions are examined.

Lemma 2.1 *The following inequality holds for a positive integer* p:

$$\|\overline{w}^{\dagger}(k)\|_{2,N} \le \beta\|\overline{e}^{\dagger}(k)\|_{2,N} \le \beta\|\overline{e}(k)\|_{2,N}. \tag{2.35}$$

Here, $\|\cdot\|_{2,N}$ *denotes an* ℓ_2 *norm, which can be defined by*

$$\|x(k)\|_{2,N} := \left(\sum_{k=1}^{N}|x(k)|^2\right)^{1/2}.$$

Proof The proof is clear from inequality (2.34). □

Lemma 2.2 *If the following inequality is satisfied with respect to the inner product of the neutral points of* (2.30) *and the backward difference*:

$$\left\langle\overline{w}^{\dagger}(k) + \beta\overline{e}^{\dagger}(k), \Delta e(k)\right\rangle_N \ge 0, \tag{2.36}$$

we can obtain the inequality

$$\|\overline{w}^{*\dagger}(k)\|_{2,N} \le \beta\|\overline{e}^{*\dagger}(k)\|_{2,N} \tag{2.37}$$

for any $q \ge 0$. *Here,* $\langle\cdot,\cdot\rangle_N$ *denotes the inner product, which is defined as*

$$\langle x_1(k), x_2(k)\rangle_N = \sum_{k=1}^{N} x_1(k)x_2(k).$$

Proof The following equation is obtained from (2.12) and (2.13):

$$\begin{aligned}\beta^2\|\overline{e}^{*\dagger}(k)\|_{2,N}^2 &- \|\overline{w}^{*\dagger}(k)\|_{2,N}^2 \\ &= \beta^2\|\overline{e}^{\dagger}(k)\|_{2,N}^2 - \|\overline{w}^{\dagger}(k)\|_{2,N}^2 + \frac{2\beta q}{h}\cdot\langle\overline{w}^{\dagger}(k) + \beta\overline{e}^{\dagger}(k), \Delta e(k)\rangle_N. \end{aligned} \tag{2.38}$$

Thus, inequality (2.37) is satisfied by using the left-side inequality of (2.35). Moreover, as for the input of $g^*(\cdot)$, the following inequality can be obtained from (2.38) and the right-side inequality of (2.35):

$$\|\overline{w}^{*\dagger}(k)\|_{2,N} \le \beta\|\overline{e}^{*}(k)\|_{2,N}. \tag{2.39}$$

□

2.6 Sum of Trapezoidal Areas

The left side of inequality (2.36) can be expressed by a sum of trapezoidal areas.

Lemma 2.3 *For any step* N, *the following equation is satisfied:*

$$\sigma(N) := \langle\,\overline{w}^{\dagger}(k) + \beta\overline{e}^{\dagger}(k), \Delta e(k)\rangle_N = \frac{1}{2}\sum_{k=1}^{N}(f(e(k)) + f(e(k-1)))\Delta e(k). \tag{2.40}$$

Proof The proof is clear from (2.33). □

In general, the sum of trapezoidal areas has the following property.

Lemma 2.4 *If inequality* (2.28) *is satisfied with respect to the discretization of the control system, the sum of trapezoidal areas becomes non-negative for any* N, *that is,*

$$\sigma(N) \ge 0. \tag{2.41}$$

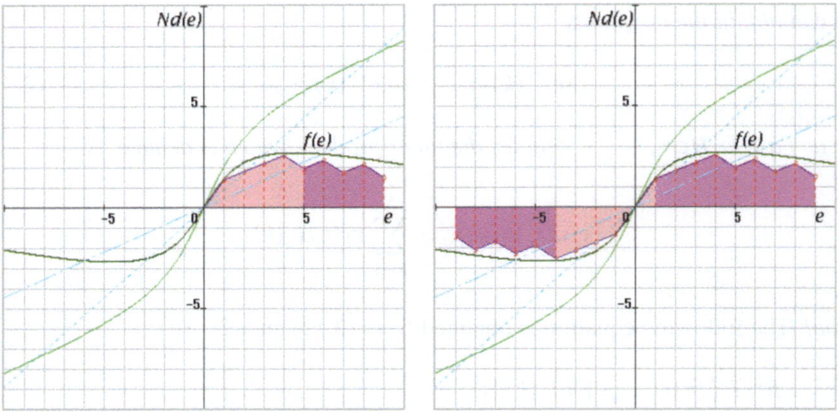

(a) (b)

Fig. 2.22 Non-negative characteristics of trapezoidal summation

Proof Since $f(e(k))$ belongs to the first and third quadrants, the area of each trapezoid

$$\tau(k) := \frac{1}{2}(f(e(k)) + f(e(k-1)))\Delta e(k) \tag{2.42}$$

is non-negative when $e(k)$ increases (decreases) in the first (third) quadrant. On the other hand, the trapezoidal area $\tau(k)$ is non-positive when $e(k)$ decreases (increases) in the first (third) quadrant.

Strictly speaking, when ($e(k) \geq 0$ and $\Delta e(k) \geq 0$) or ($e(k) \leq 0$ and $\Delta e(k) \leq 0$), $\tau(k)$ is non-negative for any k. On the other hand, when ($e(k) \geq 0$ and $\Delta e(k) \leq 0$) or ($e(k) \leq 0$ and $\Delta e(k) \geq 0$), $\tau(k)$ is non-positive for any k. Here, $\Delta e(k) \geq 0$ corresponds to $\Delta e(k) = \gamma$ or 0 (and $\Delta e(k) \leq 0$ corresponds to $\Delta e(k) = -\gamma$ or 0) for the discretized signal, when inequality (2.28) is satisfied. The sum of trapezoidal areas is given from (2.40) as:

$$\sigma(N) = \sum_{k=1}^{N} \tau(k). \tag{2.43}$$

We thus derive the following result. The sum of trapezoidal areas becomes non-negative, $\sigma(N) \geq 0$, regardless of whether $e(k)$ (and $e(k)$) increases or decreases. Since the discretized output traces the same points on the stepwise nonlinear characteristic, the sum of trapezoidal areas is canceled when $e(k)$ (and $e(k)$ decreases (increases) from a certain point $(e(k), f(e(k)))$ in the first (third) quadrant. (Here, without loss of generality, the response of discretized point $(e(k), f(e(k)))$ is assumed to commence at the origin.) Thus, the proof is concluded. □

Figures 2.22(a) and (b) show the sum of trapezoidal areas for $f(e)$ given in Fig. 2.21(a), when e is a sinusoidal input with amplitude 8.0, i.e., $e(k) = 8.0 \sin \omega k$ (ω: an arbitrary number).

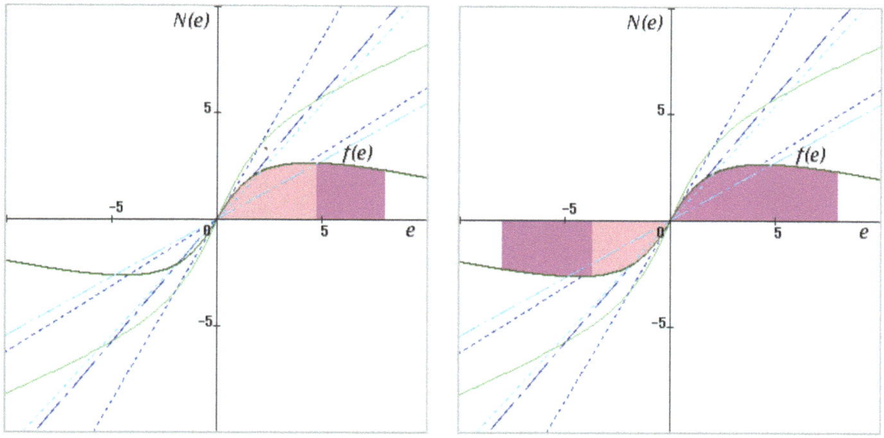

(a) (b)

Fig. 2.23 Non-negative characteristics of integrals

(a) The sinusoid starts from 0 to 8.0. Then, e decreases to $e < 5.0$.
(b) The sinusoid starts from 0, passes 8.0, 0.0, -8.0, and increases to $e > -5.0$.

In any case, the sum of trapezoids will be canceled.

On the other hand, Figs. 2.23(a) and (b) show the sum of trapezoidal areas for $f(e)$ when the sampling period h is very small (i.e., $\Delta e(k) \to 0$), in other words, the integration of $f(e)$,

$$\sigma(N) = \int_{e(0)}^{e(N)} f(e)de.$$

The latter case corresponds to the Popov stability problem for continuous control systems.[2]

For an easier understanding, examples of the sequences of continuous/discretized signals and the sum of trapezoidal areas are depicted in Figs. 2.24(a), (b) and 2.25(a), (b).

Example 2.1 The input/output characteristic shown in Fig. 2.24(a) is written as:

$$\begin{aligned} e^{\dagger} &= \gamma * (\texttt{double})(\texttt{int})(e/\gamma) \\ v &= 0.3 * e^{\dagger} + 2.7 * \texttt{atan}(0.7 * e^{\dagger}) \\ v^{\dagger} &= \gamma * (\texttt{double})(\texttt{int})(v/\gamma) \end{aligned} \tag{2.44}$$

by using a C-language expression. Here, (int) and (double) denote the conversion into an integral number (a round-down discretization) and the reconversion into a

[2]The relation to the Popov criterion will be described in Chap. 3, Sect. 3.6.

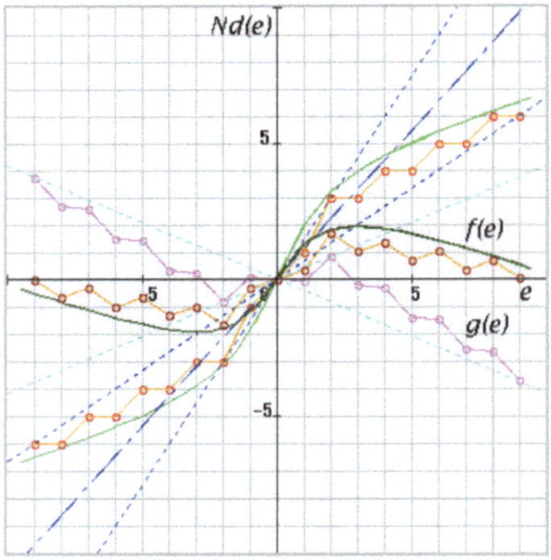

(a)

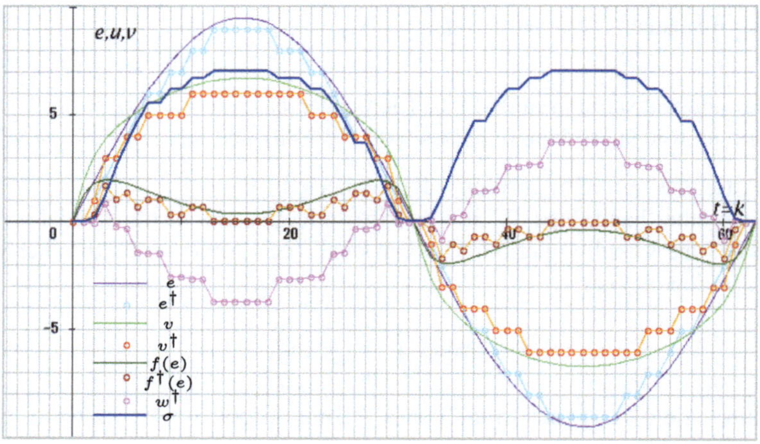

(b)

Fig. 2.24 Discretized input/output signals of a nonlinear element

double-precision real number, respectively. The second equation of (2.44) corresponds to a sigmoid (saturated) function (needless to say, $\mathtt{atan}(\cdot) = \tan^{-1}(\cdot)$).

In Fig. 2.24(b), the curve e and the sequence of circles $e^\dagger$ show the input of the nonlinear element and its discretized signal. The curve v and the sequence of circles $v^\dagger$ show the corresponding output of the nonlinear characteristic and its discretized signal, respectively. As shown in the figure, the sequences of circles $e^\dagger$ and $v^\dagger$ trace a grid pattern that is composed of integers. The sequence of circles $w^\dagger$ shows the discretized output of the nonlinear characteristic $g(\cdot)$. The curve of shifted nonlinear characteristic $f(e)$ and the sequence of circles $f^\dagger(e)$ are also shown in the figures.

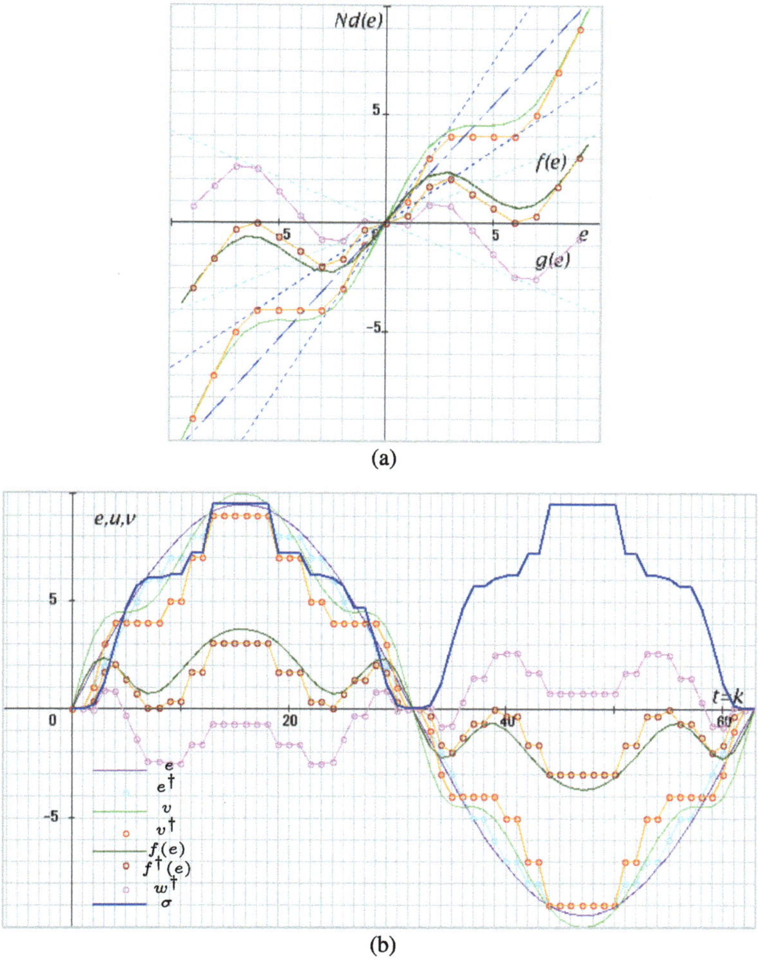

Fig. 2.25 Discretized input/output signals of a nonlinear element for Example 2.2

Example 2.2 For the example of Figs. 2.25(a) and (b), the following nonlinear characteristic is considered:

$$\begin{aligned} e^{\dagger} &= \gamma * (\texttt{double})(\texttt{int})(e/\gamma) \\ v &= 1.0 * e^{\dagger} + 1.5 * \texttt{sin}(0.7 * e^{\dagger}) \\ v^{\dagger} &= \gamma * (\texttt{double})(\texttt{int})(v/\gamma). \end{aligned} \tag{2.45}$$

The second equation of (2.45) is an inclined sinusoidal function. In either of the examples, (2.41) in Lemma 2.4 is satisfied, i.e., $\sigma(k) \geq 0$ $(k = 1, 2, \cdots)$.

Figures 2.26(a), (b) and 2.27(a), (b) show the two cases in Examples 2.1 and 2.2 with a nearly continuous characteristic, where $\gamma = 0.1$ and $h = 0.1$, that is, $1/10$ *high resolution*. As is obvious from the figures, $\sigma(t) \geq 0$ $(t = 0.1, 0.2, \cdots)$. Thus,

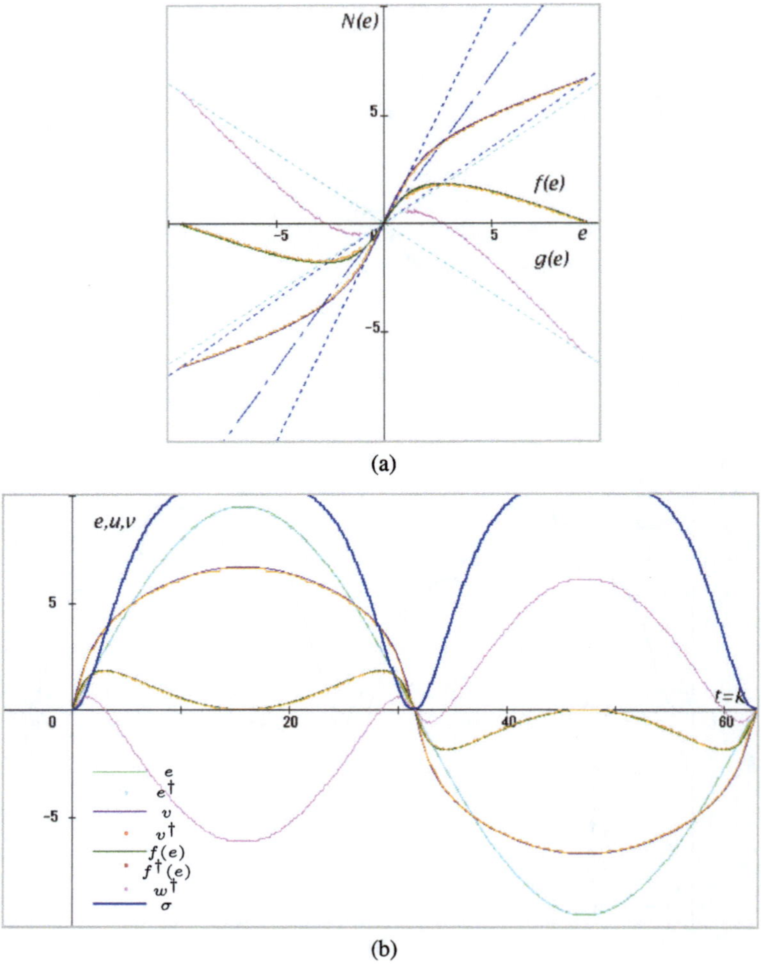

Fig. 2.26 Input/output signals of a nearly continuous characteristic for Example 2.1

the calculated results show that the input/output characteristics of nonlinear elements become similar to continuous problems, that is, Popov's criterion and other conditions in continuous time.

Example 2.3 Figures 2.28(a) and (b) illustrate the case where the following nearly nonlinear characteristic is considered:

$$\begin{aligned} e^{\dagger} &= \gamma * \texttt{(double)(int)}(e/\gamma) \\ v &= 1.0 * e^{\dagger} + 0.15 * \texttt{sin}(0.7 * e^{\dagger}) \\ v^{\dagger} &= \gamma * \texttt{(double)(int)}(v/\gamma). \end{aligned} \tag{2.46}$$

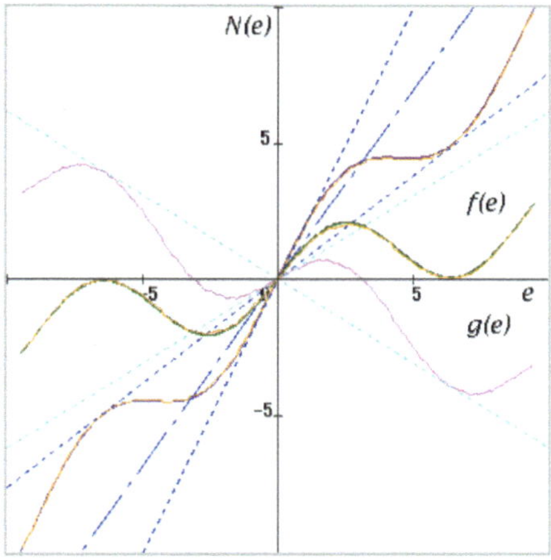

(a)

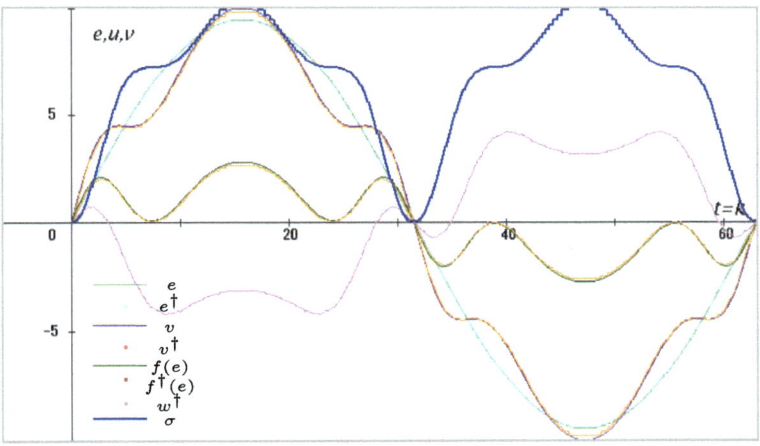

(b)

Fig. 2.27 Input/output signals of a nearly continuous characteristic for Example 2.2

In this example, the amplitude of sinusoidal function is chosen as $1/10$ for Example 2.2 (i.e., $1.5 \to 0.15$). Figures 2.29(a) and (b) illustrate the case where $1/10$ high resolution is applied (i.e., $\gamma = 0.1$ and $h = 0.1$). As is obvious from these examples, the theory of discretized static and/or dynamic nonlinear systems (in other words, discrete-time and discrete-value nonlinear systems) considered in this chapter approaches that of continuous-time and continuous-value nonlinear systems asymptotically for $\gamma \to 0$ and $h \to 0$. Of course, it includes that of continuous linear systems for $\beta \to 0$ naturally, as shown in Figs. 2.29(a) and (b).

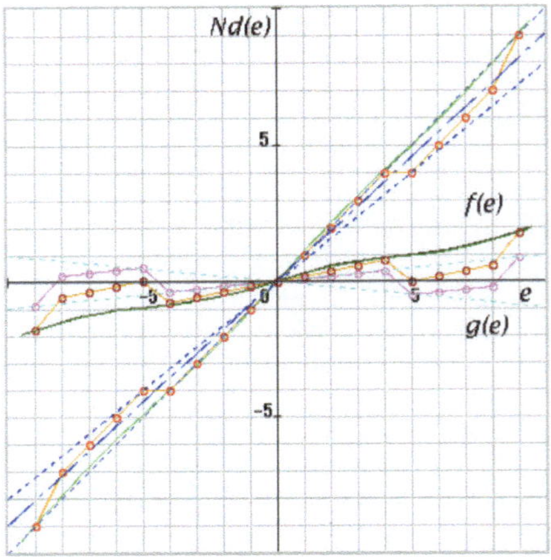

(a)

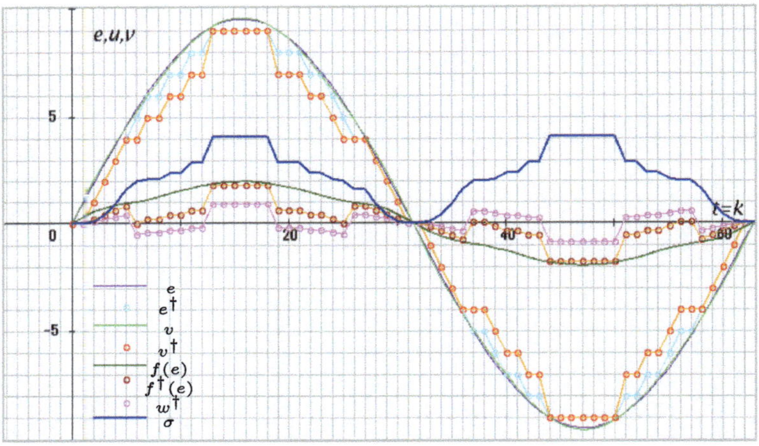

(b)

Fig. 2.28 Discretized input/output signals of a nearly continuous characteristic for Example 2.3

2.7 Exercises

(1) Prove that the sector condition in (2.5),

$$|g(e)| \leq \beta |e|,$$

is equivalently written as (2.7).

(2) Confirm that block diagram Fig. 2.16 is equivalent to Fig. 2.15.

(3) From Fig. 2.18, determine the loop transfer function $H(\beta, q, z)$ in Fig. 2.19.

(4) From (2.5) and (2.29), prove the sector inequality in (2.31), that is,

$$0 \leq \frac{f(e(k))}{e(k)} \leq 2\beta.$$

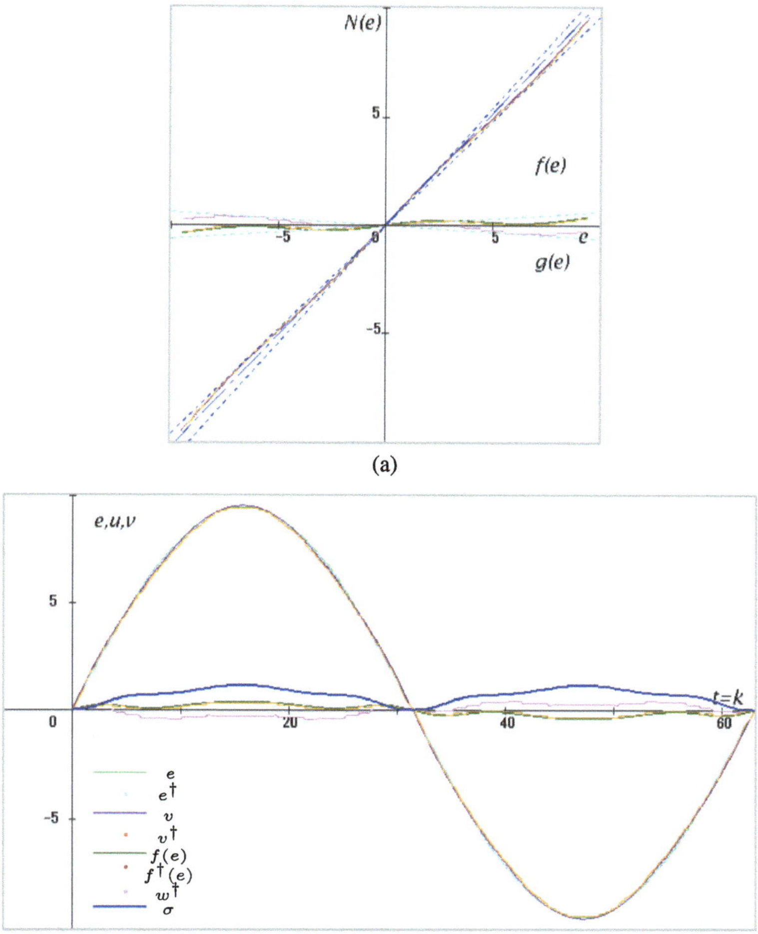

Fig. 2.29 Discretized input/output signals of a nearly continuous characteristic for Example 2.3

(5) Prove Lemma 2.1, that is,

$$\|\overline{w}^{\dagger}(k)\|_{2,N} \leq \beta\|\overline{e}^{\dagger}(k)\|_{2,N} \leq \beta\|\overline{e}(k)\|_{2,N},$$

using inequality (2.34).

(6) For $N = 2$, prove Schwarz's inequality (2.60).

(7) Using the result of (6), prove Minkowski's inequality (2.68) when $N = 2$.

Appendix A: Norms and Inner Products of L_p and ℓ_p Spaces

In this appendix, inner products and norms in an ℓ_2 space are explained for discrete-time systems. In general, norms of L_p and ℓ_p spaces are defined as follows. For a continuous-time signal $x\colon \mathbb{R}_+ \to \mathbb{R}$,

$$\|x(t)\|_p := \left(\int_0^{\infty} |x(t)|^p dt\right)^{1/p}, \quad 1 \le p < \infty, \tag{2.47}$$

$$\|x(t)\|_\infty := \operatorname{ess} \sup_{t\in[0,\infty)} |x(t)|, \tag{2.48}$$

and for a discrete-time signal $x\colon \mathbb{Z}_+ \to \mathbb{R}$ (or $\mathbb{Z}$),

$$\|x(k)\|_p := \left(\sum_{k=1}^{\infty} |x(k)|^p\right)^{1/p}, \quad 1 \le p < \infty, \tag{2.49}$$

$$\|x(k)\|_\infty := \sup_{k\ge 1} |x(k)|. \tag{2.50}$$

In the ℓ_2 space, the norm is defined as

$$\|x(k)\|_2 := \left(\sum_{k=1}^{\infty} |x(k)|^2\right)^{1/2}, \tag{2.51}$$

and the inner product is given by

$$\langle x(k), y(k)\rangle := \sum_{k=1}^{\infty} x(k)y(k). \tag{2.52}$$

The preceding definitions for finite time series $x(k)(k = 0, 1, 2, \cdots, N)$ are written as follows:

$$\|x(k)\|_{2,N} := \left(\sum_{k=1}^{N} |x(k)|^2\right)^{1/2}, \tag{2.53}$$

$$\langle x(k), y(k)\rangle_N := \sum_{k=1}^{N} x(k)y(k). \tag{2.54}$$

When $N \to \infty$, these definitions are written as:

$$\|x(k)\|_2 := \lim_{N\to\infty} \|x(k)\|_{2,N}, \tag{2.55}$$

$$\langle x(k), y(k)\rangle := \lim_{N\to\infty} \langle x(k), y(k)\rangle_N. \tag{2.56}$$

Appendix B: Hölder and Schwarz Inequalities

(1) In an L_p space, the following inequality holds:

$$\int_0^\infty |x(t)y(t)|\mathrm{d}t \le \left(\int_0^\infty |x(t)|^p \mathrm{d}t\right)^{1/p} \left(\int_0^\infty |y(t)|^q \mathrm{d}t\right)^{1/q}, \quad \frac{1}{p}+\frac{1}{q}=1. \tag{2.57}$$

As for discrete signals, the following inequality holds in an ℓ_p space:

$$\sum_{k=1}^\infty |x(k)y(k)| \le \left(\sum_{k=1}^\infty |x(k)|^p\right)^{1/p} \left(\sum_{k=1}^\infty |y(k)|^q\right)^{1/q}, \quad \frac{1}{p}+\frac{1}{q}=1. \tag{2.58}$$

These are called Hölder's inequalities [4, 8, 10]. The proof is given for $1 \le p \le \infty$ (i.e., $1 \le q \le \infty$) in, e.g., [8].

(2) An important special case of (2.58) for $p = q = 2$ is given as

$$\sum_{k=1}^\infty |x(k)y(k)| \le \left(\sum_{k=1}^\infty |x(k)|^2\right)^{1/2} \left(\sum_{k=1}^\infty |y(k)|^2\right)^{1/2}. \tag{2.59}$$

Equation (2.59) is called Schwarz's inequality. The easier proof of (2.59) is as follows. For finite sums of N steps, Schwarz's inequality (2.59) is rewritten as

$$\left(\sum_{k=1}^N |x(k)y(k)|\right)^2 \le \left(\sum_{k=1}^N |x(k)|^2\right) \left(\sum_{k=1}^N |y(k)|^2\right) \tag{2.60}$$

$$\begin{aligned} &\sum_{k=1}^N |x(k)y(k)|^2 + 2\sum_{k,l=1,k\ne l}^N |x(k)y(k)|\cdot|x(l)y(l)| \\ &\le \sum_{k=1}^N |x(k)|^2|y(k)|^2 + \sum_{k,l=1,k\ne l}^N |x(k)|^2|y(l)|^2. \end{aligned} \tag{2.61}$$

In (2.61), the following sum must be non-negative:

$$\sum_{k,l=1,k\ne l}^N |x(k)|^2|y(l)|^2 - 2\sum_{k,l=1,k\ne l}^N |x(k)y(k)|\cdot|x(l)y(l)| = \left(\sum_{k,l=1,k\ne l}^N |x(k)y(l)|\right)^2, \tag{2.62}$$

and it must hold for $N \to \infty$. Thus, inequality (2.60) has been proved.

Appendix C: Minkowski Inequalities

(1) In an L_p space, the following inequality holds:

$$\left(\int_0^\infty |x(t)+y(t)|^p \mathrm{d}t\right)^{1/p} \le \left(\int_0^\infty |x(t)|^p \mathrm{d}t\right)^{1/p} + \left(\int_0^\infty |y(t)|^p \mathrm{d}t\right)^{1/p}. \tag{2.63}$$

The norm expression based on (2.47) is given by

$$\|x(t)+y(t)\|_p \le \|x(t)\|_p + \|y(t)\|_p. \tag{2.64}$$

As for discrete signals, the following inequality holds in an ℓ_p space:

$$\left(\sum_{k=1}^{\infty} |x(k)+y(k)|^p\right)^{1/p} \le \left(\sum_{k=1}^{\infty} |x(k)|^p\right)^{1/p} + \left(\sum_{k=1}^{\infty} |y(k)|^p\right)^{1/p}. \tag{2.65}$$

The norm expression based on (2.49) is given by

$$\|x(k)+y(k)\|_p \le \|x(k)\|_p + \|y(k)\|_p. \tag{2.66}$$

These are called Minkowski's inequalities [7, 8].

(2) A special case of (2.65) for $p=2$ and finite sums of N steps is written as

$$\left(\sum_{k=1}^{N} |x(k)+y(k)|^2\right)^{1/2} \le \left(\sum_{k=1}^{N} |x(k)|^2\right)^{1/2} + \left(\sum_{k=1}^{N} |y(k)|^2\right)^{1/2} \tag{2.67}$$

$$\|x(k)+y(k)\|_{2,N} \le \|x(k)\|_{2,N} + \|y(k)\|_{2,N}. \tag{2.68}$$

In order to prove (2.65), consider the following equality:

$$(|x(k)|+|y(k)|)^p = |x(k)|(|x(k)|+|y(k)|)^{p-1} + |y(k)|(|x(k)|+|y(k)|)^{p-1}.$$

Hölder's inequality gives

$$\sum_{k=1}^{N} |x(k)|(|x(k)|+|y(k)|)^{p-1} \le \left(\sum_{k=1}^{N} |x(k)|^p\right)^{1/p} \left(\sum_{k=1}^{N} (|x(k)|+|y(k)|)^{(p-1)q}\right)^{1/q} \tag{2.69}$$

$$\sum_{k=1}^{N} |y(k)|(|x(k)|+|y(k)|)^{p-1} \le \left(\sum_{k=1}^{N} |y(k)|^p\right)^{1/p} \left(\sum_{k=1}^{N} (|x(k)|+|y(k)|)^{(p-1)q}\right)^{1/q}. \tag{2.70}$$

Since $(p-1)q = p$ and $1/q = 1 - 1/p$, the addition of (2.70) and (2.69) gives

$$\sum_{k=1}^{N}(|x(k)|+|y(k)|)^p \le \left(\sum_{k=1}^{N}(|x(k)|+|y(k)|)^p\right)^{(1-1/p)} \cdot \left[\left(\sum_{k=1}^{N}|x(k)|^p\right)^{1/p} + \left(\sum_{k=1}^{N}|y(k)|^p\right)^{1/p}\right]. \quad (2.71)$$

Moreover, $|x(k)+y(k)| \le |x(k)|+|y(k)|$. Thus, Minkowski's inequality (2.65) is obtained for $N \to \infty$.

To provide a clear understanding, the following simple equality is considered here:

$$(|x(k)|+|y(k)|)^2 = |x(k)|(|x(k)|+|y(k)|) + |y(k)|(|x(k)|+|y(k)|).$$

Schwarz's inequality gives

$$|x(1)|(|x(1)|+|y(1)|) + |x(2)|(|x(2)|+|y(2)|)$$
$$\le (|x(1)|^2+|x(2)|^2)^{1/2}[(|x(1)|+|y(1)|)^2 + (|x(2)|+|y(2)|)^2]^{1/2}$$

$$|y(1)|(|x(1)|+|y(1)|) + |y(2)|(|x(2)|+|y(2)|)$$
$$\le (|y(1)|^2+|y(2)|^2)^{1/2}[(|x(1)|+|y(1)|)^2 + (|x(2)|+|y(2)|)^2]^{1/2}.$$

By adding these inequalities, we have

$$(|x(1)|+|y(1)|)^2 + (|x(2)|+|y(2)|)^2$$
$$\le [(|x(1)|^2+|x(2)|^2)^{1/2} + (|y(1)|^2+|y(2)|^2)^{1/2}]$$
$$\cdot [(|x(1)|+|y(1)|)^2 + (|x(2)|+|y(2)|)^2]^{1/2}.$$

Thus,

$$\sqrt{|x(1)+y(1)|^2+|x(2)+y(2)|^2} \le \sqrt{|x(1)|^2+|x(2)|^2} + \sqrt{|y(1)|^2+|y(2)|^2},$$

and the equality problem is proved.

References

1. Bartoszewicz A (ed) (2011) Robust control, theory and applications. InTech, Rijeka, pp 243–260, Chap 11
2. Harris CJ, Valenca JME (1983) The stability of input-output dynamical systems. Academic Press, New York
3. Kalman RE (1956) Nonlinear aspects of sampled-data control systems. In: Proc of the symposium on nonlinear circuit analysis, vol VI, pp 273–313

4. Mashreghi J (2009) Representation theorem in Hardy spaces. Cambridge University Press, Cambridge
5. Okuyama Y (2006) Robust stability analysis for discretized nonlinear control systems in a global sense. In: Proc of the 2006 American control conference, Minneapolis, MN, USA, pp 2321–2326
6. Okuyama Y (2008) Discretized PID control and robust stabilization for continuous plants. In: Proc of the 17th IFAC world congress, Seoul, Korea, pp 1492–1498
7. Royden HL (1988) Real analysis, 3rd edn. Prentice-Hall, New York
8. Rudin W (1987) Real and complex analysis, 3rd edn. McGraw-Hill, New York
9. Vidyasagar M (1993) Nonlinear systems analysis, 2nd edn. Prentice-Hall, New York, republished by SIAM, 2002
10. Yosida K (1980) Functional analysis, 6th edn. Springer, Berlin

Chapter 3
Robust Stability Analysis

3.1 Introduction

In this chapter, the robust stability of discretized feedback systems is analyzed in the frequency domain. Throughout the theories, it is assumed that the discretization is executed on the input and output sides of a nonlinear element at equal spaces, and the sampling period is chosen of a size suitable for the discretization in the space. Based on this premise, the discretized (point-to-point) nonlinear characteristic is examined from two viewpoints: global and local.

3.2 Input-Output Stability

The basic results of input-output stability theory are presented in this section. This theory is later in origin than Lyapunov stability theory. Figure 3.1 shows the basic feedback structure of a nonlinear time-varying discrete-time system. As is obvious from the figure, the following equations are given:

$$\begin{cases} v(k) = f(e,k), \quad k = 0, 1, 2, \cdots \\ \hat{y}(z) = G(z)\hat{u}(z) \\ e(k) = r(k) - y(k), \quad u(k) = v(k) + d(k). \end{cases} \tag{3.1}$$

Here, $r(k)$ and $d(k)$ are exogenous inputs that exist in an ℓ_2 space. Moreover, in the strict sense, the nonlinear time-varying element $f(e,k)$ should be written as

$$f(e,k) = f(e(kh), kh),$$

where h is the sampling period.

Definition If $r(k) \in \ell_2$ and $d(k) \in \ell_2$ lead to $e(k) \in \ell_2$ and $y(k) \in \ell_2$, the feedback system is called bounded input-bounded output (BIBO) stable.

Y. Okuyama, *Discrete Control Systems*,
DOI 10.1007/978-1-4471-5667-3_3, © Springer-Verlag London 2014

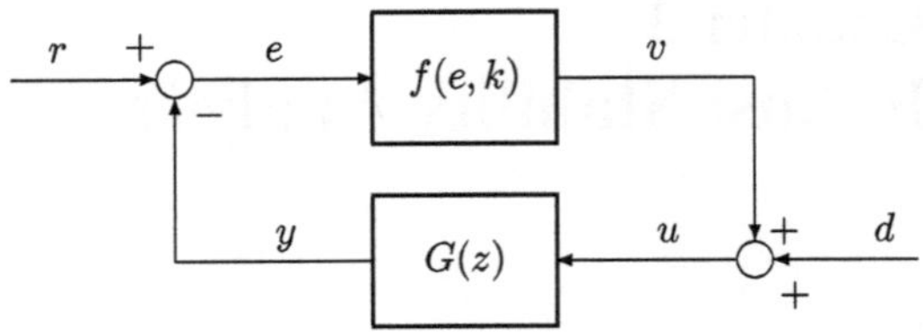

Fig. 3.1 Nonlinear time-varying feedback system

That is, if it is valid that

$$\sum_{k=0}^{\infty} |r(k)|^2 < \infty, \quad \sum_{k=0}^{\infty} |d(k)|^2 < \infty$$

lead to

$$\sum_{k=0}^{\infty} |e(k)|^2 < \infty, \quad \sum_{k=0}^{\infty} |y(k)|^2 < \infty,$$

then the nonlinear feedback system is input-output stable (in other words, ℓ_2-stable) [1, 3, 8, 13–15]. For the norm expression, the above conditions can be written as follows:

$$\|r(k)\|_2 < \infty \text{ and } \|d(k)\|_2 < \infty \Rightarrow \|e(k)\|_2 < \infty \text{ and } \|y(k)\|_2 < \infty.$$

3.3 Small Gain Theorem and Circle Criterion

From the last equations of (3.1), the norm inequalities of these variables are given by applying Minkowski's inequality[1] as follows:

$$\|e(k)\|_2 \le \|r(k)\|_2 + \|y(k)\|_2, \tag{3.2}$$

$$\|u(k)\|_2 \le \|v(k)\|_2 + \|d(k)\|_2. \tag{3.3}$$

If the nonlinear time-varying element $f(e, k)$ is written as

$$\frac{|f(e, k)|}{|e(k)|} \le \rho < \infty, \tag{3.4}$$

the norm of the output of nonlinear element $v(k)$ can be expressed as

$$\|v(k)\|_2 \le \rho \cdot \|e(k)\|_2.$$

Then, inequality (3.3) is given by

$$\|u(k)\|_2 \le \rho \cdot \|e(k)\|_2 + \|d(k)\|_2. \tag{3.5}$$

[1] See Appendix C in Chap. 2.

On the other hand, the following holds for $z = \mathrm{e}^{j\omega h}$:

$$\|\hat{y}(z)\|_2 \leq \sup_{|z|=1} |G(z)| \cdot \|\hat{u}(z)\|_2. \tag{3.6}$$

Here, $|z| = 1$ corresponds to $\omega : -\pi/h \to \pi/h$, i.e.,

$$\sup_{|z|=1} |G(z)| = \sup_{-\pi/h \leq \omega \leq \pi/h} |G(\mathrm{e}^{j\omega h})|.$$

Since open-loop system $G(z)$ is assumed to be stable, the boundary line $|z| = 1$ is considered.

By applying Parseval's identity,[2] the following inequality is obtained:

$$\|y(k)\|_2 \leq \sup_{|z|=1} |G(z)| \cdot \|u(k)\|_2. \tag{3.7}$$

By using (3.3),

$$\|e(k)\|_2 \leq \|r(k)\|_2 + \sup_{|z|=1} |G(z)| \cdot \|u(k)\|_2.$$

Thus,

$$\|e(k)\|_2 \leq \|r(k)\|_2 + \sup_{|z|=1} |G(z)|(\rho \cdot \|e(k)\| + \|d(k)\|_2. \tag{3.8}$$

Rearranging (3.8), the following inequality is obtained:

$$(1 - \rho \cdot \sup_{|z|=1} |G(z)|)\|e(k)\|_2 \leq \|r(k)\|_2 + \sup_{|z|=1} |G(z)| \cdot \|d(k)\|_2. \tag{3.9}$$

Theorem 3.1 *If* $\|r(k)\|_2 < \infty$, $\|d(k)\|_2 < \infty$, *and* $\sup_{|z|\geq 1} |G(z)| < \infty$, *then* $\|e(k)\|$ *becomes bounded when*

$$1 - \rho \cdot \sup_{|z|=1} |G(z)| > 0. \tag{3.10}$$

That is,

$$\sup_{|z|=1} |G(z)| < \frac{1}{\rho}. \tag{3.11}$$

The inequalities (3.10) *and* (3.11) *are referred to as the* small gain theorem.

Figure 3.2(a) is a graphical interpretation of (3.11), i.e.,

$$|G(\mathrm{e}^{j\omega h})| = \sqrt{U^2(\omega) + V^2(\omega)} < \frac{1}{\rho}, \tag{3.12}$$

where $U(\omega) = \Re\{G(\mathrm{e}^{j\omega h})\}$ and $V(\omega) = \Im\{G(\mathrm{e}^{j\omega h})\}$.

---

[2]See Appendix B in this chapter.

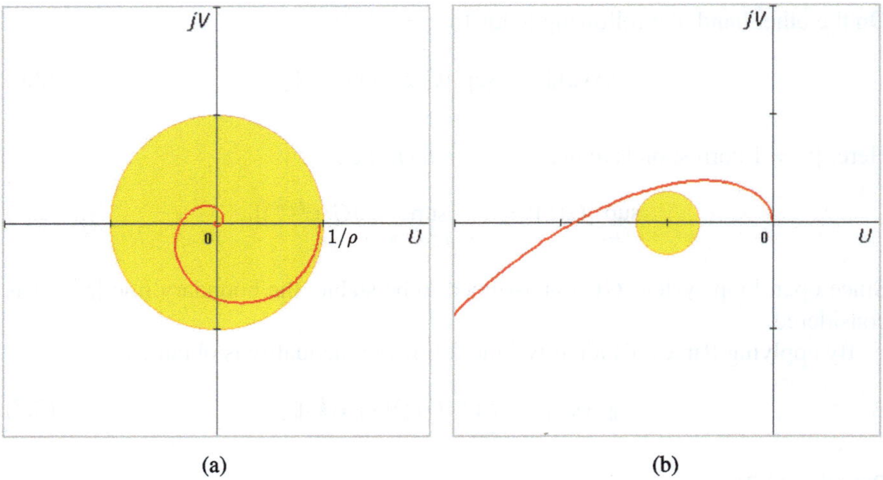

Fig. 3.2 Circle criterion for discrete-time systems

Fig. 3.3 Time-varying gain feedback system

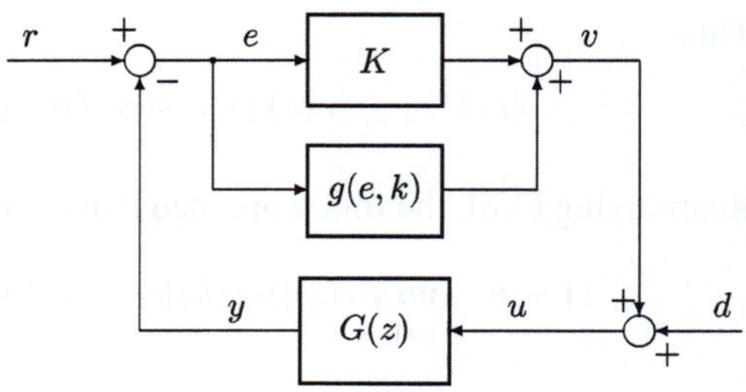

On the other hand, when the feedback system shown in Fig. 3.1 is drawn with respect to the nominal gain K as shown in Fig. 3.3, the small gain theorem (3.11) is rewritten as

$$\sup_{|z|=1} \left| \frac{G(z)}{1+KG(z)} \right| < \frac{1}{\rho}, \tag{3.13}$$

and instead of (3.4)

$$\frac{|g(e,k)|}{|e(k)|} \le \rho < \infty. \tag{3.14}$$

Inequality (3.13) can be written in regard to the inverse function of $G(z)$ as follows:

$$\left| \frac{1}{G(\mathrm{e}^{j\omega h})} + K \right| > \rho. \tag{3.15}$$

Therefore, the condition of (3.14) becomes one such that the inverse Nyquist curves of $G(\mathrm{e}^{j\omega h})$ exist in the outside of a small circle, as shown in Fig. 3.2(b). This stability criterion is also called the *circle criterion*. Note that the stability condition in the

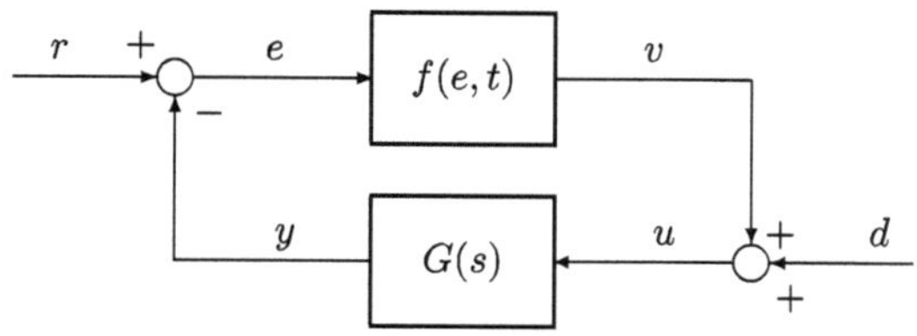

Fig. 3.4 Nonlinear continuous-time feedback system

relation between Nyquist curves and specified circles will be described in Sect. 3.7 on the Hall diagram.

Continuous-Time Systems For continuous-time systems, the nonlinear feedback systems in question should be written, of course, as shown in Fig. 3.4. The definition of input-output stability is given as follows.

Definition If $r(t) \in L_2$ and $d(t) \in L_2$ lead to $e(t) \in L_2$ and $y(t) \in L_2$, the feedback system is called BIBO stable.

That is, if it is valid that

$$\int_0^\infty |r(t)|^2 \mathrm{d}t < \infty, \quad \int_0^\infty |d(t)|^2 \mathrm{d}t < \infty$$

lead to

$$\int_0^\infty |e(t)|^2 \mathrm{d}t < \infty, \quad \int_0^\infty |y(t)|^2 \mathrm{d}t < \infty,$$

then the nonlinear feedback system is input-output stable (in other words, L_2-stable).

For the norm expression, the above conditions can be written as follows:

$$\|r(t)\|_2 < \infty \quad \text{and} \quad \|d(t)\|_2 < \infty \quad \Rightarrow \quad \|e(t)\|_2 < \infty \quad \text{and} \quad \|y(t)\|_2 < \infty.$$

As for the continuous case, Theorem 3.1 should be rewritten as

Corollary 3.2 *If $\|r(t)\|_2 < \infty$, $\|d(t)\|_2 < \infty$, and $\sup_{\Re s \geq 0} |G(s)| < \infty$, then $\|e(t)\|$ becomes bounded when*

$$1 - \rho \cdot \sup_{s=j\omega} |G(s)| > 0. \tag{3.16}$$

That is,

$$\sup_{s=j\omega} |G(s)| < \frac{1}{\rho}. \tag{3.17}$$

In recent control theory, (3.16) *and* (3.17) *may be written as*[3]

$$1 - \rho \cdot \|G(s)\|_\infty > 0 \tag{3.18}$$

and

$$\|G(s)\|_\infty < \frac{1}{\rho}. \tag{3.19}$$

Truncated Input Functions In general, even if $|x(k)| < \infty (k = 0, 1, 2, \cdots)$, the following is not always valid:

$$\|x(k)\|_2 < \infty. \tag{3.20}$$

However, if the following function x_N is considered for an arbitrary integer number $N > 0$:

$$\begin{cases} x_N(k) = x(k), & \text{for } k < N, \\ x_N(k) = 0, & \text{for } k \geq N, \end{cases} \tag{3.21}$$

inequality (3.20) is always satisfied. This operation is called *truncation*. On the other hand, if the norm of truncated function $\|x_N(k)\|_2$ is considered as[4]

$$\|x_N(k)\| = \|x(k)\|_{2,N} = \left(\sum_{k=1}^{N} |x(k)|^2 \right)^{1/2},$$

the ℓ_2 norm can be obtained as a limiting value, i.e.,

$$\|x(k)\|_2 = \lim_{N \to \infty} \|x(k)\|_{2,N}.$$

Thus, the input-output stability of discrete systems can be determined from these concepts. Figures 3.5(a) and (b) show examples of truncated discrete functions.

In the case of continuous-time systems, the truncated function can be defined as follows:

$$\begin{cases} x_T(t) = x(t), & \text{for } t < T, \\ x_T(t) = 0, & \text{for } t \geq T. \end{cases} \tag{3.22}$$

Therefore, the boundedness of norms can be clarified as

$$\|x_T(t)\|_2 = \|x(t)\|_{2,T} < \infty,$$

[3] See Appendix B in Chap. 1.

[4] See Appendix A in Chap. 2.

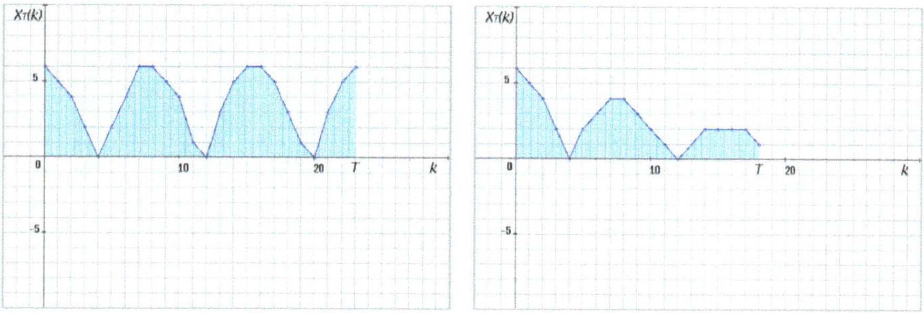

Fig. 3.5 Truncations of oscillating discrete functions

where

$$\|x(t)\|_{2,T} := \left(\int_0^T |x(t)|^2 \mathrm{d}t \right)^{1/2}.$$

Obviously,

$$\|x(t)\|_2 = \lim_{T\to\infty} \|x(t)\|_{2,T}.$$

Thus, the input-output stability of discrete systems can be determined from this expression.[5]

3.4 Discretized Nonlinear Control Systems

As was described in Chap. 2, the discretized control system shown in Fig. 3.6 is equivalently transformed into Fig. 3.7. The control system can be treated as a feedback structure, shown in Fig. 3.8.

As is obvious from these figures, the following relations are obtained:

$$w^*(k) = g^*[e^{*\dagger}(k)], \tag{3.23}$$

$$\hat{e}^{*\dagger}(z) = \hat{\epsilon}'(z) + \hat{e}^*(z). \tag{3.24}$$

Here, as is shown in (3.4) and (3.14), the following sector can be considered in regard to the nonlinear part:

$$\frac{|g^*[e^{*\dagger}(k)]|}{|e^{*\dagger}(k)|} \leq \beta < \infty. \tag{3.25}$$

[5]See Appendix A in Chap 2.

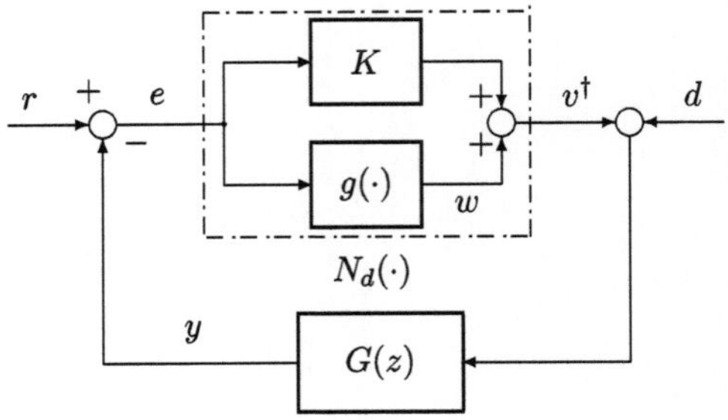

Fig. 3.6 Discretized nonlinear control system

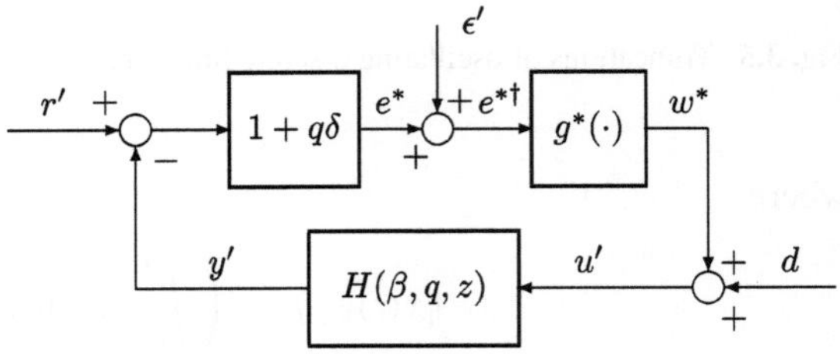

Fig. 3.7 Equivalent feedback system

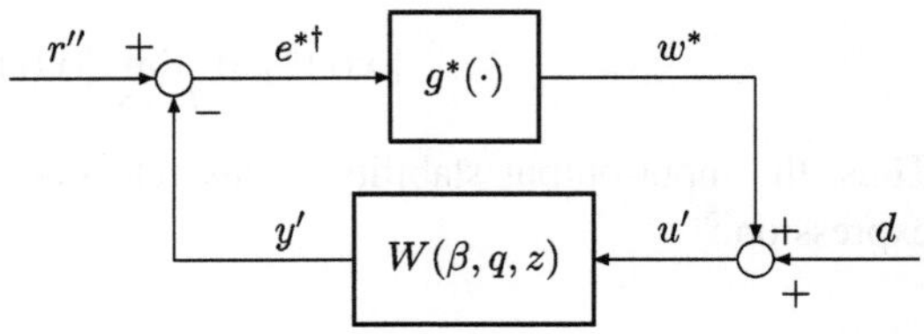

Fig. 3.8 Equivalent small-gain system

Furthermore,

$$\begin{cases} \hat{r}''(z) = \hat{\epsilon}'(z) + (1 + q\delta(z))\hat{r}'(z), \\ \hat{e}^{*\dagger}(z) = \hat{r}''(z) - \hat{y}'(z), \\ \hat{y}'(z) = H(\beta, q, z)(\hat{w}^*(z) + \hat{d}(z)). \end{cases} \tag{3.26}$$

Here,

$$\hat{r}'(z) = \frac{\hat{r}(z)}{1 + (K + \beta q\delta(z))G(z)}$$

and thus

$$H(\beta, q, z) = \frac{G(z)}{1 + (K + \beta q\delta(z))G(z)}.$$

The loop transfer function is obtained as follows:

$$W(\beta, q, z) = (1 + q\delta(z))H(\beta, q, z) = \frac{(1 + q\delta(z))G(z)}{1 + (K + \beta q\delta(z))G(z)}. \tag{3.27}$$

Therefore, the loop characteristic equation is given by

$$\hat{e}^{*\dagger}(z) = \hat{r}''(z) - W(\beta, q, z)(\hat{w}^*(z) + \hat{d}(z)), \tag{3.28}$$

where

$$\hat{r}''(z) = \hat{\epsilon}(z) + \frac{1 + q\delta(z)}{1 + (K + \beta q\delta(z))G(z)}\hat{r}(z).$$

From (3.28), the following inequalities are obtained with respect to norms in the ℓ_2 space:

$$\begin{aligned}\|e^{*\dagger}(z)\|_2 &\leq \|\hat{r}''(z)\|_2 + |W(\beta, q, z)|(\|\hat{v}^*(z)\|_2 + \|\hat{d}(z)\|_2)\\ &\leq \|\hat{r}''(z)\|_2 + \beta|W(\beta, q, z)| \cdot \|\hat{e}^{*\dagger}(z)\|_2 + |W(\beta, q, z)| \cdot \|\hat{d}(z)\|_2.\end{aligned}$$

Thus, the small gain theorem is given as follows:

$$\sup_{|z|=1} |W(\beta, q, z)| < \frac{1}{\beta}. \tag{3.29}$$

In practical physical systems, it can be written as

$$\sup_{0<\omega\leq\omega_c} |W(\beta, q, \mathrm{e}^{j\omega h})| < \frac{1}{\beta}, \tag{3.30}$$

where ω_c is a cut-off frequency. Although inequality (3.30) corresponds to a robust stability condition for discrete control systems, the allowable sector β cannot be explicitly given.

3.5 Robust Stability for Discretized Systems

In this section, based on the above result, the following robust stability condition for discretized nonlinear control systems is derived explicitly [10, 11].

Theorem 3.2 *If there exists a $q \geq 0$ in which the sector parameter β defined in* (3.25) *satisfies the following inequality, the discrete control system is robustly stable in an ℓ_2 sense*:

$$\beta < \beta_0 = K\eta(q_0, \omega_0) = \max_q \min_\omega K\eta(q, \omega), \tag{3.31}$$

when the linearized feedback system with nominal gain K is stable. (*That is, the allowable sector is given as* $[0, \beta_0]$ *from* (3.31).)

Here, the η-function is written as follows:

$$\eta(q, \omega) := \frac{-q\Omega V + \sqrt{q^2\Omega^2V^2 + (U^2 + V^2)\{(1 + U)^2 + V^2\}}}{U^2 + V^2}, \tag{3.32}$$

$$\forall\omega \in [0, \omega_c], \quad \omega_c \ : \ \text{cut-off frequency}.$$

Moreover, $\Omega(\omega)$ is the distorted frequency of angular frequency ω and is given by

$$\delta(\mathrm{e}^{j\omega h}) = j\Omega(\omega) = j\frac{2}{h}\tan\left(\frac{\omega h}{2}\right), \quad j = \sqrt{-1}. \tag{3.33}$$

Here, δ corresponds to the bilinear transformation

$$\delta(z) := \frac{2}{h}\cdot\frac{z-1}{z+1}. = \frac{2}{h}\cdot\frac{1-z^{-1}}{1+z^{-1}} \tag{3.34}$$

in the z-plane. In addition, $U(\omega)$ and $V(\omega)$ are the real and the imaginary parts of $KG(\mathrm{e}^{j\omega h})$, respectively.

Corollary 3.3 *When considering relative sector parameters $\alpha = \beta/K$, the theorem can be written simply as follows:*

$$\alpha < \frac{-q\Omega V + \sqrt{q^2\Omega^2V^2 + (U^2+V^2)\{(1+U)^2+V^2\}}}{U^2+V^2}, \quad \forall\omega \in [0, \omega_c]. \tag{3.35}$$

Proof Based on the loop characteristic in Fig. 3.8, the following inequality can be given with respect to $z = \mathrm{e}^{j\omega h}$:

$$\|\overline{e}^*(z)\|_{2,p} \le c_1\|\overline{r}'(z)\|_{2,p} + c_2\|\overline{d}(z)\|_{2,p} + \sup_{|z|=1}|W(\beta, q, z)|\cdot\|\overline{w}^{*\dagger}(z)\|_{2,p}. \tag{3.36}$$

Here, $\overline{r}'(z)$ and $\overline{d}(z)$ denote the z-transformation for the neutral points of sequences $r'(k)$ and $d(k)$, respectively. Moreover, c_1 and c_2 are positive constants.

By applying the inequality

$$\|\overline{w}^{*\dagger}(k)\|_{2,p} \le \beta\|\overline{e}^*(k)\|_{2,p}, \tag{3.37}$$

the following expression is obtained:

$$\left(1 - \beta\cdot\sup_{|z|=1}|W(\beta, q, z)|\right)\|\overline{e}^*(z)\|_{2,p} \le c_1\|\overline{r}'(z)\|_{2,p} + c_2\|\overline{d}(z)\|_{2,p}. \tag{3.38}$$

Therefore, if the following inequality (i.e., the small gain theorem with respect to ℓ_2 gains) is valid:

$$|W(\beta, q, \mathrm{e}^{j\omega h})| = \left|\frac{(1+jq\Omega(\omega))(U(\omega)+jV(\omega))}{K + (K + j\beta q\Omega(\omega))(U(\omega)+jV(\omega))}\right| < \frac{1}{\beta}, \tag{3.39}$$

the sequences $\overline{e}^*(k)$, $\overline{e}(k)$, $e(k)$, and $y(k)$ in the feedback system are restricted in finite values when exogenous inputs $r(k)$, $d(k)$ are finite and $p \to \infty$.

When $\beta = K\alpha$ is used, inequality (3.39) can be rewritten as

$$\left|\frac{(1+jq\Omega(\omega))(U(\omega)+jV(\omega))}{1 + (1 + j\alpha q\Omega(\omega))(U(\omega)+jV(\omega))}\right| < \frac{1}{\alpha}. \tag{3.40}$$

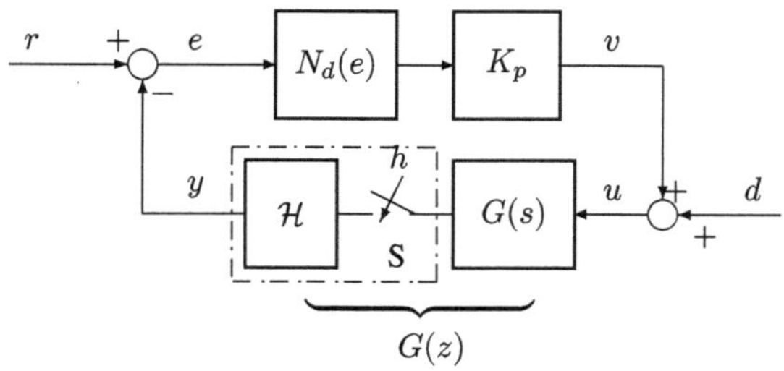

Fig. 3.9 Nonlinear sampled-data control system

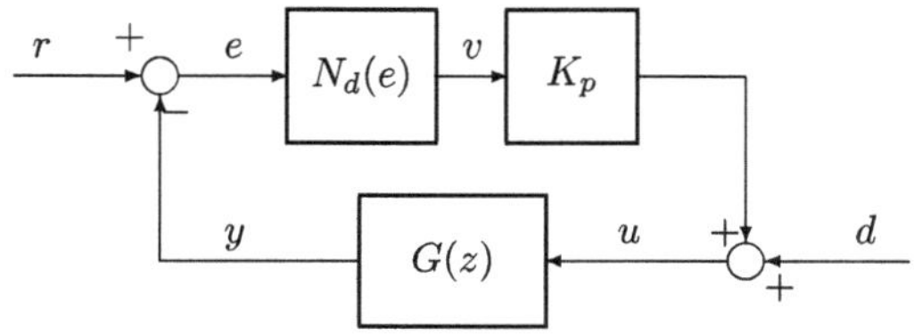

Fig. 3.10 Nonlinear discrete control system

From the square of both sides of inequality (3.40), we have

$$\alpha^2(1+q^2\Omega^2)(U^2+V^2) < (1+U-\alpha q\Omega V)^2 + (V+\alpha q\Omega U)^2.$$

The following quadratic inequality for α is obtained:

$$\alpha^2(U^2+V^2) + 2\alpha q\Omega V - \{(1+U)^2+V^2\} < 0. \tag{3.41}$$

Consequently, as a solution of inequality (3.41),

$$\alpha < \frac{-q\Omega V + \sqrt{q^2\Omega^2V^2 + (U^2+V^2)\{(1+U)^2+V^2\}}}{U^2+V^2}$$

is given. Thus Theorem 3.2 (i.e., (3.31) and (3.32)) is proved. □

Example 3.1 Consider a nonlinear feedback control system as shown in Figs. 3.9 and 3.10, which are equivalent to Figs. 2.1 and 2.2, but with a proportional controller K_p inserted in the feedback loop. Here, $N_d(\cdot)$ is a nonlinear time-invariant element, and $G(z)$ is a discrete-time linear dynamic system. When the following continuous plant is considered:

$$G(s) = \frac{K}{s(s+0.2)(s+0.4)}, \quad K = 0.02, \tag{3.42}$$

$G(z)$ is given as

$$G(z) = \frac{0.0029z^2 + 0.0099z + 0.0021}{z^3 - 2.49z^2 + 2.04z - 0.549} \tag{3.43}$$

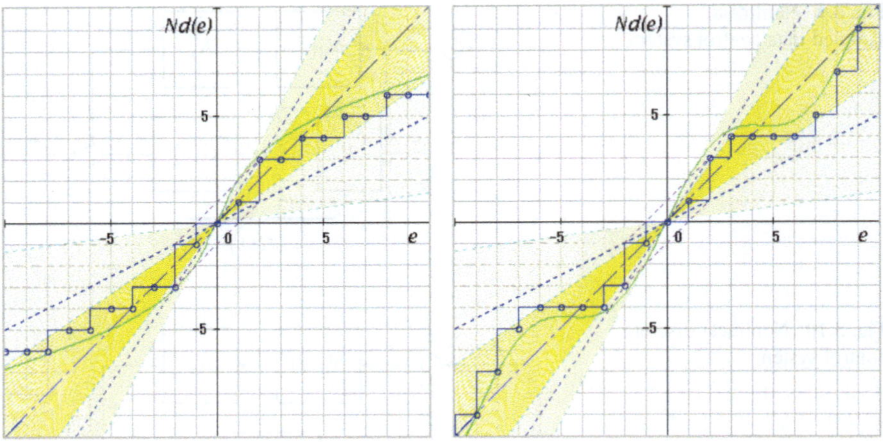

(a) (b)

Fig. 3.11 Nonlinear characteristics and allowable sectors

by using the procedure given in Sect. 1.5.2. Here, it is assumed that the discretized nonlinear characteristic to be considered is as shown in Figs. 3.11(a) and (b).[6] When proportional feedback ($K_p = 1.0$) is executed, $\eta(q, \omega)$ can be drawn as shown in Fig. 3.12 in the three-dimensional view. Figure 3.13(a) shows $\min_{\omega} K\eta(q, \omega)$ vs. ω for $K_p = 1.0, 1.2, 1.4$. For $K_p = 1.0$, $\beta_0 = \max_{q} \min_{\omega} = 0.863$ is determined. Hence, the allowable sector of β is given as $0.137 < \beta < 1.863$.[7] When $K_p = 1.4$, the

Fig. 3.12 Three-dimensional $\eta(q, \omega)$ curves

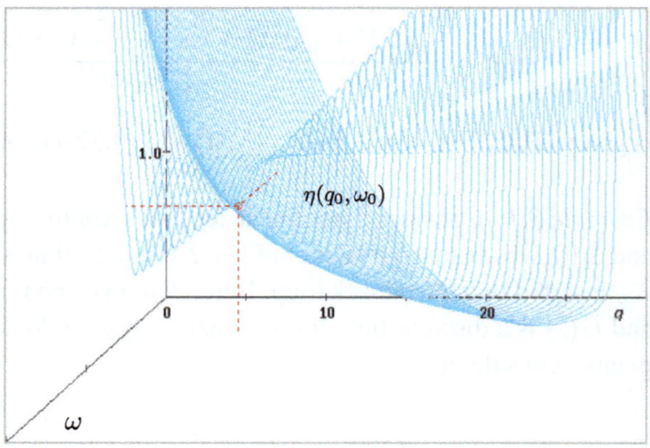

[6]In the following example, the stepwise representation is applied to the (point-to-point) nonlinear characteristic. As a result, the lower bound of the nonlinear sector is less than that of the point-transition characteristics.

[7]Actually, $0 < \beta < 1.863$ when the nominal system is stable.

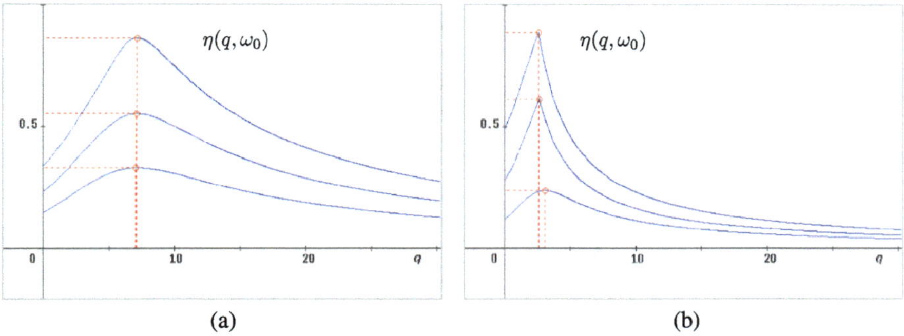

Fig. 3.13 The min-max value of $\eta(q, \omega)$ curves

stable sector, $0.669 < \beta < 1.331$, can be calculated as drawn in yellow in Fig. 3.11. Obviously, the former is guaranteed stability, while the latter is not.

Example 3.2 Next, consider the following nonminimum phase system:

$$G(s) = \frac{K(s+0.5)(-s+1.0)}{s(s+0.2)(s+1.0)}, \quad K = 0.15. \tag{3.44}$$

As described in Example 3.1, the following $G(z)$ can be obtained by using a computerized procedure:

$$G(z) = \frac{-0.0494z^2 + 0.1890z - 0.0966}{z^3 - 2.17z^2 + 1.49z - 0.301}. \tag{3.45}$$

Figure 3.13(b) shows $\min_{\omega} K\eta(q, \omega)$ vs. ω for $K_p = 1.0, 1.5, 2.0$. When $K_p = 1.0$, $\beta_0 = 0.887$ is obtained. Thus, the allowable sector of β is given as $0.113 < \beta < 1.887$. For $K_p = 1.5$, the stable sector can be calculated as $0.386 < \beta < 1.614$. In either case, the discretized feedback system with nonlinearity shown in Fig. 3.11(a) or (b) is guaranteed stability. However, when $K_p = 2.0$, $\beta_0 = 0.238$ and $0.762 < \beta < 1.238$. Obviously, the feedback system with nonlinearity as shown in Fig. 3.11 is not guaranteed in this case.

3.6 Some Relations to Traditional Theory

Aizerman's Conjecture The relationship between the robust stability condition in (3.31) and the allowable gain band (i.e., interval set parameter), in other words, the validity of Aizerman's conjecture extended into discrete-time systems, is examined.

In the following case, Theorem 3.2 becomes equal to the robust stability condition of the linear gain band that corresponds to Aizerman's conjecture.

Theorem 3.3 *If the right side of* (3.31), *i.e.*,

$$\eta(q_0, \omega_0) = \max_{\omega} \min_{q} \eta(q, \omega) \tag{3.46}$$

is satisfied at the saddle point,

$$\left(\frac{\partial \eta(q, \omega)}{\partial q} \right)_{q=q_0, \omega=\omega_0} = 0, \tag{3.47}$$

then (3.32) *of Theorem* 3.2 *becomes equal to the robust stability condition that is provided for linear discrete-time systems.*

Proof From (3.32), i.e.,

$$\eta(q, \omega) = \frac{-q\Omega V + \sqrt{q^2\Omega^2 V^2 + (U^2 + V^2)\{(1+U)^2 + V^2\}}}{U^2 + V^2},$$

the partial differential function is given by

$$\begin{aligned} \frac{\partial \eta}{\partial q} &= \frac{1}{U^2 + V^2} \left(-\Omega V + \frac{q\Omega^2 V^2}{\sqrt{(q^2\Omega^2 V^2 + (U^2 + V^2))\{(1+U)^2 + V^2\}}} \right) \\ &= \frac{-\Omega V \eta(q, \omega)}{\sqrt{q^2\Omega^2 V^2 + (U^2 + V^2)\{(1+U)^2 + V^2\}}}. \end{aligned} \tag{3.48}$$

Therefore, when (3.47) is satisfied, the following equality must hold:

$$V(\omega_0) = 0, \tag{3.49}$$

because $\Omega(\omega_0) > 0$ and $\eta(q_0, \omega_0) > 0$ for $0 < \omega_0 < \frac{\pi}{h}$. Then,

$$\eta(q_0, \omega_0) = \frac{|1 + U(\omega_0)|}{|U(\omega_0)|} = \left| 1 + \frac{1}{U(\omega_0)} \right| > \alpha, \tag{3.50}$$

that is,

$$\beta < K\eta(q_0, \omega_0) = \frac{|1 + K\Re\{G(\mathrm{e}^{j\omega h})\}|}{|\Re\{G(\mathrm{e}^{j\omega h})\}|} = \left| K + \frac{1}{\Re\{G(\mathrm{e}^{j\omega h})\}} \right|. \tag{3.51}$$

Inequalities (3.50) and (3.51) correspond to the robust stability condition which is determined for linear discrete-time systems by the linear gain band, i.e., the Nyquist and the inverse Nyquist stability criteria for discrete-time systems. □

As was shown in Fig. 3.13(a), (3.47) is satisfied in Example 3.1. However, in Example 3.2, although (3.47) is satisfied as to the small allowable sector (e.g., the case of $K_p - 2.0$ in Fig. 3.13(b)), the other cases in the figure ($K_p = 1.0, 1.5$) are not. The difference in the time responses of these cases will be described in the following section.

Popov's Criterion Inequality (3.40) can be rewritten as follows:

$$\left|\frac{\alpha\Psi(\alpha, q, \mathrm{e}^{j\omega h})}{1+\alpha\Psi(\alpha, q, \mathrm{e}^{j\omega h})}\right| < 1, \tag{3.52}$$

where

$$\Psi(\alpha, q, \mathrm{e}^{j\omega h}) = \frac{(1+jq\Omega(\omega))KG(\mathrm{e}^{j\omega h})}{1+(1-\alpha)KG(\mathrm{e}^{j\omega h})}.$$

From (3.52), the following inequality is obtained:

$$2\alpha\cdot\Re\{\Psi(\alpha, q, \mathrm{e}^{j\omega h})\}+1>0. \tag{3.53}$$

Therefore, the following robust stability condition can be given:

$$\Re\left\{\frac{1+(1+\alpha)KG(\mathrm{e}^{j\omega h})+2j\alpha q\Omega(\omega)KG(\mathrm{e}^{j\omega h})}{1+(1-\alpha)KG(\mathrm{e}^{j\omega h})}\right\}>0, \tag{3.54}$$

which is equivalent to (3.31) and (3.32). When $\alpha=1$ is chosen, (3.54) can be written as follows:

$$\frac{1}{\overline{K}_n}+\Re\{(1+jq\Omega(\omega))G(\mathrm{e}^{j\omega h})\}>0, \tag{3.55}$$

where $\overline{K}_n=2K$. In this case, the allowable sector of nonlinear characteristic $N(\cdot)$ is given by

$$0\le N(e)e\le\overline{K}_n e^2,\quad e\ne 0. \tag{3.56}$$

When h approaches zero (or ω is a low frequency), inequalities (3.55) and (3.56) are equivalent to an expression of Popov's criterion for continuous-time systems (i.e., $\delta\to s$, $\Omega\to\omega$, and $G(\mathrm{e}^{j\omega h})\to G(j\omega)$).

In the case of $q=0$, the definition of $\eta(q,\omega)$ becomes the inverse of the absolute value of complementary sensitivity function $T(j\omega)$. Then, Corollary 3.3 can be written as follows:

$$\eta(0,\omega)=\frac{\sqrt{(1+U)^2+V^2}}{\sqrt{U^2+V^2}}=\frac{1}{|T(j\omega)|}>\alpha. \tag{3.57}$$

On the other hand, from (3.54)

$$\Re\left\{\frac{1+(1+\alpha)KG(\mathrm{e}^{j\omega h})}{1+(1-\alpha)KG((\mathrm{e}^{j\omega h})}\right\}>0, \tag{3.58}$$

and thus the following criteria are obtained:

$$\frac{|1+KG(\mathrm{e}^{j\omega h})|}{|KG(\mathrm{e}^{j\omega h})|}>\alpha, \tag{3.59}$$

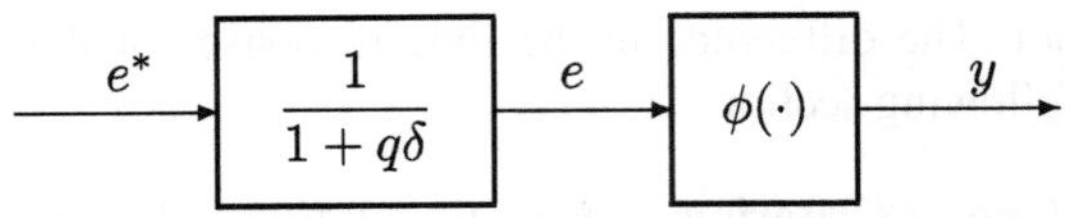

Fig. 3.14 Equivalent passivity system

that is,

$$\beta < \left| K + \frac{1}{G(\mathrm{e}^{j\omega h})} \right|. \tag{3.60}$$

Inequalities (3.57), (3.59), and (3.60) correspond to the circle criterion for nonlinear time-varying systems.

Passivity Theorem Consider a nonlinear subsystem as shown in Fig. 3.14. Here, $\phi(\cdot)$ is some nonlinear (time-invariant) element that is defined in the first and third quadrants, δ is a bilinear operator as shown in (3.34), and q is a non-negative parameter. From the left block of the diagram,

$$\hat{e}(z) + q \cdot \frac{2}{h} \cdot \frac{1 - z^{-1}}{1 + z^{-1}} \hat{e}(z) = \hat{e}^*(z). \tag{3.61}$$

Then,

$$\frac{1 + z^{-1}}{2} \hat{e}(z) + q \cdot \frac{1 - z^{-1}}{h} \hat{e}(z) = \frac{1 + z^{-1}}{2} \hat{e}^*(z). \tag{3.62}$$

Therefore, the following relation can be given with respect to time sequences:

$$\overline{e}(k) + q \cdot \frac{\Delta e(k)}{h} = \overline{e}^*(k), \tag{3.63}$$

where, as defined in Chap. 2,

$$\overline{e}(k) = \frac{e(k) + e(k-1)}{2}, \qquad \Delta e(k) = e(k) - e(k-1).$$

In regard to the inner product of the neutral point of input/output sequences, the following relation can be obtained:

$$\begin{aligned} \langle \overline{y}(k), \overline{e}^*(k) \rangle_N &= \langle \overline{\phi}(e(k)), \overline{e}^*(k) \rangle_N = \left\langle \overline{\phi}(e(k)), \overline{e}(k) + q \cdot \frac{\Delta e(k)}{h} \cdot \right\rangle_N \\ &= \langle \overline{\phi}(e(k)), \overline{e}(k) \rangle_N + \frac{q}{h} \cdot \langle \overline{\phi}(e(k)), \Delta e(k) \rangle_N, \end{aligned} \tag{3.64}$$

where

$$\overline{\phi}(e(k)) = \frac{\phi(e(k) + \phi(e(k-1))}{2}.$$

Since nonlinear characteristic $\phi(\cdot)$ belongs to the first and third quadrants, the following inequality must be satisfied for $N \to \infty$:

$$\langle \overline{\phi}(e(k)), \overline{e}(k) \rangle_N \geq 0. \tag{3.65}$$

If $q \geq 0$ and

$$\langle\overline{\phi}(e(k)), \Delta e(k)\rangle_N \geq 0, \tag{3.66}$$

the following inequality is satisfied:

$$\langle\overline{\phi}(e(k)), \overline{e}^*(k)\rangle_N \geq 0. \tag{3.67}$$

Inequality (3.66) was described in Chap. 2. Thus, the property of passivity has been elucidated with respect to Lemma 2.2.

Autonomous System Stability Consider a nonlinear control system as shown in Figs. 3.9 and 3.10. When considering continuous plant $G(s)$, $G(z)$ should be given by

$$G(z) = \mathcal{Z}\left\{\frac{1-\mathrm{e}^{-hs}}{s}G(s)\right\} = (1-z^{-1})\mathcal{Z}\left\{\frac{G(s)}{s}\right\} = (1-z^{-1})G_1(z). \tag{3.68}$$

As was described in Chap. 1, (3.68) is expressed as

$$G(z) = \frac{b_0 + b_1 z^{-1} + b_2 z^{-2} \cdots + b_n z^{-n}}{1 + a_1 z^{-1} + a_2 z^{-2} \cdots + a_n z^{-n}} = \frac{b_0 z^n + b_1 z^{n-1} + \cdots + b_{n-1} z + b_n}{z^n + a_1 z^{n-1} + \cdots + a_{n-1} z + a_n}. \tag{3.69}$$

Therefore, the input-output relation should be written as

$$\hat{y}(z) = G(z)\hat{u}(z) = \frac{b_0 z^n + b_1 z^{n-1} + \cdots + b_{n-1} z + b_n}{z^n + a_1 z^{n-1} + \cdots + a_{n-1} z + a_n} \cdot \hat{u}(z). \tag{3.70}$$

Clearly, (3.70) is rewritten as the following state-space equation:

$$\begin{cases} \boldsymbol{x}(k+1) = \boldsymbol{A}\boldsymbol{x}(k) + \boldsymbol{B}u(k), \\ y(k) = \boldsymbol{C}\boldsymbol{x}(k) + Du(k), \end{cases} \tag{3.71}$$

$$\boldsymbol{x} \in \mathbb{Z}^n, \quad u, y, D \in \mathbb{Z}, \quad \boldsymbol{A} \in \mathbb{Z}^{n\times n}, \quad \boldsymbol{B}, \boldsymbol{C} \in \mathbb{Z}^n,$$

and furthermore, from the block diagram,

$$e(k) = r(k) - y(k), \quad u(k) = v(k) + d(k), \quad v(k) = N_d(e). \tag{3.72}$$

Hence, the closed-loop characteristic can be given as:

$$\begin{cases} \boldsymbol{x}(k+1) = \boldsymbol{f}(\boldsymbol{x}(k), r(k), d(k)) \\ y(k) = \boldsymbol{g}(\boldsymbol{x}(k), r(k), d(k)) \end{cases} \tag{3.73}$$

$$\boldsymbol{x}(k) \in \mathbb{Z}^n, \quad r(k), u(k), y(k) \in \mathbb{Z}$$

$$\boldsymbol{f} : \mathbb{Z}^n \times \mathbb{Z} \to \mathbb{Z}^n, \quad \boldsymbol{g} : \mathbb{Z}^n \times \mathbb{Z} \to \mathbb{Z}, \quad k \in \mathbb{Z}_+.$$

If exogenous inputs $r(k), d(k)$ (e.g., reference and/or disturbance) are considered truncated functions:

$$\begin{cases} r_N(k) = r(k) & \text{for } k < N, \\ r_N(k) = 0 & \text{for } k \geq N, \end{cases} \qquad \begin{cases} d_N(k) = d(k) & \text{for } k < N, \\ d_N(k) = 0 & \text{for } k \geq N, \end{cases}$$

the state equation can be written in the following autonomous form:

$$\begin{cases} \boldsymbol{x}(k+1) = \tilde{\boldsymbol{f}}(\boldsymbol{x}(k)) \\ y(k) = \tilde{\boldsymbol{g}}(\boldsymbol{x}(k)) \quad \text{for } k = N, N+1, \cdots \end{cases} . \tag{3.74}$$

Therefore, if $\|y(k)\|_2 < \infty$ with respect to the above truncated inputs, the system is regarded as input-output stable in the ℓ_2 sense.

In this case, if $|y(k)| \to 0$ or $\|\boldsymbol{x}(k)\| \to 0$, the system is said to be asymptotically stable. Strictly speaking, in regard to some Lyapunov-type function $V(\boldsymbol{x})$, if $V(\boldsymbol{x}(k+1)) < V(\boldsymbol{x}(k))$ (i.e., the inclusive characteristic) holds, the system can be defined as asymptotically stable. Moreover, when $\|x(k)\| \leq X\mathrm{e}^{-at}$ $(X > 0, a > 0)$, the system may be said to be exponentially stable.

Example 3.3 We will now check the time responses of discretized feedback control for continuous plants as shown in Example 3.1, i.e.,

$$G(s) = \frac{K}{s(s+0.2)(s+0.4)}, \qquad K = 0.02.$$

The feedback control system is as shown in Figs. 3.9 and 3.10. Moreover, the discretized nonlinear characteristics are assumed to be as shown in Figs. 3.11(a) and (b).

Considering the expressions for (3.71) and (3.72) (and, e.g., (1.15) and (1.16) in Chap. 1), the following discrete-time state equation can be obtained from (3.43):

$$\begin{cases} \begin{bmatrix} x_1(k+1) \\ x_2(k+1) \\ x_3(k+1) \end{bmatrix} = \begin{bmatrix} 2.49 & 1 & 0 \\ -2.04 & 0 & 1 \\ 0.549 & 0 & 0 \end{bmatrix} \begin{bmatrix} x_1(k) \\ x_2(k) \\ x_3(k) \end{bmatrix} + \begin{bmatrix} 0.0029 \\ 0.0099 \\ 0.0021 \end{bmatrix} u(k), \\ y(k) = x_1(k), \end{cases} \tag{3.75}$$

and

$$e(k) = r_N(k) - y(k), \quad u(k) = K_p v(k) + d_N(k), \quad v(k) = N_d(e).$$

Thus, when the nonlinear characteristic is as shown in Fig. 3.11(a), the time responses for $k \geq N$ with respect to the initial states $x_1(N) = 6.0$, $x_2(N) = x_3(N) = 0.0$ become as shown in Figs. 3.15(a) and (b). On the other hand, when the nonlinear characteristic is as shown in Fig. 3.11(b), the time responses become as shown in Figs. 3.16(a) and (b). In these figures, $e(k)$, $e^{\dagger}(k)$, $\Delta e(k)$, and $\Delta e^{\dagger}(k)$

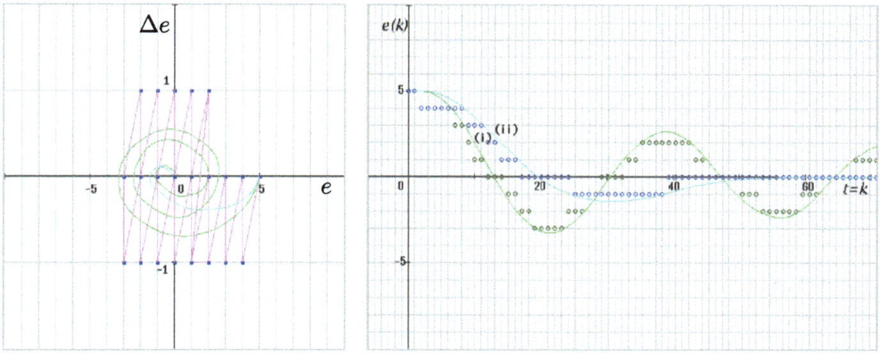

(a) (b)

Fig. 3.15 Phase traces and attenuating responses for Example 3.3 (sigmoid-type nonlinearity)

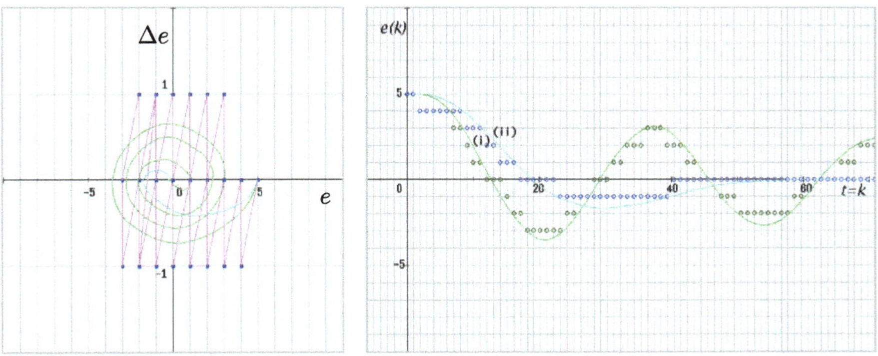

(a) (b)

Fig. 3.16 Phase traces and attenuating responses for Example 3.3 (inclined sinusoidal nonlinearity)

$(k = N, N+1, \cdots)$ are drawn when (i) $K_p = 1.0$ and (ii) $K_p = 0.5$. As is obvious in (a), $\Delta e^{\dagger}(k)$ is only 0 or ± 1.

Example 3.4 Next, consider the continuous plant as shown in Example 3.2, i.e.,

$$G(s) = \frac{K(s+0.5)(-s+1.0)}{s(s+0.2)(s+1.0)}, \quad K = 0.15.$$

In regard to the expression of (3.71) and (3.72) (and, e.g., (1.15) and (1.16) in

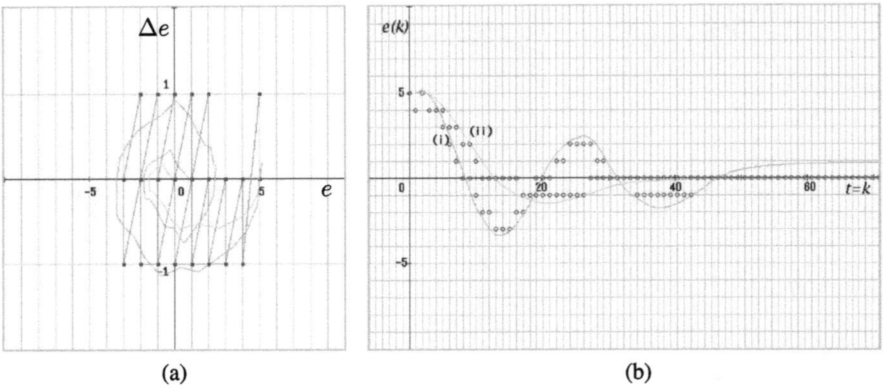

Fig. 3.17 Phase traces and attenuating responses for Example 3.4 (sigmoid-type nonlinearity)

Chap. 1), the following discrete-time state equation can be obtained from (3.43):

$$\begin{cases} \begin{bmatrix} x_1(k+1) \\ x_2(k+1) \\ x_3(k+1) \end{bmatrix} = \begin{bmatrix} 2.17 & 1 & 0 \\ -1.49 & 0 & 1 \\ 0.301 & 0 & 0 \end{bmatrix} \begin{bmatrix} x_1(k) \\ x_2(k) \\ x_3(k) \end{bmatrix} + \begin{bmatrix} -0.0494 \\ 0.1890 \\ -0.0966 \end{bmatrix} u(k), \\ y(k) = x_1(k), \end{cases} \tag{3.76}$$

and

$$e(k) = r_N(k) - y(k), \quad u(k) = K_p v(k) + d_N(k), \quad v(k) = \mathcal{N}_d(e).$$

Thus, when the nonlinear characteristic is as shown in Fig. 3.11(a), the time responses for $k \geq N$ with respect to the initial states $x_1(N) = 6.0$, $x_2(N) = x_3(N) = 0.0$ become as shown in Figs. 3.17(a) and (b). On the other hand, when the nonlinear characteristic is as shown in Fig. 3.11(b), the time responses become as shown in Figs. 3.18(a) and (b). Also in these figures, $e(k)$, $e^{\dagger}(k)$, $\Delta e(k)$, and $\Delta e^{\dagger}(k)$ $(k = N, N+1, \cdots)$ are drawn when (i) $K_p = 1.0$ and (ii) $K_p = 0.5$.

3.7 Modified Hall Diagram (Off-Axis M-Circles)

In order to interpret (3.31) and (3.32) in Theorem 3.2 graphically, a modified Hall diagram is presented in this section [10]. Here, the following inverse function of the η-function is considered:

$$\xi(q, \omega) = \frac{1}{\eta(q, \omega)}. \tag{3.77}$$

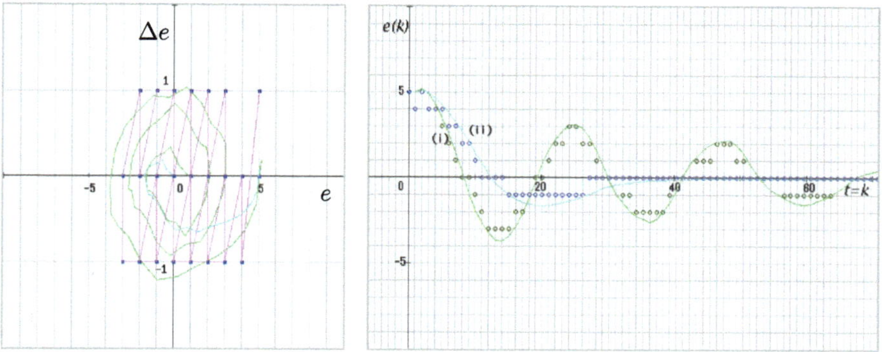

(a) (b)

Fig. 3.18 Phase traces and attenuating responses for Example 3.4 (inclined sinusoidal nonlinearity)

Using this notation, inequality (3.31) is rewritten as follows:

$$M_0 = \xi(q_0, \omega_0) = \min_q \max_\omega \xi(q, \omega) < \frac{K}{\beta}. \tag{3.78}$$

When $q = 0$, the ξ-function can be expressed as

$$\xi(0, \omega) = \frac{\sqrt{U^2 + V^2}}{\sqrt{(1+U)^2 + V^2}} = |S_c(\mathrm{e}^{j\omega h})|, \tag{3.79}$$

where $S_c(z)$ is the complementary sensitivity function for the discrete-time system. It is evident that the following curve on the complex plane,

$$\xi(0, \omega) = M, \quad (M : \text{const.}), \tag{3.80}$$

corresponds to an M-circle in the Hall diagram.[8] Figure 3.19 shows an example of a Hall diagram and Nyquist curves accompanied by performance indices, M_p, gain margin, and phase margins.

However, since (3.32) in Theorem 3.2 contains an arbitrary non-negative number q, the equality that corresponds to (3.80) should be rewritten as

$$\xi(q, \omega) := \frac{U^2 + V^2}{-q\Omega V + \sqrt{q^2\Omega^2 V^2 + (U^2 + V^2)\{(1+U)^2 + V^2\}}} = M. \tag{3.81}$$

From this expression, the following quadratic equation can be obtained:

$$(M^2 - 1)U^2 + 2M^2 U + (M^2 - 1)V^2 + M^2 - 2Mq\Omega V = 0. \tag{3.82}$$

---

[8] See Appendix D.

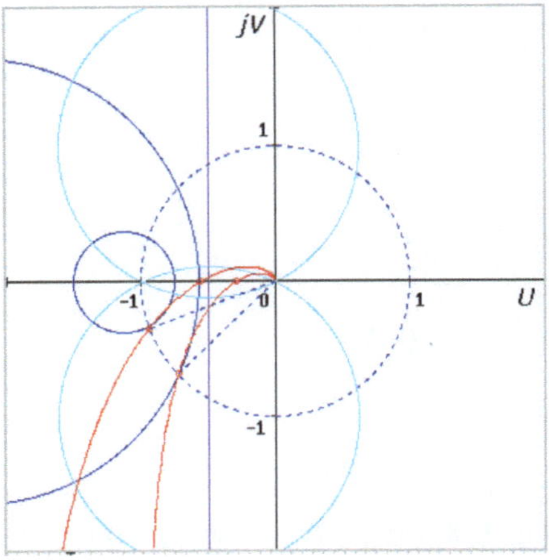

Fig. 3.19 Hall Diagram and Nyquist curves ($M_p = 1.35, 3.0$, $g_M = 11.0, 5.16$ [dB], $p_M = 43.8, 20.8$ [deg])

Obviously, when $M = 1$,

$$2U + 1 = \frac{2q\Omega}{M} \cdot V.$$

When $M > 1$, the following equation of circles is obtained from (3.82):

$$\left(U + \frac{M^2}{M^2 - 1}\right)^2 + (V - \lambda)^2 = \frac{M^2}{(M^2 - 1)^2} + \lambda^2, \tag{3.83}$$

where

$$\lambda = \frac{q\Omega M}{M^2 - 1} \geq 0. \tag{3.84}$$

Although the distorted frequency Ω is a function of ω, the term

$$c_q := q\Omega(\omega) \geq 0 \tag{3.85}$$

is assumed to be constant in this study. Hence, it can been seen that (3.83) represents off-axis circles with their center at

$$(-M^2/(M^2 - 1), \lambda)$$

and with a radius of

$$\sqrt{M^2/(M^2 - 1) + \lambda^2}.$$

Note that either of the circles crosses the real axis at

$$U = \frac{-M}{M - 1} \quad \text{or} \quad \frac{-M}{M + 1},$$

in which we will consider the latter for the robust stability problem.

The verification of robust stability using the above modified Hall diagram (off-axis M-circles) is based on the following theorem.

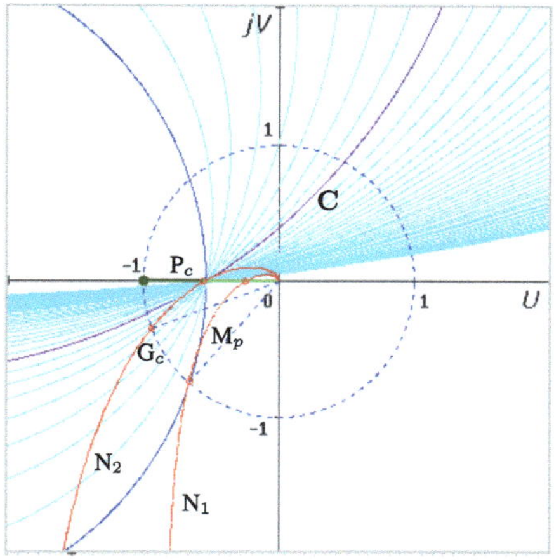

Fig. 3.20 Off-axis M-circles and Nyquist curves for Example 3.1 ($K_p = 1.0, 0.43$, $M = 1.23$, and $c_q = 0.0, \cdots, 10.0$)

Theorem 3.4 *If vector locus* $KG(\mathrm{e}^{j\omega h})$ *exists in the following area as determined by a certain* $q = q_0$ *(the outside of circle* **C** *shown in Fig.* 3.20*):*

$$\xi(q_0, \omega) = \tilde{\xi}(q_0, \omega, U, V) \leq M_0 < \frac{K}{\beta}, \tag{3.86}$$

the discretized feedback system is stable.

Proof As is evident from (3.78), the following inequality is valid in general:

$$\xi(q, \omega) \leq \xi(q, \omega_0), \quad \forall\omega \in [0, \omega_c]. \tag{3.87}$$

The right side of (3.87) is a peak value for angular frequency ω. Here, ω_0 is not always determined as only one frequency, and may not be a smooth (differentiable) point of the frequency range depending on the q-value. Nonetheless, inequality (3.38) is also satisfied for $q = q_0$. Therefore, the following inequality is obtained:

$$\xi(q_0, \omega) \leq \xi(q_0, \omega_0) = M_0, \quad \forall\omega \in [0, \omega_c]. \tag{3.88}$$

It can be shown that (3.86) in Theorem 3.4 is equivalent to (3.32). □

Figure 3.20 shows an example of the modified Hall diagram and Nyquist curves for $0 \leq c_q \leq 4.0$ and $M = 1.24$. Here, as an open-loop transfer function, we consider $G(s)$ shown in (3.42) and the proportional gain K_p. In the figure, N_1 is a vector locus that contacts with an M-circle at the peak value ($M_p = 1.24$, $K_p = 1.0, 0.43$). On the other hand, N_2 is a vector locus that contacts with a circle **C** on the real axis, where all the M-circles cross the real axis. In this case, Aizerman's conjecture extended to discrete-time systems is valid. At the continuous saddle point which is also the phase-crossover point P_c, the following equation is satisfied:

$$\left(\frac{\partial \xi(q, \omega)}{\partial q}\right)_{q=q_0, \omega=\omega_0} = 0. \tag{3.89}$$

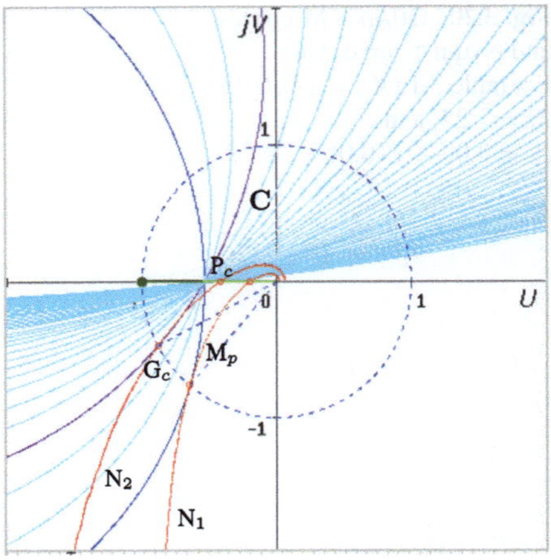

Fig. 3.21 Off-axis M-circles and Nyquist curves for Example 3.2 ($K_p = 1.0, 0.45$, $M = 1.2$, and $c_q = 0.0, \cdots, 10.0$)

Equality (3.89) is equivalent to (3.47). Incidentally, the phase margin p_M is obtained from the gain-crossover point G_c.

Figure 3.21 shows the case where the open-loop transfer function is given in (3.44). As is obvious from N_2 locus in the figure, the curve does not contact with a circle **C** on the real axis. In this case, Aizerman's conjecture is not valid; that is, the equality (3.89) is not satisfied.

3.8 Modified Nichols Diagram

Naturally, the above ideas can be applied to the Nichols diagram as well [9]. If the absolute value and the argument of $KG(e^{j\omega h})$ are defined as

$$\rho = \sqrt{U^2 + V^2} \text{ and } \theta = \tan^{-1}\left(\frac{V}{U}\right),$$

the ξ-function (3.81) can be rewritten as follows:

$$\begin{aligned}\xi(q,\omega) &= \frac{\rho^2}{-q\rho\Omega\sin\theta + \sqrt{q^2\rho^2\Omega^2\sin^2\theta + \rho^2\{(1+\rho\cos\theta)^2 + \rho^2\sin^2\theta\}}}\\ &= \frac{\rho}{-q\Omega\sin\theta + \sqrt{q^2\Omega^2\sin^2\theta + \rho^2 + 2\rho\cos\theta + 1}},\end{aligned} \quad (3.90)$$

because $U = \rho\cos\theta$ and $V = \rho\sin\theta$.

In this expression, the following robust stability condition must hold:

$$M_0 = \xi(q_0, \omega_0) = \min_q \max_\omega \xi(q,\omega) < \frac{K}{\beta}.$$

When $q = 0$, the ξ-function can be expressed as

$$\xi(0,\omega) = \frac{\rho}{\sqrt{\rho^2 + 2\rho\cos\theta + 1}} = |S_c(e^{j\omega h})|, \tag{3.91}$$

where $S_c(z)$ is the complementary sensitivity function for the discrete-time system. It is evident that the following curve on the gain-phase plane,

$$\xi(0,\omega) = M, \quad (M : \text{const.}), \tag{3.92}$$

corresponds to the contour of the constant M in the Nichols diagram.[9]

As described in the previous section, since an arbitrary non-negative number q is considered, the ξ-function that corresponds to (3.79) and (3.80) is given as follows:

$$\frac{\rho}{-q\Omega\sin\theta + \sqrt{q^2\Omega^2\sin^2\theta + \rho^2 + 2\rho\cos\theta + 1}} = M. \tag{3.93}$$

From this expression, the following quadratic equation can be obtained:

$$(M^2 - 1)\rho^2 + 2\rho M(M\cos\theta - q\Omega\sin\theta) + M^2 = 0. \tag{3.94}$$

The solution of this equation is expressed as

$$\rho = -\frac{M}{M^2 - 1}(M\cos\theta - q\Omega\sin\theta) \pm \frac{M}{M^2 - 1}\sqrt{(M\cos\theta - q\Omega\sin\theta)^2 - (M^2 - 1)}. \tag{3.95}$$

The modified contour in the gain-phase plane (θ, ρ) is drawn based on Eq. (3.95). Although the distorted frequency Ω is a function of ω, the term $q\Omega = c_q \geq 0$ is assumed to be a constant parameter. This assumption for M contours is the same as in the previous section.

Figure 3.22 shows the modified Nichols diagram for Example 3.1 ($M = 1.23$, $0 \leq c_q \leq 4.0$, and $Kp = 1.0, 0.43$. Here, GP_1 is a gain-phase curve that touches an M contour at the peak value ($M_p = 1.23$, $K_p = 0.43$). On the other hand, GP_2 is a gain-phase curve that crosses the $\theta = -180°$ line and all the M contours at the phase-crossover point P_c. That is, the gain margin g_M becomes equal to

$$g_M = -20\log_{10} M/(M + 1) = 5.4[\text{dB}].$$

Obviously, at point P_c, the gain-phase curve GP_2 touches an M contour. At this point, the following (continuous saddle point) equation is satisfied:

$$\left(\frac{\partial\xi(q,\omega)}{\partial q}\right)_{q=q_0,\omega=\omega_0} = 0. \tag{3.96}$$

[9]See Appendix E.

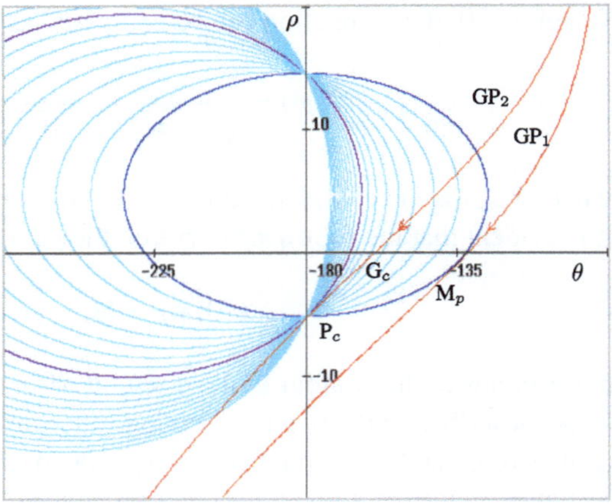

Fig. 3.22 Modified Nichols diagram for Example 3.1 ($M = 1.23$, $c_q = 0.0, 0.2, \cdots, 4.0$)

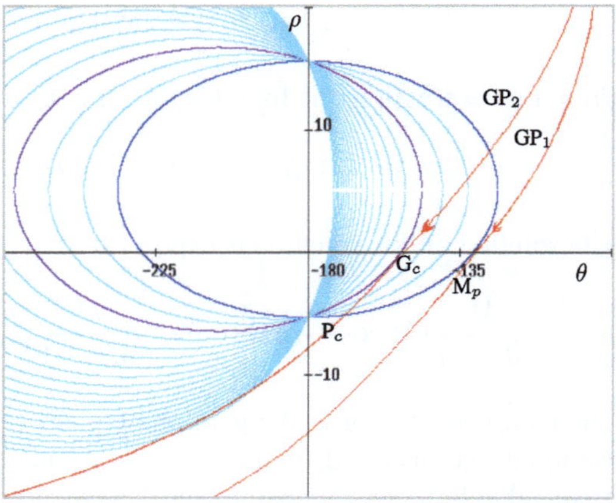

Fig. 3.23 Modified Nichols diagram for Example 3.2 ($M = 1.2$, $c_q = 0.0, 0.2, \cdots, 4.0$)

In this case, Aizerman's conjecture extended to discrete-time systems is valid. Incidentally, the phase margin p_M is obtained from the phase-crossover point P_c as follows:

$$p_M = 20.5[\text{deg}].$$

Figure 3.23 shows the modified Nichols diagram for Example 3.2 ($M = 1.2$, $0 \le c_q \le 4.0$, and $Kp = 1.0, 0.45$. Here, GP_1 is a gain-phase curve that touches an M contour at the peak value ($M_p = 1.2$, $K_p = 0.45$). Moreover, GP_2 is a gain-phase curve that crosses the $\theta = -180°$ line and all the M contours at the phase-crossover point P_c. That is, the gain margin g_M becomes equal to

$$g_M = -20 \log_{10} M/(M+1) = 14.8[\text{dB}],$$

and the phase margin p_M is obtained from G_c as

$$p_M = 49.8[\text{deg}].$$

However, the gain-phase curve does not touch an M contour at the phase-crossover point P_c. The continuous saddle point equation (3.96) is not satisfied, and thus, in this case, Aizerman's conjecture is not valid.

3.9 Exercises

(1) Show that inequality (3.15) corresponds to Fig. 3.2(b).

(2) Using the result of (3.39), prove the following inequality:

$$\beta < K \cdot \frac{-q\Omega V + \sqrt{q^2\Omega^2 V^2 + (U^2+V^2)\{(1+U)^2+V^2\}}}{U^2+V^2}.$$

(3) Show that the right side of the above inequality can be written as

$$K \cdot \frac{-q\Omega \sin\theta + \sqrt{q^2\Omega^2 \sin^2\theta + \rho^2 + 2\rho\cos\theta + 1}}{\rho},$$

where $\rho(\omega) = |KG(\mathrm{e}^{j\omega h})|$ and $\theta(\omega) = \angle KG(\mathrm{e}^{j\omega h})$.

(4) Prove the stability condition (3.11) in Theorem 3.1 based on the definition of input-output stability.

(5) Prove that

$$\delta(\mathrm{e}^{j\omega h}) = j\Omega(\omega) = j\frac{2}{h}\tan\left(\frac{\omega h}{2}\right)$$

in (3.33), where

$$\delta(z) = \frac{2}{h}\cdot\frac{z-1}{z+1}.$$

(6) Show that (3.89) is equivalent to (3.47).

(7) Derive (3.116) and (3.118) from (3.114) and (3.115).

Appendix A: Fourier-Plancherel Transform

For continuous-time signals, consider the following integral pairs:

$$\hat{x}(j\omega) = \int_{-\infty}^{\infty} x(t)\,\mathrm{e}^{-j\omega t}\,\mathrm{d}t \tag{3.97}$$

$$x(t) = \frac{1}{2\pi}\int_{-\infty}^{\infty} \hat{x}(j\omega)\,\mathrm{e}^{j\omega t}\,\mathrm{d}\omega, \tag{3.98}$$

where $\omega = 2\pi f$.[10] Usually, (3.97) is called the Fourier transform, and (3.98) is called the inverse Fourier transform [5, 6]. When evaluating at $s = j\omega$ (s: Laplace transform variable, in general, $s = \sigma + j\omega$), the preceding transforms become bilateral as follows:

$$\hat{x}(s) = \int_{-\infty}^{\infty} x(t)\,\mathrm{e}^{-st}\,\mathrm{d}t \tag{3.99}$$

$$x(t) = \frac{1}{2\pi j}\int_{-j\infty}^{j\infty} \hat{x}(s)\,\mathrm{e}^{st}\,\mathrm{d}s. \tag{3.100}$$

With respect to $t \in [0, \infty)$, the following transform is defined:

$$\hat{x}(s) = \int_{0}^{\infty} x(t)\,\mathrm{e}^{-st}\,\mathrm{d}t. \tag{3.101}$$

Definition (3.101) is the (unilateral) Laplace transform, which is well known in the field of control engineering.

The value of integrations (3.97) and (3.99) exists when the following inequality holds:

$$|\hat{x}(s)| = \left|\int_{-\infty}^{\infty} x(t)\,\mathrm{e}^{-st}\,\mathrm{d}t\right| \le \int_{-\infty}^{\infty} |x(t)|\,\mathrm{d}t < \infty.$$

In functional analysis, $x(t)$ is said to belong to the L_1 space, and is written as $x(t) \in L_1$. In general, if

$$\int_{-\infty}^{\infty} |x(t)|^p\,\mathrm{d}t < \infty,$$

$x(t)$ is said to belong to L_p and is written as $x(t) \in L_p$ or

$$x \in L_p(\mathbb{R}), \qquad \mathbb{R} := (-\infty, \infty).$$

In the L_p space, the norm is defined as follows:

$$\|x(t)\|_p := \left(\int_{-\infty}^{\infty} |x(t)|^p\,\mathrm{d}t\right)^{1/p}. \tag{3.102}$$

Obviously, in regard to the inverse Fourier (Laplace) transform, $x(s)$ must belong to the L_1 space.

Plancherel's theorem states that if $x(t) \in L_1 \cap L_2$, the above transformed function $\hat{x}(s)$ can also be determined similarly.

[10] In this book, we assume that the independent variable t represents time (expressed in the SI unit of seconds), and the transformed variables f and ω represent ordinary frequency (in Hertz) and angular frequency (in radians per second), respectively.

Plancherel Theorem *When $x \in L_2(\mathbb{R})$, the following $\hat{x}(j\omega)$ exists:*

$$\hat{x}_A(j\omega) = \int_{-A}^{A} x(t)\, \mathrm{e}^{-j\omega t} \mathrm{d}t$$

$$\int_{-\infty}^{\infty} |\hat{x}(j\omega) - \hat{x}_A(j\omega)|^2 \mathrm{d}\omega \to 0, \qquad \textit{for } A \to \infty.$$

With respect to $\hat{x}(j\omega)$, the following relation holds:

$$x_B(t) = \frac{1}{2\pi} \int_{-B}^{B} \hat{x}(j\omega)\, \mathrm{e}^{j\omega t} \mathrm{d}\omega$$

$$\int_{-\infty}^{\infty} |x(t) - x_B(t)|^2 \mathrm{d}t \to 0, \qquad \textit{for } B \to \infty.$$

Appendix B: Parseval Identity

Consider the following integral with respect to $x_1(t), x_2(t) \in L_1 \cap L_2$:

$$I = \int_{-\infty}^{\infty} \mathrm{d}t x_1(t) x_2(t). \tag{3.103}$$

By using the inverse Fourier (Laplace) transform, $x_2(t)$ is given by

$$x_2(t) = \frac{1}{2\pi j} \int_{-j\infty}^{j\infty} \mathrm{d}s \hat{x}_2(s) e^{st}.$$

Substitution of this value of $x_2(t)$ into (3.103) yields

$$I = \int_{-\infty}^{\infty} \mathrm{d}t\ x_1(t) \frac{1}{2\pi j} \int_{-j\infty}^{j\infty} \mathrm{d}s\ \mathrm{e}^{st} \hat{x}_2(s).$$

Interchange the order of the integrations,

$$I = \frac{1}{2\pi j} \int_{-j\infty}^{j\infty} \mathrm{d}s\ \hat{x}_2(s) \int_{-\infty}^{\infty} \mathrm{d}t\ \mathrm{e}^{st}\ x_1(t). \tag{3.104}$$

By applying the Fourier (two-sided Laplace) transform,

$$\hat{x}_1(s) = \int_{-\infty}^{\infty} \mathrm{d}t\ \mathrm{e}^{-st}\ x_1(t).$$

This expression can be written as

$$\hat{x}_1(-s) = \int_{-\infty}^{\infty} \mathrm{d}t\ \mathrm{e}^{st}\ x_1(t)$$

Therefore, (3.104) is expressed as

$$I = \frac{1}{2\pi j} \int_{-j\infty}^{j\infty} \mathrm{d}s \ \hat{x}_1(-s)\hat{x}_2(s). \tag{3.105}$$

For an easier understanding, (3.105) is rewritten for $s = j\omega$,

$$I = \frac{1}{2\pi} \int_{-\infty}^{\infty} \hat{x}_1(-j\omega)\hat{x}_2(j\omega) \ \mathrm{d}\omega. \tag{3.106}$$

When $x_1(t) = x_2(t) = x(t)$, (3.106) is given by

$$I = \frac{1}{2\pi} \int_{-\infty}^{\infty} \hat{x}(-j\omega)\hat{x}(j\omega) \ \mathrm{d}\omega = \frac{1}{2\pi} \int_{-\infty}^{\infty} |\hat{x}(j\omega|^2 \ \mathrm{d}\omega. \tag{3.107}$$

Then, the following equality is obtained from (3.103):

$$\int_{-\infty}^{\infty} |x(t)|^2 \ \mathrm{d}t = \frac{1}{2\pi} \int_{-\infty}^{\infty} |\hat{x}(j\omega|^2 \ \mathrm{d}\omega. \tag{3.108}$$

Here, define the following L_2 norms:

$$\|x(t)\|_2 := \left(\int_{-\infty}^{\infty} |x(t)|^2 \mathrm{d}t \right)^{1/2}$$

$$\|\hat{x}(j\omega)\|_2 := \left(\frac{1}{2\pi} \int_{-\infty}^{\infty} |\hat{x}(j\omega)|^2 \mathrm{d}\omega \right)^{1/2}.$$

Thus, we obtain

$$\|x(t)\|_2 = \|\hat{x}(j\omega)\|_2.$$

This formula is called *Parseval's identity*.

On the other hand, as for discrete-time signals, consider the following summation of discrete signals:

$$J = \sum_{k=1}^{\infty} x_1(k)x_2(k). \tag{3.109}$$

By using the inverse z-transform,

$$x_2(k) = \frac{1}{2\pi j} \int_{|z|=1} \hat{x}_2(z) z^{k-1} \mathrm{d}z, \tag{3.110}$$

where $z = \mathrm{e}^{h(\sigma + j\omega)}$. Substitution of this $x_2(k)$ into (3.109) yields

$$J = \sum_{k=0}^{\infty} x_1(k) \ \frac{1}{2\pi j} \int_{|z|=1} \hat{x}_2(z) z^{k-1} \mathrm{d}z.$$

Here, for $\sigma = 0$,

$$\mathrm{d}z = jh\mathrm{e}^{j\omega h}\,\mathrm{d}\omega.$$

Based on the z-transform,

$$\hat{x}_1(z) = \sum_{k=0}^{\infty} x_1(k)\, z^{-k},$$

the following can be defined:

$$\hat{x}_1(\overline{z}) = \sum_{k=1}^{\infty} x_1(k)\, z^{k}.$$

Thus

$$J = \frac{1}{2\pi j}\int_{|z|=1} \hat{x}_1(\overline{z})\hat{x}_2(z)z^{-1}\mathrm{d}z = \frac{h}{2\pi}\int_{-\pi}^{\pi} \hat{x}_1(\mathrm{e}^{-j\omega h})\,\hat{x}_2(\mathrm{e}^{j\omega h})\,\mathrm{d}\omega. \tag{3.111}$$

When $x_1(k) = x_2(k) = x(k)$, (3.111) is given by

$$J = \frac{h}{2\pi}\int_{-\pi}^{\pi} \hat{x}(\mathrm{e}^{-j\omega h})\,\hat{x}(\mathrm{e}^{j\omega h})\,\mathrm{d}\omega. = \frac{h}{2\pi}\int_{-\pi}^{\pi} |\hat{x}(\mathrm{e}^{-j\omega h})|^2\,\mathrm{d}\omega.$$

Then, the following equality is obtained from (3.109):

$$\sum_{k=0}^{\infty} |x(k)|^2 = \frac{h}{2\pi}\int_{-\pi}^{\pi} |\hat{x}(\mathrm{e}^{j\omega h})|^2\mathrm{d}\omega. \tag{3.112}$$

If the norm expressions in the ℓ_2 *space*,

$$\|x(k)\|_2 := \left(\sum_{k=0}^{\infty} |x(k)|^2\right)^{1/2}$$

$$\|\hat{x}(\mathrm{e}^{j\omega h})\|_2 := \left(\frac{h}{2\pi}\int_{-\pi}^{\pi} |\hat{x}(\mathrm{e}^{j\omega h})|^2\mathrm{d}\omega\right)^{1/2}$$

are used, the relation

$$\|x(k)\|_2 = \|\hat{x}(\mathrm{e}^{j\omega h})\|_2$$

can be obtained. The result is called *Parseval's identity* [7, 12].

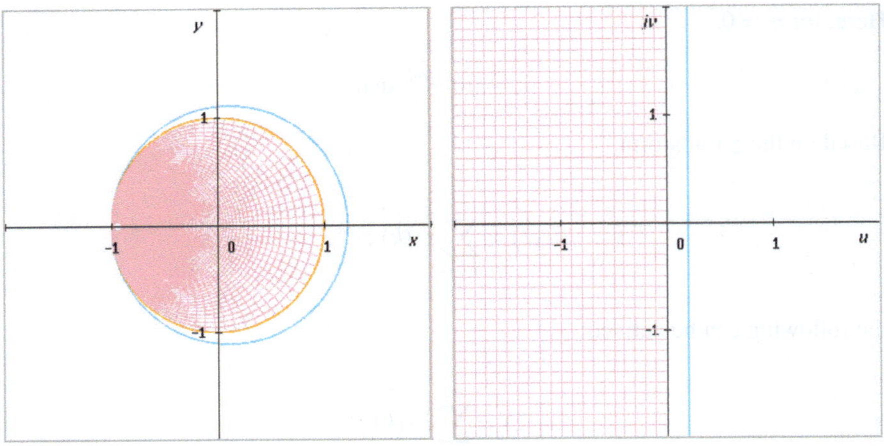

(a) (b)

Fig. 3.24 Relation between z-plane and δ-plane, and contours

Appendix C: Bilinear Transformation and Mapping

The relationship between z and δ with respect to

$$\delta = \frac{2}{h} \cdot \frac{z-1}{z+1}, \quad \text{that is,} \quad z = \frac{1+\frac{h}{2}\delta}{1-\frac{h}{2}\delta},$$

is as shown in Figs. 3.24(a) and (b). Furthermore, from the following equality:

$$\delta(\mathrm{e}^{j\omega h}) = j\Omega(\omega) = j\frac{2}{h}\tan\left(\frac{\omega h}{2}\right),$$

the relationship between ω and Ω is illustrated in Fig. 3.25.

Appendix D: The Hall Diagram

Consider a (unity feedback) closed-loop characteristic,

$$W(z) = \frac{G(z)}{1+G(z)}, \tag{3.113}$$

as shown in Fig. 3.26. When we define the following frequency characteristic for $\omega < \omega_c$ as shown in (3.32):

$$G(\mathrm{e}^{j\omega h}) = U(\omega) + jV(\omega),$$

and

$$W(\mathrm{e}^{j\omega h}) = M\mathrm{e}^{j\varphi},$$

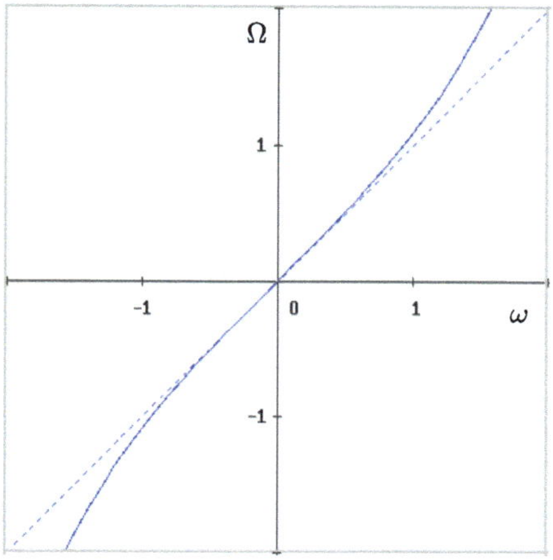

Fig. 3.25 Distorted frequency characteristic

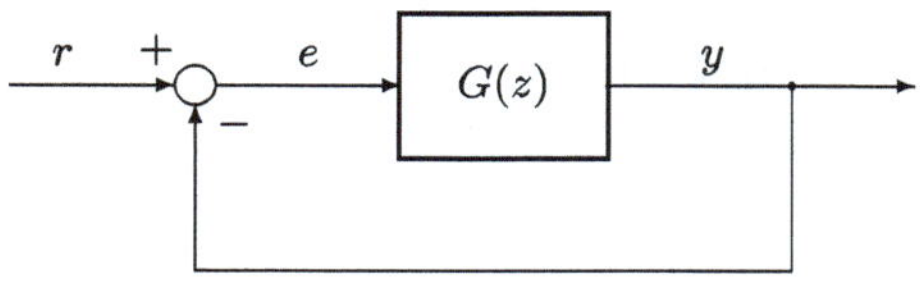

Fig. 3.26 Discrete-time unity feedback system

obviously,

$$M = |W(e^{j\omega h})| \quad \text{and} \quad \varphi = \angle W(e^{j\omega h}).$$

Therefore,

$$M = \frac{|U + jV|}{|1 + U + jV|} \tag{3.114}$$

and

$$\varphi = \tan^{-1} \frac{V}{U} - \tan^{-1} \frac{V}{1 + U}. \tag{3.115}$$

By rearranging (3.114), the following equation is obtained:

$$\left(U + \frac{M^2}{M^2 - 1}\right)^2 + V^2 = \left(\frac{M}{M^2 - 1}\right)^2, \quad \text{for} \quad M \neq 1. \tag{3.116}$$

Here, when $M = 1$, the equation is given as

$$U = -\frac{1}{2}. \tag{3.117}$$

Equation (3.117) is a (purple) line in Fig. 3.27, and (3.116) (in blue) circles the right side to (3.117) for $M < 1$ and the left side to the line for $M > 1$.

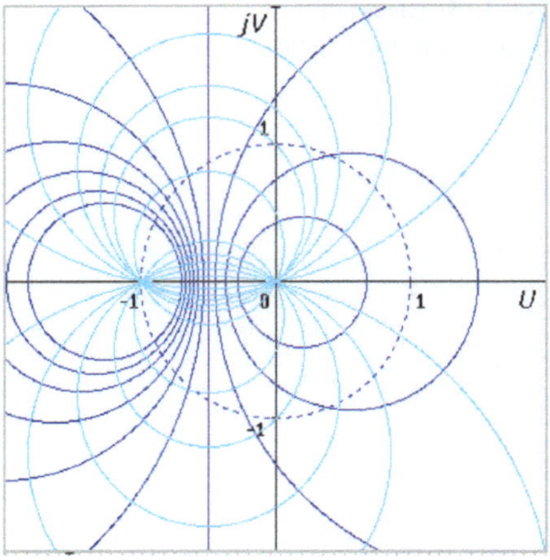

Fig. 3.27 Hall diagram

On the other hand, with respect to (3.115) the following circles equation is obtained by setting $N = \tan\varphi$:

$$\left(U+\frac{1}{2}\right)^2+\left(V-\frac{1}{2N}\right)^2=\frac{1}{4}\left(\frac{N^2+1}{N^2}\right). \tag{3.118}$$

Equation (3.118) also becomes circles, shown in light blue in Fig. 3.27. Such type of diagram is often called the Hall diagram [2].

The derivations of (3.116) and (3.118) from (3.114) and (3.115) are left for the reader.

Appendix E: The Nichols Diagram

Next, consider the open-loop characteristic $G(\mathrm{e}^{j\omega h})$ in polar coordinates as follows:

$$G(\mathrm{e}^{j\omega h})=\rho\cdot\mathrm{e}^{j\theta}. \tag{3.119}$$

Obviously, the closed-loop characteristic is given by

$$W(\mathrm{e}^{j\omega h})=\frac{\rho\cdot\mathrm{e}^{j\theta}}{1+\rho\cdot\mathrm{e}^{j\theta}}. \tag{3.120}$$

Therefore,

$$M=|W(\mathrm{e}^{j\omega h})|=\frac{\rho}{\sqrt{(1+\rho\cos\theta)^2+\rho^2\sin^2\theta}}. \tag{3.121}$$

By rearranging (3.121), the following quadratic equation is obtained for $M \neq 1$:

$$\rho^2+2\frac{M^2\cos\theta}{M^2-1}\cdot\rho+\frac{M^2}{M^2-1}=0. \tag{3.122}$$

Fig. 3.28 Nichols diagram

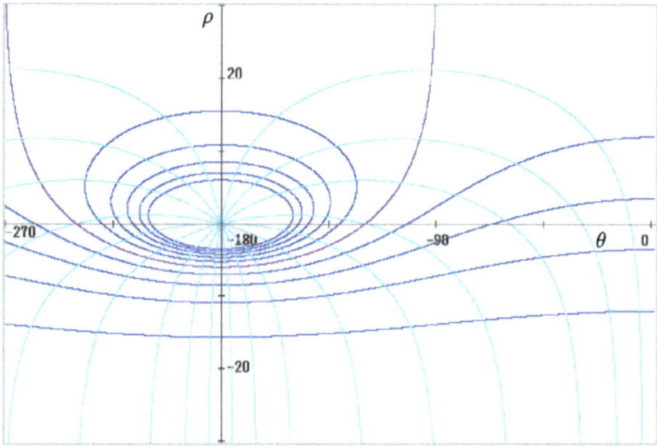

Then,

$$\rho = \frac{1}{M^2 - 1}\left(-M^2 \cos\theta \pm M\sqrt{\cos^2\theta - (M^2 - 1)}\right). \tag{3.123}$$

Curves (3.123) are drawn in blue in Fig. 3.28. When $M = 1$, note that ρ is simply written as follows:

$$\rho = -\frac{1}{2\cos\theta}. \tag{3.124}$$

This curve is drawn in purple in the figure.

On the other hand, for phase φ,

$$\varphi = \angle W(\mathrm{e}^{j\omega h}) = \theta - \tan^{-1}\left(\frac{\rho\sin\theta}{1 + \rho\cos\theta}\right). \tag{3.125}$$

From (3.125),

$$N = \tan\varphi = \frac{\sin\theta}{\cos\theta + \rho}.$$

Then,

$$\rho = \frac{\sin\theta}{\tan\varphi} - \cos\theta. \tag{3.126}$$

Curves (3.125) are drawn in sky blue in the figure. Here, the derivations of (3.123), (3.124), and (3.126) from (3.121) and (3.125) are left for the reader.

The diagram shown in Fig. 3.28 is the well-known Nichols diagram [4].

References

1. Desoer CA, Vidyasagar M (1975) Feedback systems, input-output properties. Academic Press, New York, republished by SIAM, 2009
2. Hall AC (1943) The analysis and synthesis of linear servomechanisms. MIT, Cambridge

3. Harris CJ, Valenca JME (1983) The stability of input-output dynamical systems. Academic Press, New York
4. James HM, Nichols NB, Philips RS (1947) Theory of servomechanisms. MIT, Cambridge
5. Katznelson Y (2004) Introduction to harmonic analysis, 3rd edn. Cambridge University Press, Cambridge
6. Mashreghi J (2009) Representation theorem in Hardy spaces. Cambridge University Press, Cambridge
7. Newton GC Jr, Gould LA, Kaiser JF (1957) Analytical design of linear feedback controls. Wiley, New York
8. Okuyama Y (1967) On the L_2-stability of linear systems with time-varying parameters. Trans SICE 3:252–259 (in Japanese)
9. Okuyama Y, Takemori F (1999) Robust stability evaluation for sampled-data control systems with a sector nonlinearity in a gain-phase plane. Int J Robust Nonlinear Control 9:15–32
10. Okuyama Y, Takemori F (2002) Robust stability analysis for sampled-data control systems in a frequency domain. Eur J Control 8:99–108
11. Okuyama Y, Takemori F (2002) Amplitude dependent analysis and stabilization for nonlinear sampled-data control systems. In: Proc of the 15th IFAC world congress, T-Tu-M08, Barcelona, Spain
12. Rudin W (1987) Real and complex analysis, 3rd edn. McGraw-Hill, New York
13. Sandberg IW (1964) A frequency domain condition for stability of feedback systems containing a single time-varying nonlinear element. Bell Syst Tech J 43:1601–1608
14. Vidyasagar M (1993) Nonlinear systems analysis, 2nd edn. Prentice-Hall, New York, republished by SIAM, 2002
15. Zames G (1966) On the input-output stability of nonlinear time-varying feedback systems, Pt I and II. IEEE Trans Autom Control AC-11:228–238, 465–477

Chapter 4
Model Reference Feedback and PID Control

4.1 Introduction

In the previous chapters, the robust stability of nonlinear discrete-time and discrete-value (discretized) control systems was examined in the frequency domain. In this chapter, a design problem for these discretized control systems is presented based on first a traditional but discretized PID control scheme, and then on a discrete-model reference feedback structure. The model reference feedback using a second-order continuous-value (linear) system is equivalently transformed into a traditional PID control. In the design procedure, the concepts of a modified Nyquist and Hall diagram (off-axis M-circles) and a modified Nichols chart for nonlinear control systems are applied.

4.2 Discretized PID Control

4.2.1 PID Control Scheme

The control scheme based on proportional-integral-derivative (PID) techniques has been widely used in practice and theory irrespective of whether it is continuous or discrete in time [1–4, 18, 21, 22], since it is a basic feedback control technique.

For a continuous-time classical representation,

$$u_c(t) = K_p u(t) + C_I \int u(t)\mathrm{d}t + C_D \frac{\mathrm{d}u(t)}{\mathrm{d}t}. \tag{4.1}$$

Using the Laplace transform with zero initial conditions, we have

$$\hat{u}_c(s) = K_p \hat{u}(s) + C_I \frac{\hat{u}(s)}{s} + C_D s \hat{u}(s) = C(s)\hat{u}(s). \tag{4.2}$$

Y. Okuyama, *Discrete Control Systems*,
DOI 10.1007/978-1-4471-5667-3_4, © Springer-Verlag London 2014

Thus, the controller can be written as[1]

$$C(s) = K_p\left(1 + \frac{1}{T_I s} + T_D s\right), \tag{4.3}$$

where T_I and T_D are referred to as integral and derivative time, respectively.

The PID controller is referred to as a *three-term* controller, which means that the control strategy can easily be adjusted by using only three parameters. If the stability of the control system is guaranteed, a proportional-integral (PI) controller is also used, for example, in industrial processes approximated by a first-order time-delay model [7, 8, 19]. In this case, only two parameters need to be adjusted.

In a similar representation as in the continuous case, a discrete-time PID control scheme can be given as follows:

$$\hat{u}_c(z) = K_p\hat{u}(z) + C_I \cdot \frac{\hat{u}(z)}{\delta} + C_D \cdot \delta\hat{u}(z), \tag{4.4}$$

where δ should be considered a bilinear operator as we have defined in (3.34), i.e.,

$$\delta = \frac{2}{h} \cdot \frac{1 - z^{-1}}{1 + z^{-1}}. \tag{4.5}$$

Of course, as shown in (4.3), the controller can be written as

$$C(\delta) = K_p\left(1 + \frac{1}{T_I \delta} + T_D \delta\right). \tag{4.6}$$

The operator δ has the following properties:

(1) Since

$$\delta^{-1} = \frac{h}{2} \cdot \frac{1 + z^{-1}}{1 - z^{-1}},$$

the relationship between the input and output sequences, $x(k)$, $y(k)$, $(k = 0, 1, 2, \cdots)$ can be written as

$$y(k) = y(k-1) + \frac{h}{2}(x(k) + x(k-1)). \tag{4.7}$$

This transformation corresponds to a trapezoidal summation (integration).

(2) On the other hand, with respect to the operator δ, the following relation is obtained:

$$y(k) = -y(k-1) + \frac{2}{h}(x(k) - x(k-1)). \tag{4.8}$$

[1] Since the pure differentiation cannot be realized in practice, a high-cut filter is usually added in the derivative operation.

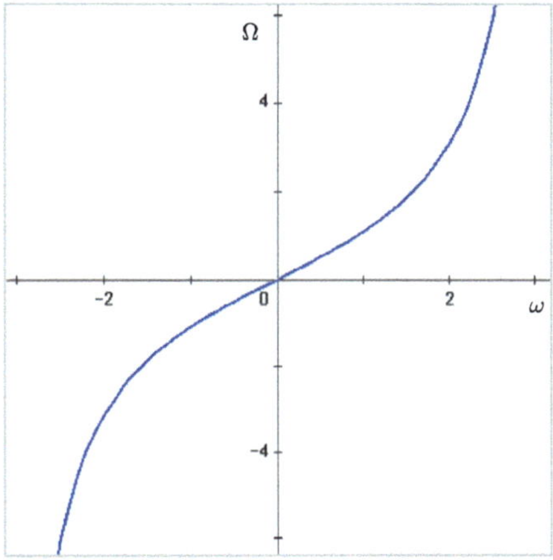

Fig. 4.1 Relationship between ω and Ω

Therefore, the PID controller can be given in a bilinear expression,

$$C(z) = K_p\left(1 + \frac{h}{2}\cdot\frac{1+z^{-1}}{T_I(1-z^{-1})} + \frac{2}{h}\cdot\frac{T_D(1-z^{-1})}{1+z^{-1}}\right). \tag{4.9}$$

In the frequency domain, (4.6) can be obtained from (3.33) as follows:

$$C(j\Omega) = K_p\left(1 + \frac{1}{jT_I\Omega} + jT_D\Omega\right) = K_p\left[1 + j\left(T_D\Omega - \frac{1}{T_I\Omega}\right)\right], \tag{4.10}$$

where Ω is the distorted frequency of ω that is defined in (3.33), i.e.,

$$\Omega(\omega) = \frac{2}{h}\tan\left(\frac{\omega h}{2}\right).$$

The relationship between ω and $\Omega(\omega)$ is drawn as shown in Fig. 4.1.

We can also apply the direct difference method to the discrete-time PID controller as follows:

$$u_c(k) = K_p u(k) + C_I\sum_{j=0}^{k} u(j) + C_D\Delta u(k), \tag{4.11}$$

where $\Delta u(k) = u(k) - u(k-1)$ is a backward difference of the input signal. Using the z-transform expression, (4.11) can be written as

$$\hat{u}_c(z) = K_p\hat{u}(z) + C_I(1 + z^{-1} + z^{-2} + \cdots)\hat{u}(z) + C_D(1 - z^{-1})\hat{u}(z). \tag{4.12}$$

Therefore, when considering $|z| > 1$, (4.12) can be written in closed form,

$$\hat{u}_c(z) = K_p\hat{u}(z) + C_I\cdot\frac{1}{1-z^{-1}}\hat{u}(z) + C_D(1 - z^{-1})\hat{u}(z). \tag{4.13}$$

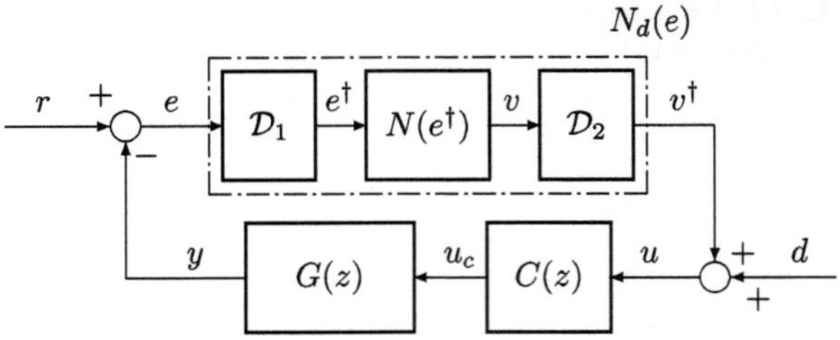

Fig. 4.2 Discretized PID control system

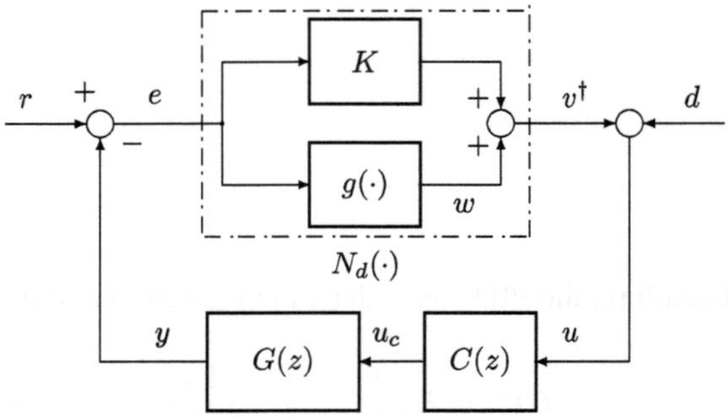

Fig. 4.3 Discretized nonlinear PID control system

Thus, the controller can be given as follows:

$$C(z) = K_p \left(1 + \frac{h}{T_I(1 - z^{-1})} + \frac{T_D}{h}(1 - z^{-1}) \right). \tag{4.14}$$

By using the above PID controllers, the discretized control systems can be represented as shown in Fig. 4.2 and equivalently in Fig. 4.3.

4.2.2 Controller Algorithm

The PID control algorithm based on (4.13) and (4.14) is easily realized in a computer (or a microprocessor). It can also be represented by a block diagram, as shown in Fig. 4.4. On the other hand, the controller algorithm using a bilinear approximation method based on (4.9) becomes a little complicated, as shown in Fig. 4.5.

However, the difference in the controller performance between these algorithms is not so large in the frequency response for $\omega \ll \omega_s/2 = \pi/h$. For example, the gain curves in the frequency domain become as shown in Figs. 4.6(a) and (b), where $K_p = C_I = C_D = 1.0$ and $h = 1.0$ are chosen. Here, (i) and (ii) are frequency responses (vs. ω) of the PID controller using the direct difference and bilinear approximation methods, respectively.

In addition, if the controller algorithm without division as shown in (4.11) and (4.12) is used, all variables can be realized in integers when parameters K_p,

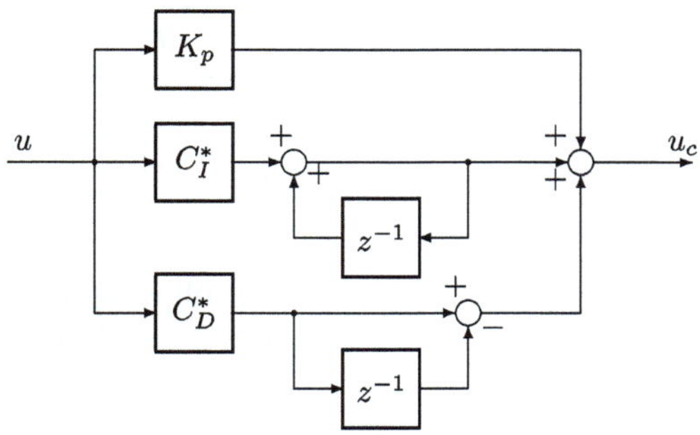

Fig. 4.4 PID controller using a direct difference method, where $C_I^* = C_I h = K_p h/T_I$ and $C_D^* = C_D/h = K_P T_D/h$

C_I, and C_D are also chosen in integers. Therefore, the direct difference method (4.11) will be used in the following examples.

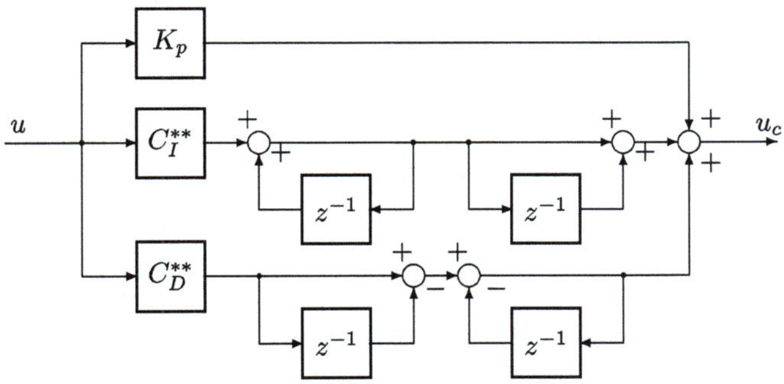

Fig. 4.5 PID controller using a bilinear approximation method, where $C_I^{**} = C_I h/2 = K_p h/2T_I$ and $C_D^{**} = 2C_D/h = 2K_P T_D/h$

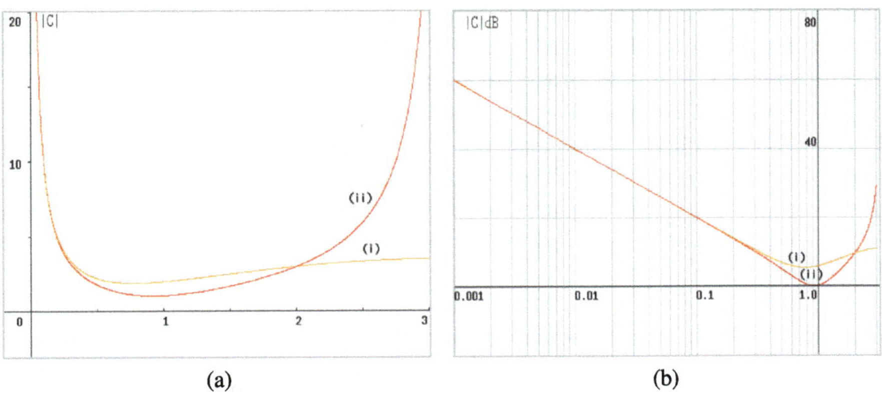

Fig. 4.6 Frequency-domain characteristics of discretized PID controllers (gain curves)

Table 4.1 PID parameters and control performances for Example 4.1

Cases	K_p	C_I	C_D	M_p	g_M [dB]	p_M [deg]
(i)	1.0	1.0	0.0	1.70	13.6	34.5
(ii)	1.0	1.0	0.5	1.67	16.3	35.2

Example 4.1 (A) Consider the following continuous plant without a pole on the origin (compare with Examples 3.1 and 3.2):

$$G(s) = \frac{K}{(s+0.2)(s+0.4)}, \qquad K = 0.02. \tag{4.15}$$

Here, the sampling period is assumed to be $h = 1.0$. Including the integration of the zero-order hold as shown in (1.66), the following partial fraction is obtained:

$$G_I(s) = \frac{G(s)}{s} = \frac{0.25}{s} - \frac{0.5}{s+0.2} + \frac{0.25}{s+0.4}. \tag{4.16}$$

Then,

$$G_I(z) = \frac{0.25}{1-z^{-1}} - \frac{0.5}{1-\mathrm{e}^{-0.2}z^{-1}} + \frac{0.25}{1-\mathrm{e}^{-0.4}z^{-1}}.$$

The z-transform of (4.15) with the zero-order hold is given as

$$G(z) = (1-z^{-1})G_I(z) = \frac{0.0082z + 0.0067}{z^2 - 1.489z + 0.549}. \tag{4.17}$$

Using a direct difference method as shown in (4.14), the controller can be written as

$$C(z) = K_p + \frac{C_I}{1-z^{-1}} + C_D(1-z^{-1}), \tag{4.18}$$

where $C_I = K_p/T_I$ and $C_D = K_pT_D$. The PID parameters and control performances are given in Table 4.1. Here, M_p, g_M, and p_M are the peak values, gain margins, and phase margins, respectively. Figure 4.7 shows the discretized nonlinear characteristics for this example, and Fig. 4.8 the Hall diagram. Figures 4.9(a) and (b) show the phase traces and step responses for cases (i) and (ii). The figures clearly show that the control system responses are well stabilized by using these PI and PID controllers. In this case, the modified Hall diagram is drawn as shown in Figs. 4.8(i) and (ii). Obviously, the robust stability is satisfied as $\beta > 1.0$ in either case. In this first example, the discretized nonlinear characteristic considered in the feedback loop is assumed to be a sigmoid (saturated) function as shown in Fig. 4.7(a) [11, 12].

Example 4.1 (B) Consider the case in which the discretized nonlinear characteristic is given as shown in Fig. 4.7(b) (i.e., the inclined sine function, as was shown in Figs. 2.6–2.8 (b).) In this case, similar phase traces and step responses are obtained, as shown in Figs. 4.10(a) and (b).

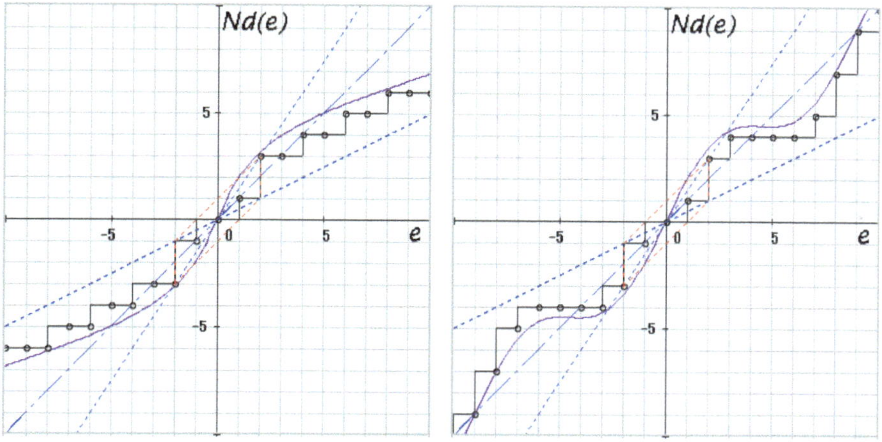

(a) (b)

Fig. 4.7 Discretized nonlinear characteristics for Example 4.1

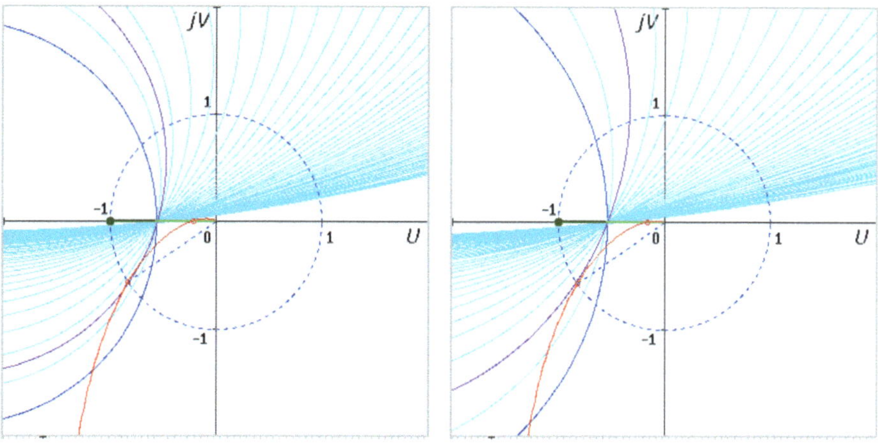

(i) (ii)

Fig. 4.8 Modified Hall diagram for Example 4.1

Example 4.2 Consider the following continuous plant without a pole on the origin (with a slightly quicker response than in Example 4.1):

$$G(s) = \frac{K}{(s+0.5)(s+1.0)}, \quad K = 0.3. \tag{4.19}$$

Including the integration of the zero-order hold, we have

$$G_I(s) = \frac{G(s)}{s} = \frac{0.6}{s} - \frac{1.2}{s+0.5} + \frac{0.6}{s+1.0}. \tag{4.20}$$

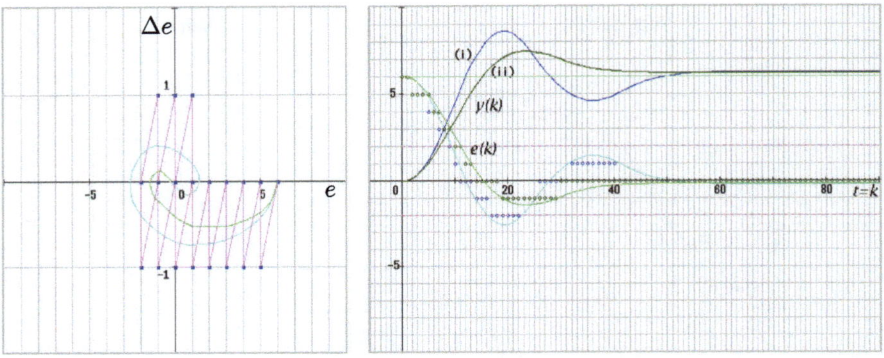

(a) (b)

Fig. 4.9 Phase traces and step responses for Example 4.1 (A)

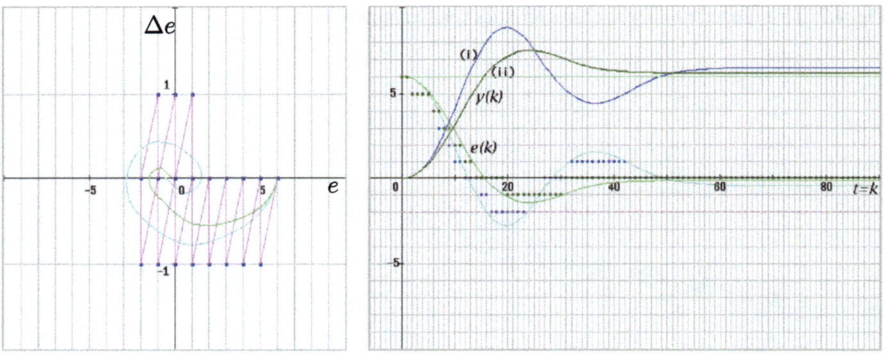

(a) (b)

Fig. 4.10 Phase traces and step responses for Example 4.1 (B)

Here, the sampling period is chosen to be the same as in Example 4.1, i.e., $h = 1.0$. Therefore,

$$G_I(z) = \frac{0.6}{1 - z^{-1}} - \frac{1.2}{1 - e^{-0.5}z^{-1}} + \frac{0.6}{1 - e^{-1.0}z^{-1}}.$$

The z-transform of (4.19) with the zero-order hold is given as follows:

$$G(z) = (1 - z^{-1})G_I(z) = \frac{0.093z + 0.056}{z^2 - 0.974z + 0.223}. \tag{4.21}$$

In the case of (i) in Table 4.1 (i.e., PI control), the discrete PI controller is written as

$$C(z) = K_p + \frac{C_I}{1 - z^{-1}}.$$

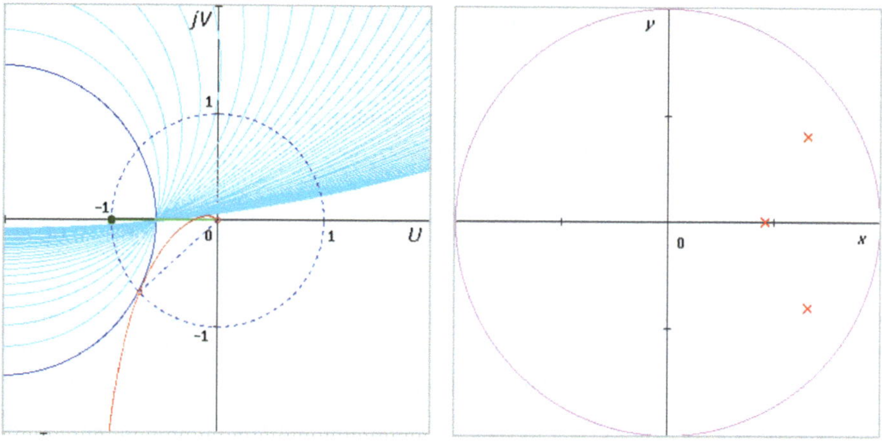

(a) (b)

Fig. 4.11 Modified Hall diagram and pole location for Example 4.2

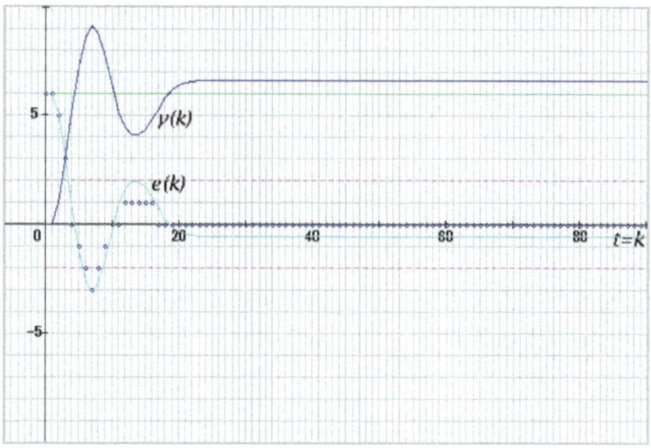

Fig. 4.12 Step responses for Example 4.2

If the controller parameters are chosen as $K_p = 1.0$ and $C_I = 1.0$, the characteristic equation is approximately given as

$$\hat{f}(z) = z^3 - 1.79z^2 + 1.22z - 0.28 = 0. \tag{4.22}$$

In this case, the modified Hall diagram is drawn as shown in Fig. 4.11(a). Figure 4.11(b) shows the location of the characteristic roots of the control system in the z-plane. The step responses of the PI control system are shown in Fig. 4.12. The robust performance of (4.22) will be discussed in Chap. 6.

4.2.3 Controlled Systems with Time Delay

The transfer function of a continuous plant (controlled system) with a time delay can be written based on the form of (1.65) as

$$G(s) = G_0(s)\mathrm{e}^{-L_p s} = \frac{N_p(s)}{(s-p_1)(s-p_2)\cdots(s-p_n)} \cdot \mathrm{e}^{-L_p s}, \tag{4.23}$$

irrespective of whether the time delay exists in the input or output side of the plant. Here, L_p is the time delay and $N_p(s)$ is a numerator polynomial. The order of the numerator is less than that of the denominator in general. From the derivation of $G(z)$ in Sect. 1.5.2, the z-transform transfer function of the continuous plant and holding circuit is given by

$$G(z) = (1 - z^{-1})G_I(z)z^{-d_p}, \tag{4.24}$$

where $d_p = L_p/h$. Thus, with respect to time-delayed systems, the following expression can be obtained; see (1.73) and (1.74):

$$G(z) = \frac{b_0 + b_1 z^{-1} + b_2 z^{-2} \cdots + b_n z^{-n}}{1 + a_1 z^{-1} + a_2 z^{-2} \cdots + a_n z^{-n}} \cdot z^{-d_p} \tag{4.25}$$

$$= \frac{b_0 z^n + b_1 z^{n-1} + \cdots + b_{n-1} z + b_n}{z^n + a_1 z^{n-1} + \cdots + a_{n-1} z + a_n} \cdot z^{-d_p}. \tag{4.26}$$

Example 4.3 Consider a plant with time delay L_p (or a transmission delay in the control action), i.e.,

$$G(s) = \frac{K}{(s+0.2)(s+0.4)} \cdot \mathrm{e}^{-L_p s}, \quad K = 0.02, \quad L_p = 2.0. \tag{4.27}$$

Since $h = 1.0$, from (4.17) the z-transform of $G(s)$ with a zero-order hold can be written as

$$G(z) = \frac{0.0082z + 0.0067}{z^2 - 1.489z + 0.549} \cdot z^{-2}. \tag{4.28}$$

Therefore, when applying PID algorithm (4.18), the calculation results for the phase traces and step responses are given as shown in Figs. 4.13(a) and (b). In this case, allowable sectors β with respect to the robust stability are calculated as shown in Table 4.2. In case (ii), the robust stability is satisfied with regard to the discretized nonlinearities as shown in Figs. 3.11(a) and (b) (i.e., sector [0.5, 1.5]). The modified Hall and Nichols diagrams are shown in Figs. 4.14(a) and (b).

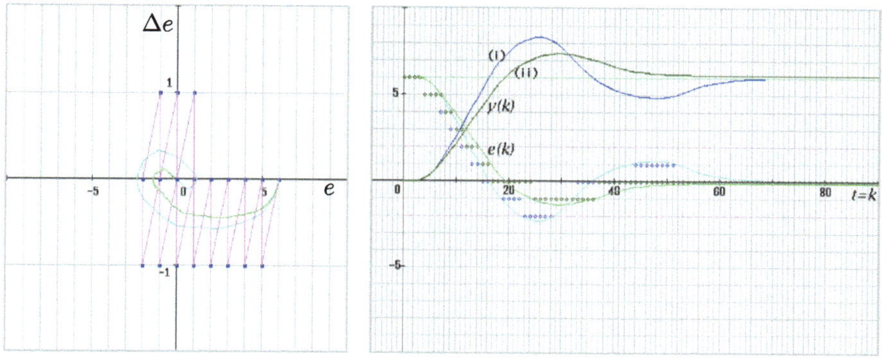

(a) (b)

Fig. 4.13 Phase traces and step responses for Example 4.3

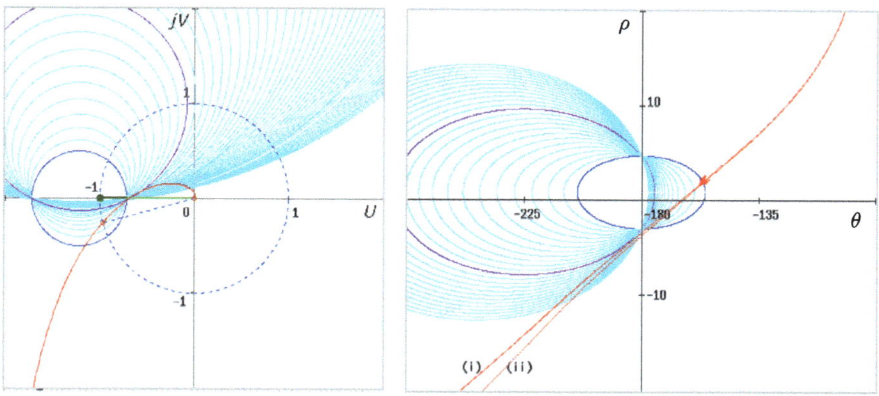

(a) (b)

Fig. 4.14 Modified Hall and Nichols diagrams for Example 4.3

Table 4.2 PID parameters and control performances for Example 4.3

Cases	K_p	C_I	C_D	M_p	g_M [dB]	p_M [deg]	β
(i)	1.0	1.0	0.5	4.3	3.5	15.1	0.49
(ii)	1.0	1.0	1.0	4.0	3.8	16.1	0.55

4.3 Model Reference Feedback Control

4.3.1 Discrete Model Reference and Observer

A model reference feedback structure for a robust control system was proposed by the author in 1964 [9]. A discrete-time version of the model reference feedback was presented in [20]. In this chapter, a discretized model reference control system (see

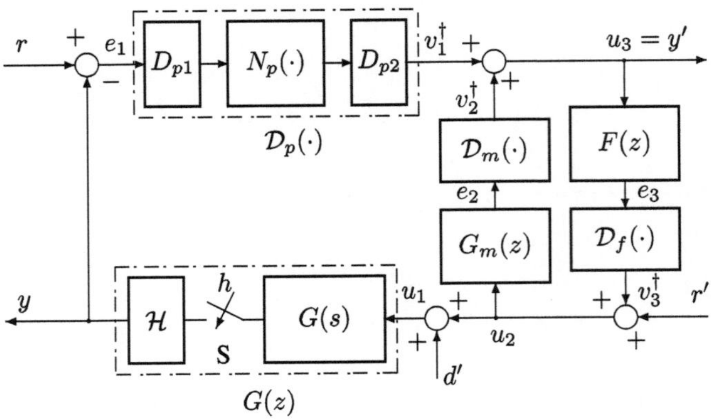

Fig. 4.15 Model reference feedback system

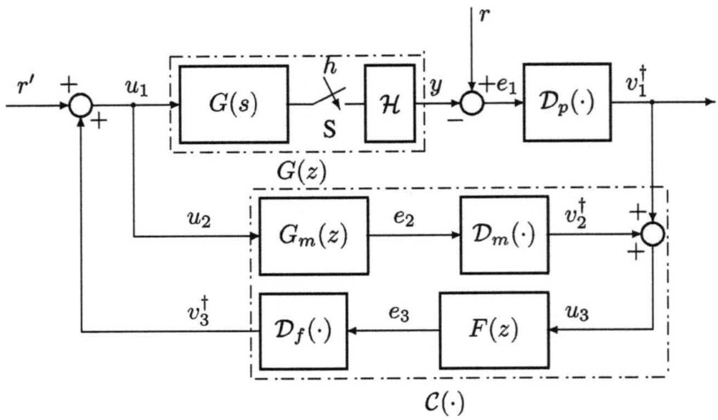

Fig. 4.16 Model reference feedback structure

Fig. 4.15) is examined [13–15]. The control system is redrawn as shown in Fig. 4.16. In these figures, $G(s)$ is a continuous-time and continuous-value plant, and $G(z)$ is the z-transform of $G(s)$ together with the zero-order hold $\mathcal{H}$. Also, $N_p(\cdot)$, $\mathcal{D}_{p1}$, and $\mathcal{D}_{p2}$ are, respectively, a nonlinear element and the input and output discretizing units which are determined by sensor and actuator elements. Moreover, $G_m(z)$ is a plant model and $F(z)$ is a feedback compensator (from another point of view, a discrete observer). $\mathcal{D}_m(\cdot)$ and $\mathcal{D}_f(\cdot)$ are the nonlinear discretized elements of the model and the compensator, respectively.

In the figures, each symbol e, v, $\cdots$ indicates the sequence $e(k)$, $v(k)$, $\cdots$, ($k = 0, 1, 2, \cdots$) in discrete time, but for continuous values. Each symbol $e^{\dagger}$, $v^{\dagger}$, $\cdots$ indicates a discrete value that can be assigned to an integer number, e.g.,

$$\begin{aligned} e_1^{\dagger} &\in \{\cdots, -2\gamma, -\gamma, 0, \ \gamma, \ 2\gamma, \cdots\}, \\ v_1^{\dagger} &\in \{\cdots, -2\gamma, -\gamma, 0, \ \gamma, \ 2\gamma, \cdots\}, \end{aligned} \tag{4.29}$$

where γ is the resolution of each variable. An example of the discretization characteristics D_{p1}, D_{p2} and the discretized characteristic $v_1^{\dagger} = \mathcal{D}_p(e_1)$ is depicted as shown in Fig. 4.7(A). Here, without loss of generality, the resolutions in D_{p1} and D_{p2} are assumed to be $\gamma = 1.0$, and the continuous nonlinear curve $N_p(\cdot)$ is chosen as a sigmoid function. Note that nonlinear discretized characteristic $\mathcal{D}_p(\cdot)$ corresponds to $N_d(\cdot)$ in Fig. 4.2.

4.3.2 Bilinear Transformation and Discrete Model

The model system $K_m G_m(z)$ is assumed to be a second-order lag system, e.g.,

$$K_m G_m(z) = K_m \tilde{G}_m(\delta) = \frac{K_m}{1 + C_1\delta + C_2\delta^2}, \tag{4.30}$$

where δ is the following bilinear transformation as shown in (4.5):

$$\delta = \frac{2}{h} \cdot \frac{z-1}{z+1}. \tag{4.31}$$

Here, K_m is the nominal gain of $\mathcal{D}_m$, which is usually chosen as the nominal gain of discretized nonlinear characteristic $\mathcal{D}_p$ (i.e., K), and C_1 and C_2 are the design parameters of the model system.

This type of model system will correspond to a discrete observer, although it does not clearly observe the states of the plant. The z-transform expression of (4.30) is given by

$$K_m G_m(z) = \frac{K_m h^2(z+1)^2}{h^2(z+1)^2 + 2C_1 h(z+1)(z-1) + 4C_2(z-1)^2}. \tag{4.32}$$

Obviously, the δ-function approaches Laplace transform variable s, when the sampling period is $h \to 0$.

In this chapter, hereafter, δ will be used instead of the z-transform operator. Therefore, the feedback compensator $K_f F(z)$ as shown in Figs. 4.15 and 4.16 is defined as

$$K_f F(z) = K_f \tilde{F}(\delta) = \frac{1 + C_1\delta + C_2\delta^2}{K_m(1 + c_1\delta + c_2\delta^2)}. \tag{4.33}$$

Here, K_f is the nominal gain of $\mathcal{D}_f$ that will be substituted by $1/K_f$, and C_1 and C_2 are the design parameters of the feedback compensator. Thus, the z-transform expression of (4.33) is written as

$$F(z) = \frac{h^2(z+1)^2 + 2C_1 h(z+1)(z-1) + 4C_2(z-1)^2}{h^2(z+1)^2 + 2c_1 h(z+1)(z-1) + 4c_2(z-1)^2}. \tag{4.34}$$

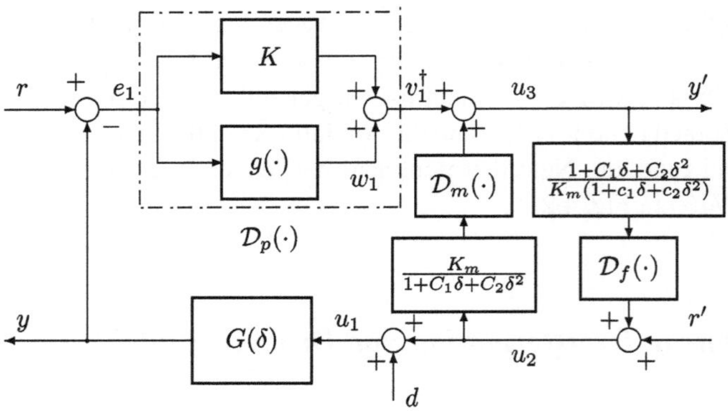

Fig. 4.17 Second-order model reference feedback

In the frequency domain, as was described in (3.33), δ can be expressed as

$$\delta(\mathrm{e}^{j\omega h}) = j\Omega(\omega) = j\frac{2}{h}\tan\left(\frac{\omega h}{2}\right), \tag{4.35}$$

where Ω is a distorted frequency of ω. By using expression (4.35), (4.30) and (4.33) can be written as follows:

$$G_m(\mathrm{e}^{j\omega h}) = \tilde{G}_m(j\Omega) = \frac{1}{1 - C_2\Omega^2 + jC_1\Omega}, \tag{4.36}$$

and

$$F(\mathrm{e}^{j\omega h}) = \tilde{F}(j\Omega) = \frac{1 - C_2\Omega^2 + jC_1\Omega}{1 - c_2\Omega^2 + jc_1\Omega}. \tag{4.37}$$

By applying a second-order model, the model reference control systems can be redrawn as shown in Fig. 4.17.

When the controllers are in high resolution (i.e., $\gamma \to 0$), the model reference control system as shown in Fig. 4.17 can be transformed into Fig. 4.18. Here, d' is a disturbance signal generated by the discretization of controllers. The equivalent controller $C(\delta)$ and the pre-compensator $D(\delta)$ are given by

$$C(\delta) = \frac{K_f F(\delta)}{1 - K_m K_f G_m(\delta) F(\delta)} = \frac{1 + C_1\delta + C_2\delta^2}{K_m(c_1\delta + c_2\delta^2)}, \tag{4.38}$$

$$D(\delta) = \frac{1}{K_f F(\delta)} = \frac{K_m(1 + c_1\delta + c_2\delta^2)}{1 + C_1\delta + C_2\delta^2}. \tag{4.39}$$

Then, the block diagram of the equivalent PID control system is drawn simply as shown in Fig. 4.19. Here, $C(\delta)$ can be considered to be a controller when $c_2 \ll c_1$. If $c_2 \to 0$, the controller is approximately written as

$$C(\delta) = \frac{1}{\kappa}\delta^{-1} + \frac{C_1}{\kappa} + \frac{C_2}{\kappa}\delta, \tag{4.40}$$

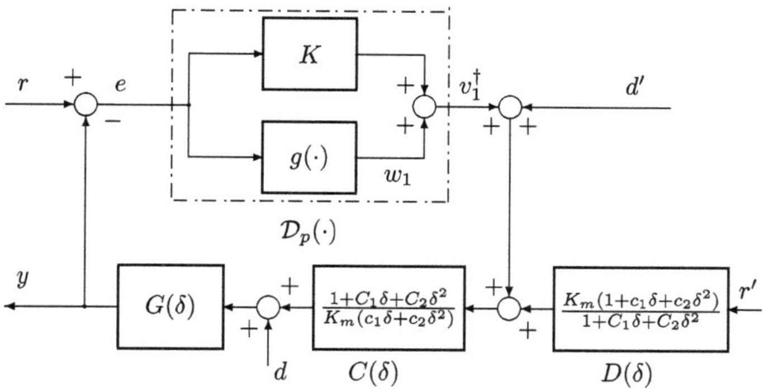

Fig. 4.18 Approximate PID control system

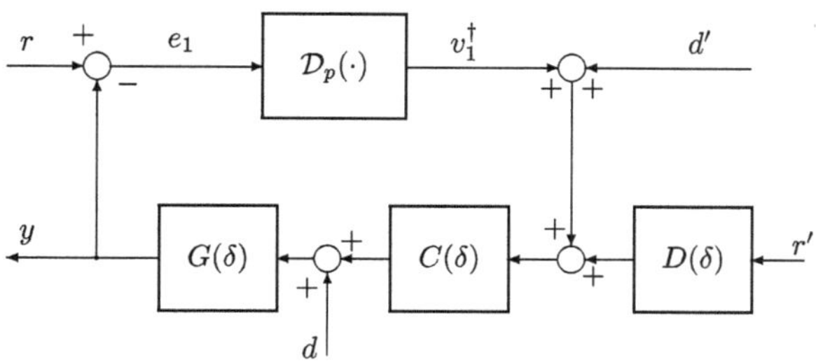

Fig. 4.19 Equivalent PID control system

where $\kappa = K_m c_1$. We can see that (4.40) is a three-term controller based on the bilinear transform expression. When the sampling period is $h \to 0$, the δ-function approaches Laplace transform variable s. Therefore, the scheme given in (4.40) will correspond to a traditional continuous PID control.

The z-transform of (4.40) is written as

$$C(z) = \frac{h}{2\kappa} \cdot \frac{z+1}{z-1} + \frac{C_1}{\kappa} + \frac{2C_2}{\kappa h} \cdot \frac{z-1}{z+1}. \tag{4.41}$$

In the distorted frequency domain, it can be expressed as

$$C(j\Omega) = \frac{C_1}{\kappa} + j \cdot \frac{1}{\kappa}\left(C_2\Omega - \frac{1}{\Omega}\right). \tag{4.42}$$

These algorithms (4.40) and (4.41) can be regarded as quasi-PID control algorithms.

With respect to higher order and time-delay plants, the model system $KG_m(z)$ can be improved as follows:

$$K_m G_m(\delta) = \frac{K_m}{1 + C_1\delta + C_2\delta^2} \cdot \mathrm{e}^{-L_m s}, \tag{4.43}$$

where L_m is the inserted time delay of the model system. Since a feedback compensator that has a time-lead characteristic cannot be realized, the quasi-PID controller should be written as

$$C(\delta) = \frac{1 + C_1\delta + C_2\delta^2}{K_m(1 + c_1\delta + c_2\delta^2) - z^{-d_m}}, \qquad d_m = L_m/h. \tag{4.44}$$

4.4 Discretized Nonlinear Characteristics and Inequality Conditions

4.4.1 Partition of Nonlinear Characteristics

The input/output discretization processes and the discretized nonlinear characteristics are illustrated in Fig. 4.20. The nonlinear characteristics can be partitioned as follows:

$$v_1^\dagger = \mathcal{D}_p(e_1^\dagger) = K e_1^\dagger + g_1(e_1^\dagger), \;\; 0 < K < \infty, \tag{4.45}$$

$$|w_1^\dagger| = |g_1(e_1^\dagger)| < \infty, \quad \text{for } |e_1^\dagger| < \varepsilon_1, \tag{4.46}$$

$$|w_1^\dagger| = |g_1(e_1^\dagger)| \le \beta_1 |e_1^\dagger|, \quad \text{for } |e_1^\dagger| \ge \varepsilon_1. \tag{4.47}$$

Since the input-side discretization is equivalently represented as shown in Fig. 4.21, the input signal is considered as an integer, $e_1 = e_1^\dagger$. Here, $w_1^\dagger$ is not always an integer, although † is attached to the symbol. In these inequalities, when analyzing the robust stability in a global sense, it is sufficient to consider nonlinear term (4.47), because nonlinear term (4.46) can be treated as a disturbance signal. In this study, since the nonlinear characteristic (4.45) is assumed to exist in the first and third quadrants, the sector parameter β_1 should be considered in $0 \le \beta_1 \le K$.

In regard to the model reference feedback system, the nonlinear characteristic in the discrete model should be considered:

$$v_2^\dagger = \mathcal{D}_m(e_2^\dagger) = K_m e_2^\dagger + g_2(e_2^\dagger), \;\; 0 < K_m < \infty, \tag{4.48}$$

$$|w_2^\dagger| = |g_2(e_2^\dagger)| < \infty, \quad \text{for } |e_2^\dagger| < \varepsilon_2, \tag{4.49}$$

$$|w_2^\dagger| = |g_2(e_2^\dagger)| \le \beta_2 |e_2^\dagger|, \quad \text{for } |e_2^\dagger| \ge \varepsilon_2, \tag{4.50}$$

and $0 \le \beta_2 \le K_m$. Moreover, in regard to the feedback compensator, the following expression will be given:

$$v_3^\dagger = \mathcal{D}_f(e_3^\dagger) = K_f e_3^\dagger + g_3(e_3^\dagger), \;\; 0 < K_f < \infty, \tag{4.51}$$

$$|w_3^\dagger| = |g_3(e_3^\dagger)| < \infty, \quad \text{for } |e_3^\dagger| < \varepsilon_3, \tag{4.52}$$

$$|w_3^\dagger| = |g_3(e_3^\dagger)| \le \beta_3 |e_3^\dagger|, \quad \text{for } |e_3^\dagger| \ge \varepsilon_3, \tag{4.53}$$

and $0 \le \beta_3 \le K_f$.

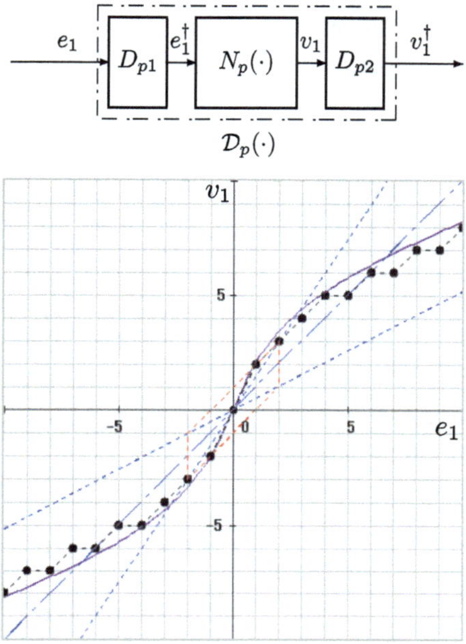

Fig. 4.20 Discretization for a nonlinear characteristic

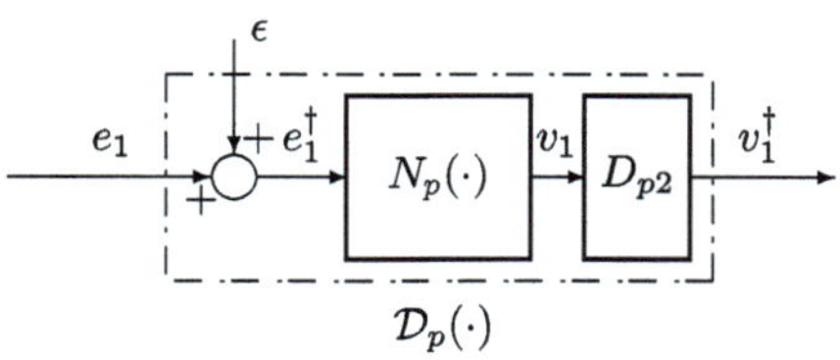

Fig. 4.21 Equivalent expression ($\epsilon = e_1^\dagger - e_1 = d_1$: a sawtooth signal)

Therefore, the robust stability of model reference feedback systems with multi-nonlinearity is analyzed based on the inner product and norm analysis in the ℓ_2 space. In regard to (4.47), the following new nonlinear function can be defined:[2]

$$f_i(e_i) := g_i(e_i) + \beta_i \cdot e_i, \quad i = 1, 2, 3. \tag{4.54}$$

When considering the discretized output of the nonlinear characteristic, $w_i^\dagger = g_i(e_i^\dagger)$, the following expression is given:

$$f_i(e_i^\dagger(k)) = w_i^\dagger(k) + \beta_i \cdot e_i^\dagger(k). \tag{4.55}$$

From inequality (4.47), it can be seen that (4.55) belongs to the first and third quadrants.

[2]Hereafter, only $i = 1$ is considered.

For the neutral points of $e_i^\dagger(k)$ and $w_i^\dagger(k)$, the following expression can be given from (4.55):

$$\frac{1}{2}(f_i(e_i^\dagger(k)) + f_i(e_i^\dagger(k-1))) = \overline{w}_i^\dagger(k) + \beta_i \cdot \overline{e}_i^\dagger(k), \tag{4.56}$$

where

$$\overline{w}_i^\dagger(k) = \frac{w_i^\dagger(k) + w_i^\dagger(k-1)}{2}, \quad \overline{e}_i^\dagger(k) = \frac{e_i^\dagger(k) + e_i^\dagger(k-1)}{2}. \tag{4.57}$$

Then, the trapezoidal area of the one-step transition in integer grid coordinates, $f_i(e)$, is written as

$$\begin{aligned}\tau_i(k) &:= \frac{1}{2}(f_i(e^\dagger(k)) + f_i(e_i^\dagger(k-1)))\Delta e_i^\dagger(k) \\ &= (\overline{w}_i^\dagger(k) + \beta_i \overline{e}_i^\dagger(k))\Delta e_i^\dagger(k).\end{aligned} \tag{4.58}$$

Here, $\Delta e_i^\dagger(k)$ is the backward difference of sequence $e_i^\dagger(k)$.

$$\Delta e_i^\dagger(k) = e_i^\dagger(k) - e_i^\dagger(k-1).$$

Since $f(e_i^\dagger(k))$ belongs to the first and third quadrants, the area of each trapezoid $\tau_i(k)$ is non-negative when $e_i(k)$ increases (decreases) in the first (third) quadrant. On the other hand, the trapezoidal area $\tau_i(k)$ is non-positive when $e_i(k)$ decreases (increases) in the first (third) quadrant.

As described in Chap. 3, the following assumption is provided with respect to the discretized responses on the integer grid coordinates.

Assumption The absolute value of the backward difference of sequence $e(k)$ does not exceed γ, i.e.,

$$|\Delta e_i(k)| = |e_i(k) - e_i(k-1)| \le \gamma. \tag{4.59}$$

If condition (4.59) is satisfied, $\Delta e^\dagger(k)$ is exactly $\pm\gamma$ or 0 because of the discretization. That is, the absolute value of the backward difference can be given as

$$|\Delta e_i^\dagger(k)| = |e_i^\dagger(k) - e_i^\dagger(k-1)| = \gamma \;\text{ or }\; 0.$$

This assumption states that each point of the response traces on adjacent points in the integer grid coordinates.

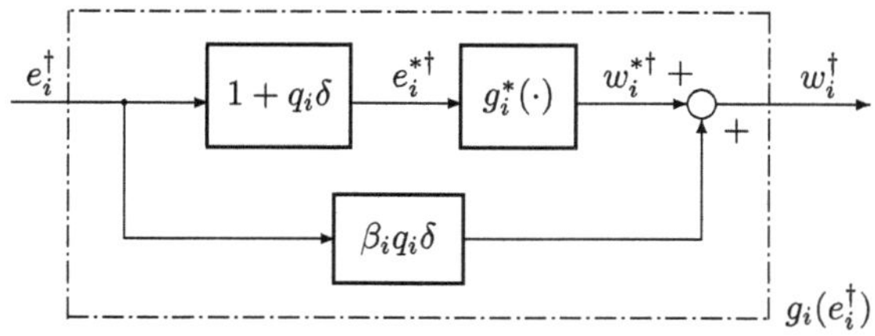

Fig. 4.22 Nonlinear subsystem ($i = 1, 2, 3$)

4.4.2 Sum of Trapezoidal Areas

Consider the following sum of trapezoidal areas:

$$\sigma_i(p) = \sum_{k=1}^{p} \tau_i(k). \tag{4.60}$$

If the above assumption is satisfied with respect to the discretization of the control system, the sum of trapezoidal areas, $\sigma(p)$, becomes non-negative for any p. Since the discretized output traces the same points on the stepwise nonlinear characteristic, the sum of trapezoidal areas is canceled when $e_i(k)$ (and $e_i^\dagger(k)$) decreases (increases) from a certain point $(e_i^\dagger(k), f_i(e_i^\dagger(k)))$ in the first (third) quadrant. (Here, without loss of generality, the response of discretized point $(e_i^\dagger(k), f_i(e_i^\dagger(k)))$ is assumed to commence at the origin.)

From Eq. (4.58), the sum of trapezoidal areas can be expressed as follows:

$$\begin{aligned}\sigma_i(p) &:= \frac{1}{2}\sum_{k=1}^{p}(f_i(e_i^\dagger(k)) + f_i(e_i^\dagger(k-1)))\Delta e_i^\dagger(k)\\ &= \langle\, \overline{w}_i^\dagger(k) + \beta_i \overline{e}_i^\dagger(k), \Delta \overline{e}_i^\dagger(k)\,\rangle_p.\end{aligned} \tag{4.61}$$

Here, $\langle\cdot,\cdot\rangle_p$ denotes the inner product in the ℓ_2 space,

$$\langle\, x(k), y(k)\,\rangle_p = \sum_{k=1}^{p} x(k)y(k).$$

In order to derive the robust stability condition, the following new sequences are considered:

$$\overline{e}_i^{*\dagger}(k) = \overline{e}_i^\dagger(k) + q_i \cdot \frac{\Delta e_i^\dagger(k)}{h}, \tag{4.62}$$

$$\overline{w}_i^{*\dagger}(k) = \overline{w}_i^\dagger(k) - \beta_i q_i \cdot \frac{\Delta e_i^\dagger(k)}{h}, \tag{4.63}$$

where q_i is a non-negative number. The relationship between Eqs. (4.62) and (4.63) is as shown in Fig. 4.22.

Based on these sequences, we give the following lemma.

Lemma 4.1 *If the following inequality is satisfied with respect to the inner product of the neutral points of* (4.55) *and the backward difference*:

$$\langle\, \overline{w}_i^{\dagger}(k) + \beta_i \overline{e}_i^{\dagger}(k),\, \Delta e_i^{\dagger}(k)\,\rangle_p \geq 0, \tag{4.64}$$

we can obtain

$$\|\overline{w}_i^{*\dagger}(k)\|_{2,p} \leq \beta_i \|\overline{e}_i^{*\dagger}(k)\|_{2,p} \leq \beta_i \|\overline{e}_i^{*}(k)\|_{2,p} \tag{4.65}$$

for any $q_i \geq 0$ *and* $p \to \infty$. *Here,* $\|\cdot\|_{2,p}$ *denotes the Euclidean norm, which can be defined as*

$$\|x(k)\|_{2,p} := \left(\sum_{k=1}^{p} |x(k)|^2\right)^{1/2}.$$

Proof The following equation is obtained from (4.62) and (4.63):

$$\begin{aligned}&\beta_i^2 \|\overline{e}_i^{*\dagger}(k)\|_{2,p}^2 - \|\overline{w}_i^{*\dagger}(k)\|_{2,p}^2 \\ &\quad = \beta_i^2 \|\overline{e}_i^{\dagger}(k)\|_{2,p}^2 - \|\overline{w}_i^{\dagger}(k)\|_{2,p}^2 + \frac{2\beta_i q_i}{h} \cdot \langle \overline{w}_i^{\dagger}(k) + \beta_i \overline{e}_i^{\dagger}(k), \Delta e_i^{\dagger}(k)\rangle_p.\end{aligned}$$

From (4.47), (4.50), and (4.53), it holds that

$$\|\overline{w}_i^{\dagger}(k)\|_{2,p} \leq \beta_i \|\overline{e}_i^{\dagger}(k)\|_{2,p}. \tag{4.66}$$

Thus, from (4.64) we can obtain

$$\|\overline{w}_i^{*\dagger}(k)\|_{2,p} \leq \beta_i \|\overline{e}_i^{*\dagger}(k)\|_{2,p}. \tag{4.67}$$

Since $\|\overline{e}_i^{\dagger}(k)\|_{2,p} \leq \|\overline{e}_i(k)\|_{2,p}$ for $p \to \infty$, the following inequality holds in the frequency domain based on Parseval's formula:[3]

$$\|\overline{e}_i^{*\dagger}(\delta)\|_2 = |1 + q_i\delta| \cdot \|\overline{e}_i^{\dagger}(\delta)\|_2 \leq |1 + q_i\delta| \cdot \|\overline{e}_i(\delta)\|_2 = \|\overline{e}_i^{*}(\delta)\|_2.$$

Then, $\|\overline{e}_i^{*\dagger}(k)\|_{2,p} \leq \|\overline{e}_i^{*}(k)\|_{2,p}$, and thus the right side of inequality (4.65) is satisfied. □

[3]Hereafter, δ is considered $j\Omega(\omega)$ as shown in (4.35).

4.5 Vector-Matrix Expression for Model Reference Control

The model reference control system as shown in Fig. 4.15 is given by the following vector-matrix expression:[4]

$$\begin{bmatrix} e_1(\delta) \\ e_2(\delta) \\ e_3(\delta) \end{bmatrix} = \begin{bmatrix} 1 \\ 0 \\ 0 \end{bmatrix} r(\delta) + \begin{bmatrix} -G(\delta) & 0 & 0 \\ 0 & G_m(\delta) & 0 \\ 0 & 0 & F(\delta) \end{bmatrix} \begin{bmatrix} u_1(\delta) \\ u_2(\delta) \\ u_3(\delta) \end{bmatrix}, \tag{4.68}$$

where control inputs u_1, u_2, and u_3 are given by

$$\begin{bmatrix} u_1(\delta) \\ u_2(\delta) \\ u_3(\delta) \end{bmatrix} = \begin{bmatrix} u_2(\delta) \\ v_3^\dagger(\delta) \\ v_1^\dagger(\delta) + v_2^\dagger(\delta) \end{bmatrix} + \begin{bmatrix} d'(\delta) \\ r'(\delta) \\ 0 \end{bmatrix}.$$

As shown in Fig. 4.22, the nonlinear parts (point-to-point characteristics) of the system can be written as

$$\begin{bmatrix} v_1^\dagger \\ v_2^\dagger \\ v_3^\dagger \end{bmatrix} = \begin{bmatrix} \Gamma_1(\delta) & 0 & 0 \\ 0 & \Gamma_2(\delta) & 0 \\ 0 & 0 & \Gamma_3(\delta) \end{bmatrix} \begin{bmatrix} e_1^\dagger \\ e_2^\dagger \\ e_3^\dagger \end{bmatrix} + \begin{bmatrix} w_1^{*\dagger} \\ w_2^{*\dagger} \\ w_3^{*\dagger} \end{bmatrix}, \tag{4.69}$$

where $\Gamma_1(\delta) = K + \beta_1 q_1 \delta$, $\Gamma_2(\delta) = K_m + \beta_2 q_2 \delta$, $\Gamma_3(\delta) = K_f + \beta_3 q_3 \delta$.

If the exogenous inputs are $r' = d' = 0$, u_1 is equal to u_2. Moreover, since $e_i^\dagger = e_i - d_i$ $(i = 1, 2, 3)$, the following closed-loop system equation can be given:

$$\begin{aligned} \begin{bmatrix} e_1^\dagger \\ e_2^\dagger \\ e_3^\dagger \end{bmatrix} &= \begin{bmatrix} d_1 + r \\ d_2 \\ d_3 \end{bmatrix} + \begin{bmatrix} -G & 0 \\ G_m & 0 \\ 0 & F \end{bmatrix} \begin{bmatrix} u_1 \\ u_3 \end{bmatrix} \\ &= \begin{bmatrix} d_1 + r \\ d_2 \\ d_3 \end{bmatrix} + \begin{bmatrix} -G & 0 \\ G_m & 0 \\ 0 & F \end{bmatrix} \begin{bmatrix} 0 & 0 & \Gamma_3 \\ \Gamma_1 & \Gamma_2 & 0 \end{bmatrix} \begin{bmatrix} e_1^\dagger \\ e_2^\dagger \\ e_3^\dagger \end{bmatrix} \\ &\quad + \begin{bmatrix} -G & 0 \\ G_m & 0 \\ 0 & F \end{bmatrix} \begin{bmatrix} 0 & 0 & 1 \\ 1 & 1 & 0 \end{bmatrix} \begin{bmatrix} w_1^{*\dagger} \\ w_2^{*\dagger} \\ w_3^{*\dagger} \end{bmatrix}, \end{aligned} \tag{4.70}$$

where d_i $(|d_i| \leq \gamma)$ are discretized/quantized errors.

---

[4]In the following, the transformed variables, e.g., $e(\delta)$, $e(z)$, $u(\delta)$, $u(z)$, $\cdots$ will be used without the "hat" symbol.

For the neutral points defined in (4.57), the following expression can be given:

$$\begin{bmatrix} 1 & 0 & \Gamma_3 G \\ 0 & 1 & -\Gamma_3 G_m \\ -\Gamma_1 F & -\Gamma_2 F & 1 \end{bmatrix} \begin{bmatrix} \overline{e}_1^{\dagger} \\ \overline{e}_2^{\dagger} \\ \overline{e}_3^{\dagger} \end{bmatrix} = \begin{bmatrix} \overline{d}_1 + \overline{r} \\ \overline{d}_2 \\ \overline{d}_3 \end{bmatrix} \begin{bmatrix} 0 & 0 & -G \\ 0 & 0 & G_m \\ F & F & 0 \end{bmatrix} \begin{bmatrix} \overline{w}_1^{*\dagger} \\ \overline{w}_2^{*\dagger} \\ \overline{w}_3^{*\dagger} \end{bmatrix}. \tag{4.71}$$

The inverse matrix of the left side of the equation is written as

$$\begin{aligned} &\begin{bmatrix} 1 & 0 & \Gamma_3 G \\ 0 & 1 & -\Gamma_3 G_m \\ -\Gamma_1 F & -\Gamma_2 F & 1 \end{bmatrix}^{-1} \\ &= \frac{1}{1 + (\Gamma_1 G - \Gamma_2 G_m)\Gamma_3 F} \cdot \begin{bmatrix} 1 - \Gamma_2 \Gamma_3 G_m F & -\Gamma_2 \Gamma_3 G F & -\Gamma_3 G \\ \Gamma_1 \Gamma_3 G_m F & 1 + \Gamma_1 \Gamma_3 G F & \Gamma_3 G_m \\ \Gamma_1 F & \Gamma_2 F & 1 \end{bmatrix} \\ &= \begin{bmatrix} \Psi_{11} & \Psi_{12} & \Psi_{13} \\ \Psi_{21} & \Psi_{22} & \Psi_{23} \\ \Psi_{31} & \Psi_{32} & \Psi_{33} \end{bmatrix}. \end{aligned} \tag{4.72}$$

Hereafter, in order to simplify the equation, we will use the symbols

$$\Lambda_i(\delta) = 1 + q_i \delta, \quad i = 1, 2, 3. \tag{4.73}$$

Thus,

$$\begin{bmatrix} \overline{e}_1^{*\dagger}(\delta) \\ \overline{e}_2^{*\dagger}(\delta) \\ \overline{e}_3^{*\dagger}(\delta) \end{bmatrix} = \begin{bmatrix} \Lambda_1(\delta) & 0 & 0 \\ 0 & \Lambda_2(\delta) & 0 \\ 0 & 0 & \Lambda_3(\delta) \end{bmatrix} \begin{bmatrix} \overline{e}_1^{\dagger}(\delta) \\ \overline{e}_2^{\dagger}(\delta) \\ \overline{e}_3^{\dagger}(\delta) \end{bmatrix}. \tag{4.74}$$

Hence, the vector-matrix expression for the control system is written by functions of δ as follows:

$$\begin{aligned} \begin{bmatrix} \overline{e}_1^{*\dagger} \\ \overline{e}_2^{*\dagger} \\ \overline{e}_3^{*\dagger} \end{bmatrix} &= \begin{bmatrix} \Lambda_1 & 0 & 0 \\ 0 & \Lambda_2 & 0 \\ 0 & 0 & \Lambda_3 \end{bmatrix} \begin{bmatrix} \Psi_{11} & \Psi_{12} & \Psi_{13} \\ \Psi_{21} & \Psi_{22} & \Psi_{23} \\ \Psi_{31} & \Psi_{32} & \Psi_{33} \end{bmatrix} \begin{bmatrix} \overline{d}_1 + \overline{r} \\ \overline{d}_2 \\ \overline{d}_3 \end{bmatrix} \\ &+ \begin{bmatrix} \Lambda_1 & 0 & 0 \\ 0 & \Lambda_2 & 0 \\ 0 & 0 & \Lambda_3 \end{bmatrix} \begin{bmatrix} \Psi_{11} & \Psi_{12} & \Psi_{13} \\ \Psi_{21} & \Psi_{22} & \Psi_{23} \\ \Psi_{31} & \Psi_{32} & \Psi_{33} \end{bmatrix} \begin{bmatrix} 0 & 0 & -P \\ 0 & 0 & P_m \\ F & F & 0 \end{bmatrix} \begin{bmatrix} \overline{w}_1^{*\dagger} \\ \overline{w}_2^{*\dagger} \\ \overline{w}_3^{*\dagger} \end{bmatrix} \\ &= \begin{bmatrix} \Lambda_1 \Psi_{11} & \Lambda_1 \Psi_{12} & \Lambda_1 \Psi_{13} \\ \Lambda_2 \Psi_{21} & \Lambda_2 \Psi_{22} & \Lambda_2 \Psi_{23} \\ \Lambda_3 \Psi_{31} & \Lambda_3 \Psi_{32} & \Lambda_3 \Psi_{33} \end{bmatrix} \begin{bmatrix} \overline{d}_1 + \overline{r} \\ \overline{d}_2 \\ \overline{d}_3 \end{bmatrix} \end{aligned}$$

$$+\begin{bmatrix} \Lambda_1\Psi_{13}F & \Lambda_1\Psi_{13}F & -\Lambda_1(\Psi_{11}G-\Psi_{12}G_m) \\ \Lambda_2\Psi_{23}F & \Lambda_2\Psi_{23}F & -\Lambda_2(\Psi_{21}G-\Psi_{22}G_m) \\ \Lambda_3\Psi_{33}F & \Lambda_3\Psi_{33}F & -\Lambda_3(\Psi_{31}G-\Psi_{32}G_m) \end{bmatrix}\begin{bmatrix} \overline{w}_1^{*\dagger} \\ \overline{w}_2^{*\dagger} \\ \overline{w}_3^{*\dagger} \end{bmatrix}.$$

In regard to each norm of the equation, we obtain the inequality

$$\begin{bmatrix} \|\overline{e}_1^{*\dagger}\|_2 \\ \|\overline{e}_2^{*\dagger}\|_2 \\ \|\overline{e}_3^{*\dagger}\|_2 \end{bmatrix} \leqq \begin{bmatrix} |\Lambda_1\Psi_{11}| & |\Lambda_1\Psi_{12}| & |\Lambda_1\Psi_{13}| \\ |\Lambda_2\Psi_{21}| & |\Lambda_2\Psi_{22}| & |\Lambda_2\Psi_{23}| \\ |\Lambda_3\Psi_{31}| & |\Lambda_3\Psi_{32}| & |\Lambda_3\Psi_{33}| \end{bmatrix}\begin{bmatrix} \|\overline{d}_1\|_2+\|\overline{r}\|_2 \\ \|\overline{d}_2\|_2 \\ \|\overline{d}_3\|_2 \end{bmatrix}$$

$$+\begin{bmatrix} |\Lambda_1\Psi_{13}F| & |\Lambda_1\Psi_{13}F| & |\Lambda_1(\Psi_{11}G-\Psi_{12}G_m)| \\ |\Lambda_2\Psi_{23}F| & |\Lambda_2\Psi_{23}F| & |\Lambda_2(\Psi_{21}G-\Psi_{22}G_m)| \\ |\Lambda_3\Psi_{33}F| & |\Lambda_3\Psi_{33}F| & |\Lambda_3(\Psi_{31}G-\Psi_{32}G_m)| \end{bmatrix}\begin{bmatrix} \|\overline{w}_1^{*\dagger}\|_2 \\ \|\overline{w}_2^{*\dagger}\|_2 \\ \|\overline{w}_3^{*\dagger}\|_2 \end{bmatrix}. \quad (4.75)$$

Here, the symbol $\leqq$ denotes a set of inequalities for each element.

4.6 Robust Stability Condition for Multi-Nonlinearity Systems

By using the result of Lemma 4.1, the following inequality is derived:

$$\begin{bmatrix} 1-\beta_1|\Lambda_1\Psi_{13}F| & -\beta_2|\Lambda_1\Psi_{13}F| & -\beta_3|\Lambda_1(\Psi_{11}G-\Psi_{12}G_m)| \\ -\beta_1|\Lambda_2\Psi_{23}F| & 1-\beta_2|\Lambda_2\Psi_{23}F| & -\beta_3|\Lambda_2(\Psi_{21}G-\Psi_{22}G_m)| \\ -\beta_1|\Lambda_3\Psi_{33}F| & -\beta_2|\Lambda_3\Psi_{33}F| & 1-\beta_3|\Lambda_3(\Psi_{31}G-\Psi_{32}G_m)| \end{bmatrix}$$

$$\cdot\begin{bmatrix} \|\overline{e}_1^{*\dagger}\|_2 \\ \|\overline{e}_2^{*\dagger}\|_2 \\ \|\overline{e}_3^{*\dagger}\|_2 \end{bmatrix} \leqq \begin{bmatrix} |\Lambda_1\Psi_{11}| & |\Lambda_1\Psi_{12}| & |\Lambda_1\Psi_{13}| \\ |\Lambda_2\Psi_{21}| & |\Lambda_2\Psi_{22}| & |\Lambda_2\Psi_{23}| \\ |\Lambda_3\Psi_{31}| & |\Lambda_3\Psi_{32}| & |\Lambda_3\Psi_{33}| \end{bmatrix}\begin{bmatrix} \|\overline{d}_1\|_2+\|\overline{r}\|_2 \\ \|\overline{d}_2\|_2 \\ \|\overline{d}_3\|_2 \end{bmatrix}. \quad (4.76)$$

When the matrix of the left side of this inequality is written as

$$\boldsymbol{A}=\begin{bmatrix} a_{11} & a_{12} & a_{13} \\ a_{21} & a_{22} & a_{23} \\ a_{31} & a_{32} & a_{33} \end{bmatrix}, \quad (4.77)$$

all non-diagonal elements are obviously non-positive. In addition, if all diagonal elements are positive and if all principal minors of all orders are positive, this matrix $\boldsymbol{A}$ is called an M-matrix [5, 6, 16, 17].

Theorem 4.1 *If there exists a $q_i \geq 0$ in which matrix* (4.77) *becomes an Ostrowski's M-matrix, the discretized model reference control system with sector nonlinearities* (4.47), (4.50), *and* (4.53) *is robustly stable in an ℓ_2 sense, when the linearized system with nominal gains K, K_m, and K_f is asymptotically stable* [14, 15].

Proof From (4.77), inequality (4.76) can be written as

$$\begin{bmatrix} a_{11} & a_{12} & a_{13} \\ a_{21} & a_{22} & a_{23} \\ a_{31} & a_{32} & a_{33} \end{bmatrix} \begin{bmatrix} x_1 \\ x_2 \\ x_3 \end{bmatrix} = \begin{bmatrix} \tilde{y}_1 \\ \tilde{y}_2 \\ \tilde{y}_3 \end{bmatrix} \leqq \begin{bmatrix} y_1 \\ y_2 \\ y_3 \end{bmatrix}, \tag{4.78}$$

where $x_i \geq 0$, $y_j \geq 0$, and $a_{ij} \leq 0$ for $i \neq j$ $(i, j = 1, 2, 3)$. Here, $\tilde{y}_j$ are arbitrary values that satisfy $0 < \tilde{y}_j \leq y_j < \infty$ $(j = 1, 2, 3)$.

The equality part of (4.78) can be rewritten as follows:

$$\begin{bmatrix} a_{11}^{(1)} & a_{12}^{(1)} & a_{13}^{(1)} \\ 0 & a_{22}^{(2)} & a_{23}^{(2)} \\ 0 & 0 & a_{33}^{(3)} \end{bmatrix} \begin{bmatrix} x_1^{(1)} \\ x_2^{(1)} \\ x_3^{(1)} \end{bmatrix} = \begin{bmatrix} \tilde{y}_1^{(1)} \\ \tilde{y}_2^{(2)} \\ \tilde{y}_3^{(3)} \end{bmatrix}, \tag{4.79}$$

where $x_i^{(1)} = x_i$, $\tilde{y}_j^{(1)} = \tilde{y}_j$, and $a_{ij}^{(1)} = a_{ij}$ $(i, j = 1, 2, 3)$, and furthermore,

$$\begin{cases} a_{22}^{(2)} = \dfrac{1}{a_{11}^{(1)}} \begin{vmatrix} a_{11}^{(1)} & a_{12}^{(1)} \\ a_{21}^{(1)} & a_{22}^{(1)} \end{vmatrix}, & a_{23}^{(2)} = \dfrac{1}{a_{11}^{(1)}} \begin{vmatrix} a_{11}^{(1)} & a_{13}^{(1)} \\ a_{21}^{(1)} & a_{23}^{(1)} \end{vmatrix}, \\ a_{32}^{(2)} = \dfrac{1}{a_{11}^{(1)}} \begin{vmatrix} a_{11}^{(1)} & a_{12}^{(1)} \\ a_{31}^{(1)} & a_{32}^{(1)} \end{vmatrix}, & a_{33}^{(2)} = \dfrac{1}{a_{11}^{(1)}} \begin{vmatrix} a_{11}^{(1)} & a_{13}^{(1)} \\ a_{31}^{(1)} & a_{33}^{(1)} \end{vmatrix}, \end{cases}$$

and

$$a_{33}^{(3)} = \frac{1}{a_{22}^{(2)}} \begin{vmatrix} a_{22}^{(2)} & a_{23}^{(2)} \\ a_{32}^{(2)} & a_{33}^{(2)} \end{vmatrix}.$$

Then, the right side of (4.79) can be written as

$$\tilde{y}_1^{(1)} = \tilde{y}_1, \quad \tilde{y}_2^{(2)} = \tilde{y}_2^{(1)} - \frac{a_{21}^{(1)}}{a_{11}^{(1)}} \tilde{y}_1^{(1)}, \quad \tilde{y}_3^{(3)} = \tilde{y}_3^{(2)} - \frac{a_{32}^{(2)}}{a_{22}^{(2)}} \tilde{y}_2^{(2)},$$

provided $a_{11}^{(1)} > 0$ and $a_{22}^{(2)} > 0$, where $0 < \tilde{y}_j^{(j)} \leq y_j^{(j)}$ $(j = 1, 2, 3)$. Thus, we can see that these values are non-negative and bounded if each norm of the exogenous inputs is bounded (i.e., $\|\bar{r}\|_2 < \infty$, $\|\bar{d}_j\|_2 < \infty$). In addition, if $a_{33}^{(3)} > 0$ is satisfied, $0 < x_3^{(1)} < \infty$, $0 < x_2^{(1)} < \infty$, and $0 < x_1^{(1)} < \infty$ are obtained in reverse order. Note that the above sweep-out procedure for inequalities is similar to the method shown in Appendix A of Chap. 1.[5]

On the other hand, (4.79) can be rewritten as

$$\begin{bmatrix} a_{11}^{(1)} x_1^{(1)} \\ a_{22}^{(2)} x_2^{(1)} \\ a_{33}^{(3)} x_3^{(1)} \end{bmatrix} = \begin{bmatrix} \tilde{y}_1^{(1)} \\ \tilde{y}_2^{(2)} \\ \tilde{y}_3^{(3)} \end{bmatrix} + \begin{bmatrix} 0 & -a_{12}^{(1)} & -a_{13}^{(1)} \\ 0 & 0 & -a_{23}^{(2)} \\ 0 & 0 & 0 \end{bmatrix} \begin{bmatrix} x_1^{(1)} \\ x_2^{(1)} \\ x_3^{(1)} \end{bmatrix}. \tag{4.80}$$

[5]The author proposed a method for inequalities in [10].

Therefore, if $0 < x_3 < \infty$, $0 < x_2 < \infty$, and $0 < x_1 < \infty$, then $a_{33}^{(3)} > 0$, $a_{22}^{(2)} > 0$, and $a_{11}^{(1)} > 0$ are obtained in reverse order.

Noted that the above conditions $a_{11}^{(1)} > 0$, $a_{22}^{(2)} > 0$, and $a_{33}^{(3)} > 0$ can be rewritten as follows:

$$a_{11}^{(1)} = \Delta_1 = a_{11} > 0$$

$$a_{22}^{(2)} = \frac{\Delta_2}{\Delta_1} = \frac{\begin{vmatrix} a_{11} & a_{12} \\ a_{21} & a_{22} \end{vmatrix}}{a_{11}} > 0$$

$$a_{33}^{(3)} = \frac{\Delta_3}{\Delta_2} = \frac{\begin{vmatrix} a_{11} & a_{12} & a_{13} \\ a_{21} & a_{22} & a_{23} \\ a_{31} & a_{32} & a_{33} \end{vmatrix}}{\begin{vmatrix} a_{11} & a_{12} \\ a_{21} & a_{22} \end{vmatrix}} > 0.$$

The conditions say that all principal minors of matrix $\boldsymbol{A}$ are positive, which means that the matrix becomes an M-matrix.[6] Thus, it can be proven that

$$\|e_i^*\| < \infty \quad \text{and} \quad \|e_i\| < \infty, \quad i = 1, 2, 3,$$

for a nominal control system with gains K, K_m, and K_f. Thus, the proof of Theorem 4.1 based on the concept of bounded input-bounded output (BIBO) stability of model reference control systems is completed. □

Example 4.4 Consider the following continuous plant:

$$G(s) = \frac{K_1}{(s+0.1)(s+0.3)(s+0.5)}, \tag{4.81}$$

where the gain constant is $K_1 = 0.03$. The sampling period and the resolution value are assumed to be $h = 1.0$ and $\gamma = 1.0$. That is, the responses of the control systems trace on integer grid coordinates. The discretized nonlinear characteristic (discretized sigmoid, i.e., arctangent) is as shown in Fig. 4.20. The Hall diagram is shown in Fig. 4.23. The input/output characteristic of the discretization process can be written by, e.g., a C-language expression as follows:

$$\begin{aligned} e_1^\dagger &= \gamma * \texttt{(double)(int)}(e_1/\gamma) \\ v_1 &= 0.4 * e_1^\dagger + 3.0 * \texttt{atan}(0.6 * e_1^\dagger) \\ v_1^\dagger &= \gamma * \texttt{(double)(int)}(v_1/\gamma). \end{aligned} \tag{4.82}$$

[6] The derivation of the result in general form is written in Chap. 5, and graphical representations are presented in Appendix B of Chap. 5.

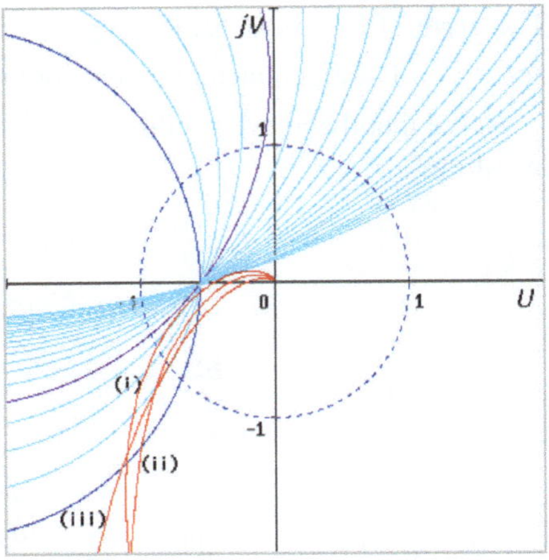

Fig. 4.23 Modified Hall diagram for Example 4.4

Table 4.3 Performances of model reference control system for Example 4.4

Cases	C_1	M_p	g_M [dB]	p_M [deg]
(i)	16.0	2.2	7.1	28.3
(ii)	12.0	2.1	10.0	35.9
(iii)	8.0	2.1	14.9	37.7

When the nominal gain $K = 1.0$ and the threshold $\varepsilon_1 = 2.0$ are considered, the sectorial area of the point-to-point characteristic for $\varepsilon_1 \leq |e_1| \leq 40.0$ can be determined as [0.5, 1.5].

The model system is chosen as

$$G_m(\delta) = \frac{1}{1 + C_1\delta + 8.0\delta^2}, \quad K_m = 1.0, \tag{4.83}$$

and the feedback compensator is chosen as

$$F(\delta) = \frac{1 + C_1\delta + 8.0\delta^2}{1 + 8.0\delta + \delta^2}, \quad K_f = 1.0. \tag{4.84}$$

For the three cases in Table 4.3, the phase traces and the step responses are depicted as shown in Figs. 4.24(a) and (b). By applying the model reference feedback structure, the control system can be sufficiently stabilized. The control performances are obtained as shown in Table 4.3.

In order to check the robust stability of the discrete control system, the loop characteristics and the stability margins,

$$|H| = |\Lambda_1 \Psi_{13} F| \tag{4.85}$$

and

$$\Delta_1 = 1 - \beta_1 |\Lambda_1 \Psi_{13} F|, \tag{4.86}$$

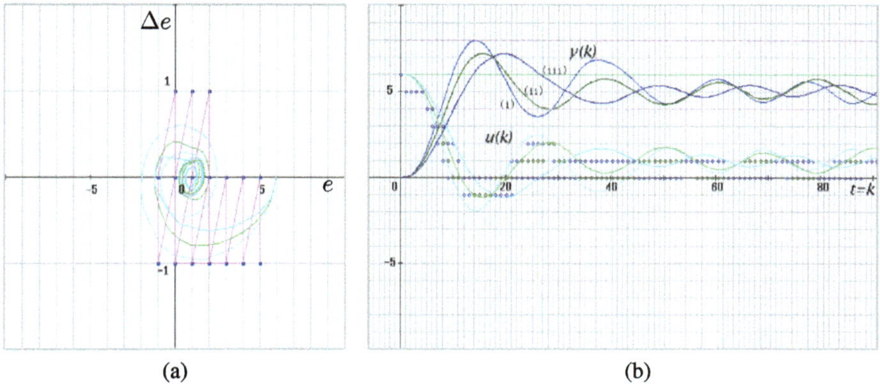

Fig. 4.24 Phase traces and step responses for Example 4.4 ((i) $C_1 = 16.0$, (ii) $C_1 = 12.0$, (iii) $C_1 = 8.0$)

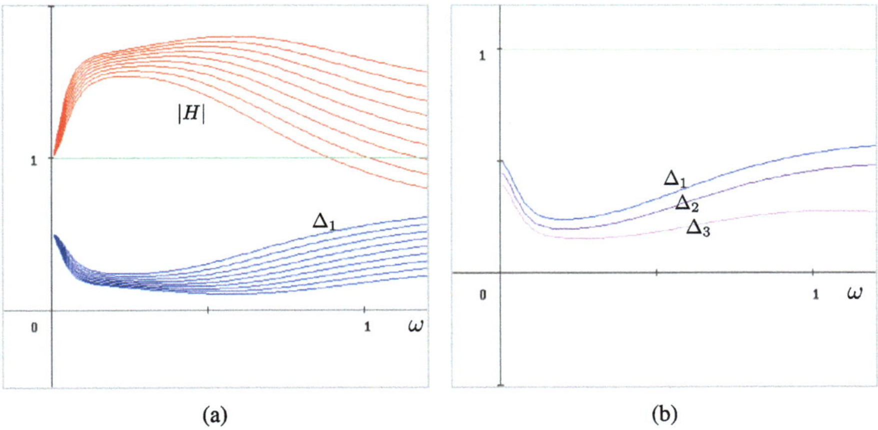

Fig. 4.25 Checking of robust stability margins for Example 4.4 when $\beta_1 = 0.5$, $\beta_2 = 0.1$, $\beta_3 = 0.1$, and $C_1 = 5.0$

in (4.77) for $q_1 = 20.0 \sim 36.0$ are calculated as shown in Fig. 4.25(a). Figure 4.25(b) shows calculated results of the following principal minors for $q_1 = 20.0$, $q_2 = 5.0$, and $q_3 = 5.0$:

$$\Delta_1 = a_{11}, \quad \Delta_2 = \begin{vmatrix} a_{11} & a_{12} \\ a_{21} & a_{22} \end{vmatrix}, \quad \Delta_3 = \begin{vmatrix} a_{11} & a_{12} & a_{13} \\ a_{21} & a_{22} & a_{23} \\ a_{31} & a_{32} & a_{33} \end{vmatrix}.$$

It is obvious from these figures that the stability margins for the discrete control systems are satisfied. For reference, the modified Hall diagram and Nyquist curves are shown in Fig. 4.23 when the controllers are in high resolution.

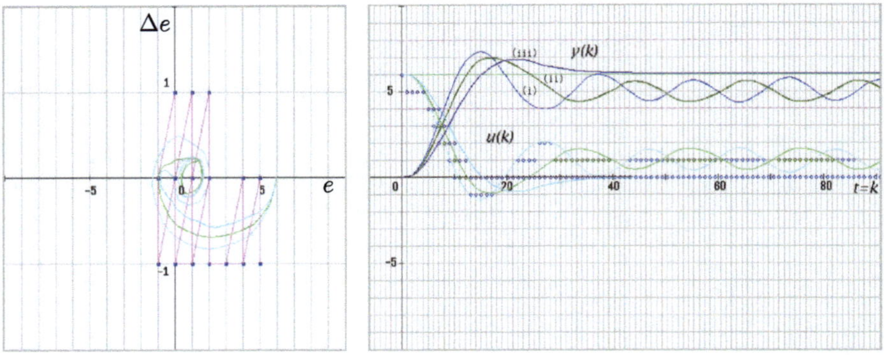

(a) (b)

Fig. 4.26 Phase traces and step responses for Example 4.5 ((i) $C_1 = 16.0$ and $L_m = 2.0$, (ii) $C_1 = 12.0$ and $L_m = 2.0$, (iii) $C_1 = 8.0$ and $L_m = 2.0$)

Example 4.5 In this example, the model system is assumed to be the following time-delay system:

$$G_m(\delta) = \frac{1}{1 + C_1\delta + 8.0\delta^2} \cdot \mathrm{e}^{-L_m s}, \quad L_m = 2.0, \quad K_m = 1.0. \tag{4.87}$$

As for the three cases described in Example 4.4, the phase traces and the step responses are depicted as shown in Figs. 4.26(a) and (b). By using a model system with time delay, the responses of the control system can be well stabilized.

4.7 Model Reference Control with Transmission Delay

Since the discretized model reference and quasi-PID control techniques in the previous sections are analyzed in the frequency domain, the stabilization and design of control systems with some transmission delay can also be applied. As was described in Sect. 4.4, the transfer function of a continuous plant with time delay can be written as

$$G(s) = \frac{N_p(s)}{(s - p_1)(s - p_2)\cdots(s - p_n)} \cdot \mathrm{e}^{-L_p s}, \tag{4.88}$$

irrespective of whether the time delay exists in the input or the output side of the plant. Here, L_p is the transmission delay (time delay) and the order of the numerator polynomial $N_p(s)$ is less than that of the denominator. With respect to time-delayed systems, the following expression can be given:

$$\begin{aligned} G(z) &= \frac{b_0 + b_1 z^{-1} + b_2 z^{-2} \cdots + b_n z^{-n}}{1 + a_1 z^{-1} + a_2 z^{-2} \cdots + a_n z^{-n}} \cdot z^{-d_p} \\ &= \frac{b_0 z^n + b_1 z^{n-1} + \cdots + b_{n-1} z + b_n}{z^n + a_1 z^{n-1} + \cdots + a_{n-1} z + a_n} \cdot z^{-d_p}. \end{aligned}$$

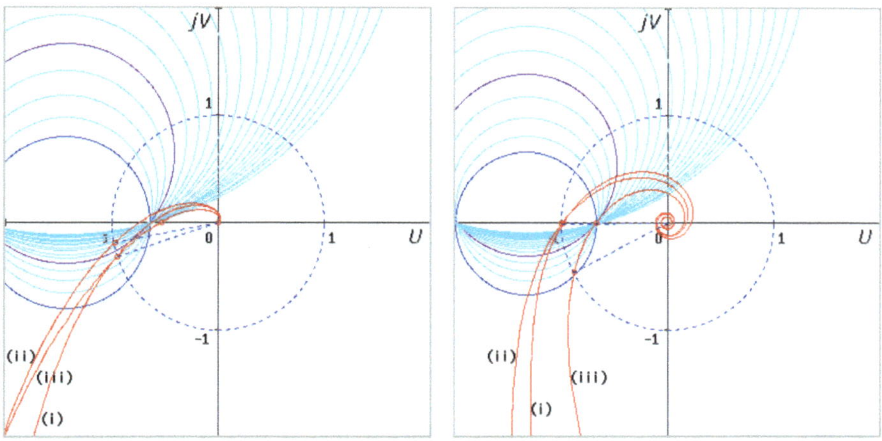

(a) (b)

Fig. 4.27 Modified Hall diagram for Examples 4.6 and 4.7

Table 4.4 Performances of model reference control system for Example 4.6

Cases	C_1	L_p	L_m	M_p	g_M [dB]	p_M [deg]
(i)	8.0	2.0	0.0	3.38	4.43	17.2
(ii)	8.0	2.0	2.0	5.49	2.66	11.4
(iii)	8.0	2.0	4.0	3.18	5.84	19.1

Example 4.6 Consider the following continuous plant with transmission delay:

$$G(s) = \frac{K_p}{(s+0.1)(s+0.3)(s+0.5)} \cdot \mathrm{e}^{-L_p s}, \quad K_p = 0.03, \quad L_p = 2.0. \tag{4.89}$$

Here, the model system is assumed to be

$$G_m(\delta) = \frac{1}{1 + C_1\delta + 8.0\delta^2} \cdot \mathrm{e}^{-L_m s}, \quad L_m = 2.0, \quad K_m = 1.0. \tag{4.90}$$

The feedback compensator $F(\delta)$ is the same as (4.84). The Hall diagram and Nyquist curves are shown in Fig. 4.27. Also in this example, the phase trace and the step responses are well stabilized and depicted as shown in Figs. 4.28(a) and (b). Table 4.4 shows the model and compensator parameters and the control performances. The relationship between the step responses and the performance indices is not as clear as it appeared in linear control systems.

Example 4.7 The model reference control technique can be applied to controlled systems with long transmission delay. In this last example, consider the following

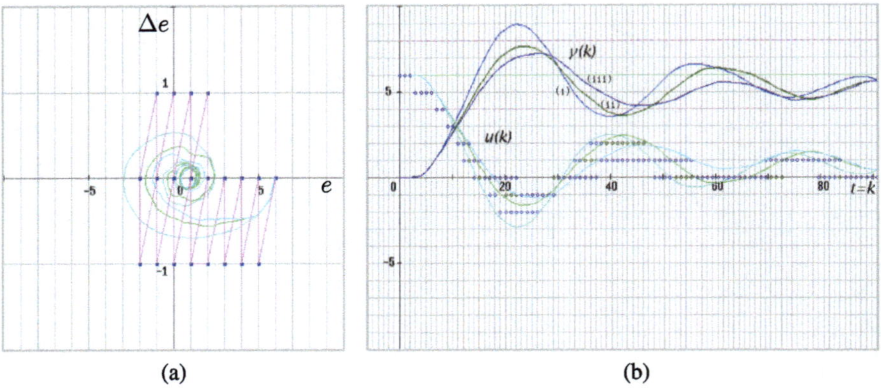

Fig. 4.28 Phase traces and step responses for Example 4.6 ($C_1 = 8.0$, (i) $L_m = 0.0$, (ii) $L_m = 2.0$, (iii) $L_m = 4.0$)

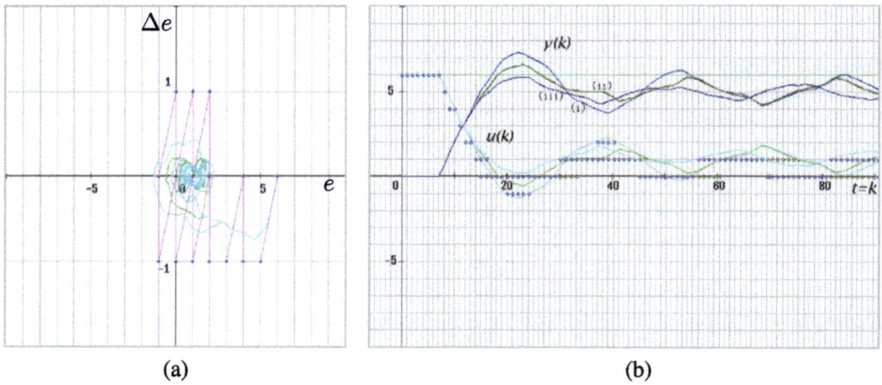

Fig. 4.29 Phase traces and step responses for Example 4.7 ($C_1 = 8.0$, (i) $L_m = 2.0$, (ii) $L_m = 4.0$, (iii) $L_m = 6.0$)

Table 4.5 Performances of model reference control system for Example 4.7

Cases	C_1	L_p	L_m	M_p	g_M [dB]	p_M [deg]
(i)	8.0	6.0	2.0	2.1	0.20	0.29
(ii)	8.0	6.0	4.0	1.8	0.06	0.69
(iii)	8.0	6.0	6.0	1.6	3.78	25.6

continuous plant:

$$G(s) = \frac{K_p}{(s+0.2)(s+0.4)} \cdot \mathrm{e}^{-L_p s}, \quad K_p = 0.1, \quad L_p = 6.0. \tag{4.91}$$

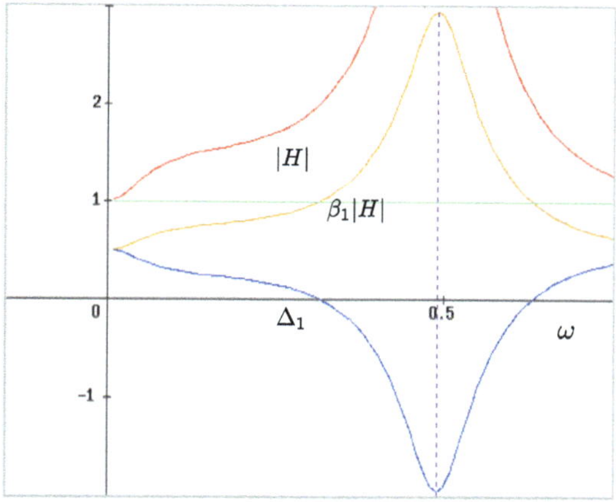

Fig. 4.30 Robust performance measures Δ_1 ($C_1 = 8.0$, $L_p = 6.0$, and $L_m = 6.0$)

Figures 4.29(a) and (b) show the phase traces and the step responses when using the model system given as

$$G_m(\delta) = \frac{1}{1 + C_1\delta + 8.0\delta^2} \cdot e^{-L_m s}. \ K_m = 1.0, \tag{4.92}$$

where $L_m = 2.0, 4.0$, and 6.0. Although there are some deformations in the responses, the stabilization of the control system is achieved well. However, the relationship between the step responses and the performance indices is considerably different from those in (usual) linear control systems. The relationship between the model and compensator parameters and the control performances becomes as shown in Table 4.5.

For reference, the robust performance measures $|H|$, $\beta_1|H|$, and Δ_1 that correspond to (4.85) and (4.86) in Example 4.4 are depicted in Fig. 4.30. The modified Hall diagram and Nyquist curves for Examples 4.6 and 4.7 are as shown in Figs. 4.27(a) and (b) when the controllers are in high resolution.

4.8 Exercises

(1) Confirm that the following equality holds in regard to (4.3) and (4.6):

$$\lim_{h \to 0} C(\delta) = C(s).$$

(2) Determine the z-transform of the following plant with a zero-order hold:

$$G(s) = \frac{K}{(s+1)(s+2)}, \quad K = 1.0.$$

Assume the sampling period $h = 0.1$, and use the result of Example 4.1, i.e., Eqs. (4.15)–(4.17).

(3) Determine the characteristic equation $\mathcal{F}(z) = 0$ for Example 4.1 (A) when a PI controller is used (i.e., case (i), $K_p = 1.0$, $C_I = 1.0$, and $C_D = 0.0$).
(4) Show that the approximate PID control system in Fig. 4.18 is obtained from the model reference feedback system in Fig. 4.17, when $\mathcal{D}_m(\cdot)$ and $\mathcal{D}_f(\cdot)$ are K_m and $K_f = 1/K_m$, respectively.
(5) Regarding the simultaneous (linear) inequalities,

$$\begin{cases} a_{11}x_1 + a_{12}x_2 + a_{13}x_3 \leq y_1 \\ a_{21}x_1 + a_{22}x_2 + a_{23}x_3 \leq y_2 \\ a_{31}x_1 + a_{32}x_2 + a_{33}x_3 \leq y_3, \end{cases}$$

show that the following operations are valid (a similar concept to (1), (2), and (3) in Appendix A of Chap. 1):

(1) interchanging two inequalities,
(2) multiplying each term in one inequality by a positive constant,
(3) adding a positive multiple of one inequality to another.

(6) In regard to a vector-matrix expression, $\boldsymbol{Ax} \leqq \boldsymbol{y}$, where

$$\boldsymbol{A} = \begin{bmatrix} a_{11} & a_{12} & a_{13} \\ a_{21} & a_{22} & a_{23} \\ a_{31} & a_{32} & a_{33} \end{bmatrix}, \quad \boldsymbol{x} = \begin{bmatrix} x_1 \\ x_2 \\ x_3 \end{bmatrix}, \quad \boldsymbol{y} = \begin{bmatrix} y_1 \\ y_2 \\ y_3 \end{bmatrix},$$

show that the following operations are valid (a similar concept to (a), (b), and (c) in Appendix A of Chap. 1):

(a) interchanging two rows,
(b) multiplying each term in one row by a positive constant,
(c) adding a positive multiple of one row to another,

considering the influence on variables x_i and y_j $(i, j = 1, 2, 3)$.
(7) Show that the condition of the M-matrix is not influenced by the above operations (a), (b), and (c).

Appendix

In this chapter, C-language functions for polynomial and complex variable operations are provided as a reference for the reader.

A.1 Polynomial Operations

```
/* polynomial multiplication */
void    p_mul(na,nb,a,b,c)
```

```
int      na,nb;
double   a[20],b[20],c[20];
{
         int i,ia,ib,nc;
         nc=na+nb;
         for(i=0;i<=nc;i++){
                  c[i]=0.0;
         }
         for(ia=0;ia<=na;ia++){
                  for(ib=0;ib<=nb;ib++){
                           c[ia+ib]=c[ia+ib]+a[ia]*b[ib];
                  }
         }
}

/* polynomial addition */
void     p_add(na,nb,a,b,c)
int      na,nb;
double   a[20],b[20],c[20];
{
         int i,ia,ib,nc;
         if(na>=nb){
                  nc=na;
                  for(ib=nb+1;ib<=nc;ib++){
                           b[ib]=0.0;
                  }
         }
         else{
         nc=nb;
         for(ia=na+1;ia<=nc;ia++){
                           a[ia]=0.0;
                  }
         }
         for(i=0;i<=nc;i++){
                  c[i]=a[i]+b[i];
         }
}

/* polynomial subtraction */
void     p_sub(na,nb,a,b,c)
int      na,nb;
double   a[20],b[20],c[20];
{
         int i,ia,ib,nc;
         if(na>=nb){
         nc=na;
         for(ib=nb+1;ib<=nc;ib++){
                  b[ib]=0.0;
                  }
         }
         else{
         nc=nb;
         for(ia=na+1;ia<=nc;ia++){
                  a[ia]=0.0;
```

```
                }
        }
        for(i=0;i<=nc;i++){
                c[i]=a[i]-b[i];
        }
}

/* derivative of ploynomial */
void    p_der(nd,ad,cd)
int     nd;
double  ad[20],cd[19];
{
        int i;
        for(i=0;i<=nd;i++){
                cd[i]=0.0;
        }
        for(i=0;i<=nd-1;i++){
                cd[i]=(double)(i+1)*ad[i+1];
        }
}

/* values of polynomial */
complex p_val(nv,av,s)
int     nv;
double  av[20];
complex s;
{
        int i;
        complex z,f;
        f.re=av[nv];f.im=0.0;
        for(i=1;i<=nv;i++){
        z.re=av[nv-i]+f.re*s.re-f.im*s.im;
                z.im=f.re*s.im+f.im*s.re;
                f=z;
        }
        return(f);
}
```

A.2 Complex Variable Functions

```
/* complex variable subroutine functions */
#define TINY    0.000001
typedef struct{
                double  re;
                double  im;
}complex;
/* complex addition */
complex cadd(a,b)
complex a,b;
{
        complex   z;
```

```
        z.re=a.re+b.re;
        z.im=a.im+b.im;
        return(z);
}

/* complex subtraction */
complex csub(a,b)
complex a,b;
{
        complex   x;

        x.re=a.re-b.re;
        x.im=a.im-b.im;
        return(x);
}

/* complex multiplication */
complex cmul(a,b)
complex a,b;
{
        complex   x;

        x.re=a.re*b.re-a.im*b.im;
        x.im=a.re*b.im+a.im*b.re;
        return(x);
}

/* complex division */
complex cdiv(a,b)
complex a,b;
{
        complex   x;
        double    d;

        d=b.re*b.re+b.im*b.im;
        if(d==0){
               printf("\n\n divided by zero error !!!\n");
               d=TINY;
        }
        printf("d=%lf\n",d);
        x.re=(a.re*b.re+a.im*b.im)/d;
        x.im=(a.im*b.re-a.re*b.im)/d;
        return(x);
}
```

A.3 Partial Fraction Expansions

```
/* partial fraction expansion */
/* K*(s-z1)....(s-zm)/(s-p1)....(s-pn) */
void    p_pf(n,m,p,z,k,ap)
int     n,m;
```

```
double  p[20],z[20],k,ap[20][3];
{
        int     i,j;
        double  pd=1.0;
        double  zd=1.0;
        for(i=1;i<=n;i++){
                pd=1.0;
                for(j=1;j<=n;j++){
                        if(j!=i) pd=pd*(p[i]-p[j]);
                }
                zd=1.0;
                for(j=1;j<=m;j++){
                        zd=zd*(p[i]-z[j]);
                }
                ap[i][0]=k*zd/pd;
        }
}

/* partial fraction expansion-1 */
/* K*(s-z1)....(s-zm)/s(s-p1)....(s-pn) */
void    p_pf1(n,m,p,z,k,ap)
int     n,m;
double  p[20],z[20],k,ap[20][3];
{
        int    i,j;
        double  pn=1.0;
        double  pd=1.0;
        double  zn=1.0;
        double  zd=1.0;
        for(i=1;i<=n;i++){
                pn=pn*(-p[i]);
        }
        for(j=1;j<=m;j++){
                zn=zn*(-z[j]);
        }
        ap[0][0]=k*zn/pn;

        for(i=1;i<=n;i++){
                pd=p[i];
                for(j=1;j<=n;j++){
                        if(j!=i) pd=pd*(p[i]-p[j]);
                }
                zd=1.0;
                for(j=1;j<=m;j++){
                        zd=zd*(p[i]-z[j]);
                }
                ap[i][0]=k*zd/pd;
        }
}
```

References

1. Åstrom KJ, Hägglund T (1995) PID controllers: theory, design, and tuning, 2nd edn. Instrument Society of America, Research Triangle Park
2. Åstrom KJ, Hägglund T (2006) Advanced PID controllers: theory, design, and tuning. Instrument Society of America, Research Triangle Park
3. Brown GS, Campbell DP (1948) Principles of servomechanisms. Wiley, New York
4. Datta A, Ho MT, Bhattacharyya SP (2000) Structure and synthesis of PID controllers. Springer, Berlin
5. Fan M (1960) Note on M-matrices. Q J Math, 43–49
6. Fiedler M, Pták V (1962) On matrices with non-positive off-diagonal elements and positive principal minors. Czechoslov Math J, 382–400
7. Johnson MA, Moradi MH (eds) (2005) PID control—new identification and design methods. Springer, Berlin
8. Krajewski W, Lepschy A, Miani S, Viaro U (2005) Frequency-domain approach to robust control. J Franklin Inst 342:674–687
9. Okuyama Y (1964) Estimation of parameter sensitivity of linear control systems and synthesis of passive adaptive control systems. J SICE 3:759–767 (in Japanese)
10. Okuyama Y (1967) On the L_2-stability of linear systems with time-varying parameters. Trans SICE 3:252–259 (in Japanese)
11. Okuyama Y (2009) Discretized PID control and robust stabilization for continuous plants on an integer-grid pattern. In: Proc of the European control conference, Budapest, Hungary, pp 514–519
12. Okuyama Y (2009) Robust stabilization for discretized PID control systems with transmission delay. In: Proc of the IEEE int conf on decision and control, Shanghai, PR China, pp 5120–5126
13. Okuyama Y (2011) Model-reference discretized PID control and robust stabilization for continuous plants. In: Proc of the 18th IFAC world congress, Milano, Italy, pp 5813–5818
14. Okuyama Y (2012) Discretized model-reference feedback and robust stabilization on integer grid coordinates. In: Proc of the 7th IFAC symp on robust control design, Aalborg, Denmark, pp 400–405
15. Okuyama Y (2013) Robust stabilization of discrete model-reference control systems on integer grid coordinates. In: Proc of the 2013 American control conference, Washington DC, USA, pp 5613–5620
16. Ostrowski AM (1937) Über die Determinanten mit überwiegender Hauptdiagonale. Comment Math Helv 22:69–96 (in German)
17. Ostrowski AM (1955) Note on bounds for some determinants. Duke Math J 22:95–102
18. Silva GJ, Datta A, Bhattacharyya SP (2002) New results on the synthesis of PID controllers. IEEE Trans Autom Control 47:241–252
19. Silva GJ, Datta A, Bhattacharyya SP (2005) PID controllers for time-delay systems. Birkhäuser, Boston
20. Takemori F, Okuyama Y (2000) Discrete-time model reference feedback and PID control for interval plants. In: Digital control 2000: past, present and future of PID control. Pergamon Press, Oxford, pp 260–265
21. Zhuang M, Atherton DP (1993) Automatic tuning of optimum PID controllers. Proc IEE, Part D 140:216–224
22. Ziegler JG, Nichols NB (1942) Optimum settings for automatic controller. Trans Am Soc Mech Eng 64:759–768

Chapter 5
Multi-Loop Feedback Systems

5.1 Introduction

In this chapter, the concepts described in Chaps. 2 and 3 are extended to multivariable and multi-loop feedback systems. For the input-output stability of multivariable systems, matrix representations and inequalities based on the norms in the ℓ_2 space will be defined. Usually, multivariable control systems are analyzed in a state space and are designed by using quadratic forms; see, e.g., [8]. However, a design method on the basis of vector-matrix representations (the linear quadratic, LQ, form) is not practical, because controlled systems are accompanied by many nonlinearities [3, 9]. Therefore, in this book, for multiple nonlinearities, inequality conditions and Ostrowski's M-matrix [7] are applied to the stability problem.

5.2 Input-Output Stability for Multi-Loop Systems

First, a multi-loop control system as shown in Fig. 5.1 is considered. Here, each variable is given in vector form as follows:[1]

$$
\begin{aligned}
&\boldsymbol{r}(k) = (r_1(k), r_2(k), \cdots, r_n(k))^T, \quad \boldsymbol{e}(k) = (e_1(k), e_1(k), \cdots, e_n(k))^T \\
&\boldsymbol{v}(k) = (v_1(k), v_2(k), \cdots, v_n(k))^T, \quad \boldsymbol{u}(k) = (u_1(k), u_1(k), \cdots, u_n(k))^T \\
&\boldsymbol{d}(k) = (d_1(k), d_2(k), \cdots, d_n(k))^T, \quad \boldsymbol{y}(k) = (y_1(k), y_1(k), \cdots, y_n(k))^T \\
&k = 0, 1, 2, \cdots .
\end{aligned}
$$

[1] In this chapter, the symbol "n" is used for the number of loops regardless of the system order and the signal resolution.

Y. Okuyama, *Discrete Control Systems*,
DOI 10.1007/978-1-4471-5667-3_5, © Springer-Verlag London 2014

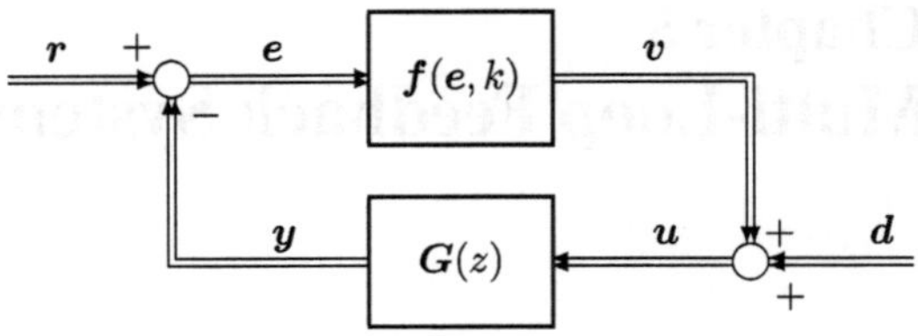

Fig. 5.1 Nonlinear time-varying multi-loop feedback system

As is obvious from the figure, the following equations are given:

$$\begin{cases} \boldsymbol{v}(k) = \boldsymbol{f}(\boldsymbol{e}, k), \quad k = 0, 1, 2, \cdots \\ \hat{\boldsymbol{y}}(z) = \boldsymbol{G}(z)\hat{\boldsymbol{u}}(z) \\ \boldsymbol{e}(k) = \boldsymbol{r}(k) - \boldsymbol{y}(k), \quad \boldsymbol{u}(k) = \boldsymbol{v}(k) + \boldsymbol{d}(k). \end{cases} \tag{5.1}$$

Here, it is assumed that the nonlinear time-varying element $\boldsymbol{f}(\boldsymbol{e}, k)$ is given by the following diagonal matrix:

$$\boldsymbol{f}(\boldsymbol{e}, k) = \text{diag}\{f_1(e_1, k), f_2(e_2, k), \cdots, f_n(e_n, k)\}, \tag{5.2}$$

and the linear (discrete-time) dynamical system $\boldsymbol{G}(z)$ can be written as the following $n \times n$ matrix with z-transformed elements:

$$\boldsymbol{G}(z) = \begin{bmatrix} G_{11}(z) & G_{12}(z) & \dots & G_{1n}(z) \\ G_{21}(z) & G_{22}(z) & \dots & G_{2n}(z) \\ \vdots & \vdots & \ddots & \vdots \\ G_{n1}(z) & G_{n2}(z) & \dots & G_{nn}(z) \end{bmatrix}. \tag{5.3}$$

Moreover, $r_i(k)$ and $d_i(k)$ $(i = 1, 2, \cdots, n)$ are exogenous inputs that exist in the ℓ_2 space. Here, in the strict sense, (5.2) should be written as

$$\boldsymbol{f}(\boldsymbol{e}, k) = \text{diag}\{f_1(e_1(kh), kh), f_2(e_2(kh), kh), \cdots, f_n(e(kh), kh)\},$$

where h is the sampling period.

Definition If $r_i(k) \in \ell_2$ and $d_i(k) \in \ell_2$ $(i = 1, 2, \cdots, n)$ lead to $e_i(k) \in \ell_2$ and $y_i(k) \in \ell_2$ $(i = 1, 2, \cdots, n)$, the feedback system is called *bounded input-bounded output stable* (BIBO stable).

If it is valid that

$$\sum_{k=0}^{\infty} |r_i(k)|^2 < \infty, \quad \sum_{k=0}^{\infty} |d_i(k)|^2 < \infty, \qquad i = 1, 2, \cdots, n$$

lead to

$$\sum_{k=0}^{\infty} |e_i(k)|^2 < \infty, \quad \sum_{k=0}^{\infty} |y_i(k)|^2 < \infty, \qquad i = 1, 2, \cdots, n,$$

then the nonlinear time-varying multi-loop system is input-output stable. For a norm expression, the above conditions can be written as follows:

$$\|r_i(k)\|_2 < \infty \quad \text{and} \quad \|d_i(k)\|_2 < \infty \quad \Rightarrow \quad \|e_i(k)\|_2 < \infty \quad \text{and} \quad \|y_i(k)\|_2 < \infty, \quad (\forall i).$$

5.3 Stability Condition for Multi-Loop Systems

From the last equations of (5.1), the norm inequalities of these variables are given by applying Minkowski's inequality:

$$\|e_i(k)\|_2 \leq \|r_i(k)\|_2 + \|y_i(k)\|_2,, \qquad i = 1, 2, \cdots, n \tag{5.4}$$

$$\|u_i(k)\|_2 \leq \|v_i(k)\|_2 + \|d_i(k)\|_2., \qquad i = 1, 2, \cdots, n. \tag{5.5}$$

In this chapter, these inequalities are rewritten in vector form as follows:[2]

$$\begin{bmatrix} \|e_1(k)\|_2 \\ \|e_2(k)\|_2 \\ \vdots \\ \|e_n(k)\|_2 \end{bmatrix} \leqq \begin{bmatrix} \|r_1(k)\|_2 \\ \|r_2(k)\|_2 \\ \vdots \\ \|r_n(k)\|_2 \end{bmatrix} + \begin{bmatrix} \|y_1(k)\|_2 \\ \|y_2(k)\|_2 \\ \vdots \\ \|y_n(k)\|_2 \end{bmatrix}, \tag{5.6}$$

$$\begin{bmatrix} \|u_1(k)\|_2 \\ \|u_2(k)\|_2 \\ \vdots \\ \|u_n(k)\|_2 \end{bmatrix} \leqq \begin{bmatrix} \|v_1(k)\|_2 \\ \|v_2(k)\|_2 \\ \vdots \\ \|v_n(k)\|_2 \end{bmatrix} + \begin{bmatrix} \|d_1(k)\|_2 \\ \|d_2(k)\|_2 \\ \vdots \\ \|d_n(k)\|_2 \end{bmatrix}. \tag{5.7}$$

In this book, Euclidean norms of these vectors, e.g.,

$$\|\boldsymbol{x}\|_2 = \left(\sum_{i=1}^{n} |x_i|^2 \right)^{1/2}$$

are not considered, because the stability analysis of nonlinear feedback systems using those norms may become conservative.

If each component of the nonlinear time-varying element, $\boldsymbol{f}(\boldsymbol{e}, k)$, is given by

$$\frac{|f_i(e_i, k)|}{|e_i(k)|} \leq \rho_i < \infty, \qquad i = 1, 2, \cdots, n, \tag{5.8}$$

[2]The inequality symbol $\leqq$ denotes that each component of the left-side vector is less than or equal to each component of the right-side one.

the norm of the output of the ith nonlinear element $v_i(k)$ can be expressed in vector-matrix form as follows:

$$\begin{bmatrix} \|v_1(k)\|_2 \\ \|v_2(k)\|_2 \\ \vdots \\ \|v_n(k)\|_2 \end{bmatrix} \leq \begin{bmatrix} \rho_1 & 0 & \dots & 0 \\ 0 & \rho_2 & \dots & 0 \\ \vdots & \vdots & \ddots & \vdots \\ 0 & 0 & \dots & \rho_n \end{bmatrix} \begin{bmatrix} \|e_1(k)\|_2 \\ \|e_2(k)\|_2 \\ \vdots \\ \|e_n(k)\|_2 \end{bmatrix},$$

where $0 < \rho_i < \infty$ $(i = 1, 2, \cdots, n)$. Then, inequality (5.5) can be written as

$$\begin{bmatrix} \|u_1(k)\|_2 \\ \|u_2(k)\|_2 \\ \vdots \\ \|u_n(k)\|_2 \end{bmatrix} \leq \begin{bmatrix} \rho_1 & 0 & \dots & 0 \\ 0 & \rho_2 & \dots & 0 \\ \vdots & \vdots & \ddots & \vdots \\ 0 & 0 & \dots & \rho_n \end{bmatrix} \begin{bmatrix} \|e_1(k)\|_2 \\ \|e_2(k)\|_2 \\ \vdots \\ \|e_n(k)\|_2 \end{bmatrix} + \begin{bmatrix} \|d_1(k)\|_2 \\ \|d_2(k)\|_2 \\ \vdots \\ \|d_n(k)\|_2 \end{bmatrix}. \tag{5.9}$$

On the other hand, the following relation holds for $z = e^{j\omega h}$:

$$\begin{bmatrix} \|\hat{y}_1(z)\|_2 \\ \|\hat{y}_2(z)\|_2 \\ \vdots \\ \|\hat{y}_n(z)\|_2 \end{bmatrix} \leq \begin{bmatrix} \sup_{|z|=1}|G_{11}(z)| & \sup_{|z|=1}|G_{12}(z)| & \dots & \sup_{|z|=1}|G_{1n}(z)| \\ \sup_{|z|=1}|G_{21}(z)| & \sup_{|z|=1}|G_{22}(z)| & \dots & \sup_{|z|=1}|G_{2n}(z)| \\ \vdots & \vdots & \ddots & \vdots \\ \sup_{|z|=1}|G_{n1}(z)| & \sup_{|z|=1}|G_{n2}(z)| & \dots & \sup_{|z|=1}|G_{nn}(z)| \end{bmatrix} \cdot \begin{bmatrix} \|\hat{u}_1(z)\|_2 \\ \|\hat{u}_2(z)\|_2 \\ \vdots \\ \|\hat{u}_n(z)\|_2 \end{bmatrix}. \tag{5.10}$$

Here, $|z| = 1$ corresponds to $\omega : -\pi/h \to \pi/h$, i.e.,

$$\sup_{|z|=1} |G_{ij}(z)| = \sup_{-\pi/h \leq \omega \leq \pi/h} |G_{ij}(e^{j\omega h})|, \qquad i, j = 1, 2, \cdots, n.$$

By applying Parseval's identity,[3] the following inequality is obtained:

$$\begin{bmatrix} \|y_1(k)\|_2 \\ \|y_2(k)\|_2 \\ \vdots \\ \|y_n(k)\|_2 \end{bmatrix} \leq \begin{bmatrix} |G_{11}(e^{j\omega h})| & |G_{12}(e^{j\omega h})| & \dots & |G_{1n}(e^{j\omega h})| \\ |G_{21}(e^{j\omega h})| & |G_{22}(e^{j\omega h})| & \dots & |G_{2n}(e^{j\omega h})| \\ \vdots & \vdots & \ddots & \vdots \\ |G_{n1}(e^{j\omega h})| & |G_{n2}(e^{j\omega h})| & \dots & |G_{nn}(e^{j\omega h})| \end{bmatrix} \cdot \begin{bmatrix} \|u_1(k)\|_2 \\ \|u_2(k)\|_2 \\ \vdots \\ \|u_n(k)\|_2 \end{bmatrix}. \tag{5.11}$$

[3] See Appendix B in Chap. 3.

Using (5.6),

$$\begin{bmatrix} \|e_1(k)\|_2 \\ \|e_2(k)\|_2 \\ \vdots \\ \|e_n(k)\|_2 \end{bmatrix} \leq \begin{bmatrix} \|r_1(k)\|_2 \\ \|r_2(k)\|_2 \\ \vdots \\ \|r_n(k)\|_2 \end{bmatrix} + \begin{bmatrix} |G_{11}(\mathrm{e}^{j\omega h})| & \dots & |G_{1n}(\mathrm{e}^{j\omega h})| \\ |G_{21}(\mathrm{e}^{j\omega h})| & \dots & |G_{2n}(\mathrm{e}^{j\omega h})| \\ \vdots & \ddots & \vdots \\ |G_{n1}(\mathrm{e}^{j\omega h})| & \dots & |G_{nn}(\mathrm{e}^{j\omega h})| \end{bmatrix} \cdot \begin{bmatrix} \|u_1(k)\|_2 \\ \|u_2(k)\|_2 \\ \vdots \\ \|u_n(k)\|_2 \end{bmatrix}. \tag{5.12}$$

Thus,

$$\begin{bmatrix} \|e_1(k)\|_2 \\ \|e_2(k)\|_2 \\ \vdots \\ \|e_n(k)\|_2 \end{bmatrix} \leq \begin{bmatrix} \|r_1(k)\|_2 \\ \|r_2(k)\|_2 \\ \vdots \\ \|r_n(k)\|_2 \end{bmatrix} + \begin{bmatrix} |G_{11}(\mathrm{e}^{j\omega h})| & |G_{12}(\mathrm{e}^{j\omega h})| & \dots & |G_{1n}(\mathrm{e}^{j\omega h})| \\ |G_{21}(\mathrm{e}^{j\omega h})| & |G_{22}(\mathrm{e}^{j\omega h})| & \dots & |G_{2n}(\mathrm{e}^{j\omega h})| \\ \vdots & \vdots & \ddots & \vdots \\ |G_{n1}(\mathrm{e}^{j\omega h})| & |G_{n2}(\mathrm{e}^{j\omega h})| & \dots & |G_{nn}(\mathrm{e}^{j\omega h})| \end{bmatrix}$$
$$\cdot \left\{ \begin{bmatrix} \rho_1 & 0 & \dots & 0 \\ 0 & \rho_2 & \dots & 0 \\ \vdots & \vdots & \ddots & \vdots \\ 0 & 0 & \dots & \rho_n \end{bmatrix} \begin{bmatrix} \|e_1(k)\|_2 \\ \|e_2(k)\|_2 \\ \vdots \\ \|e_n(k)\|_2 \end{bmatrix} + \begin{bmatrix} \|d_1(k)\|_2 \\ \|d_2(k)\|_2 \\ \vdots \\ \|d_n(k)\|_2 \end{bmatrix} \right\}. \tag{5.13}$$

Rearranging inequality (5.13), we obtain:

$$\begin{bmatrix} 1-\rho_1|G_{11}(\mathrm{e}^{j\omega h})| & -\rho_2|G_{12}(\mathrm{e}^{j\omega h})| & \dots & -\rho_n|G_{1n}(\mathrm{e}^{j\omega h})| \\ -\rho_1|G_{21}(\mathrm{e}^{j\omega h})| & 1-\rho_2|G_{22}(\mathrm{e}^{j\omega h})| & \dots & -\rho_n|G_{2n}(\mathrm{e}^{j\omega h})| \\ \vdots & \vdots & \ddots & \vdots \\ -\rho_1|G_{n1}(\mathrm{e}^{j\omega h})| & -\rho_2|G_{n2}(\mathrm{e}^{j\omega h})| & \dots & 1-\rho_n|G_{nn}(\mathrm{e}^{j\omega h})| \end{bmatrix} \begin{bmatrix} \|e_1(k)\|_2 \\ \|e_2(k)\|_2 \\ \vdots \\ \|e_n(k)\|_2 \end{bmatrix}$$
$$\leq \begin{bmatrix} \|r_1(k)\|_2 \\ \|r_2(k)\|_2 \\ \vdots \\ \|r_n(k)\|_2 \end{bmatrix} + \begin{bmatrix} |G_{11}(\mathrm{e}^{j\omega h})| & |G_{12}(\mathrm{e}^{j\omega h})| & \dots & |G_{1n}(\mathrm{e}^{j\omega h})| \\ |G_{21}(\mathrm{e}^{j\omega h})| & |G_{22}(\mathrm{e}^{j\omega h})| & \dots & |G_{2n}(\mathrm{e}^{j\omega h})| \\ \vdots & \vdots & \ddots & \vdots \\ |G_{n1}(\mathrm{e}^{j\omega h})| & |G_{n2}(\mathrm{e}^{j\omega h})| & \dots & |G_{nn}(\mathrm{e}^{j\omega h})| \end{bmatrix} \begin{bmatrix} \|d_1(k)\|_2 \\ \|d_2(k)\|_2 \\ \vdots \\ \|d_n(k)\|_2 \end{bmatrix}. \tag{5.14}$$

When the matrix on the left side of the above inequality is written as

$$\boldsymbol{A} = \begin{bmatrix} a_{11} & a_{12} & \dots & a_{1n} \\ a_{21} & a_{22} & \dots & a_{2n} \\ \vdots & \vdots & \ddots & \vdots \\ a_{n1} & a_{n2} & \dots & a_{nn} \end{bmatrix}, \tag{5.15}$$

all non-diagonal elements are obviously non-positive. In addition, if all principal minors of (5.15) are positive,[4] matrix $\boldsymbol{A}$ is called an M-matrix [6, 7].

Based on the above premise, the following theorem for multi-loop discretized nonlinear systems is obtained.

Theorem 5.1 *If* $\|r_i(k)\|_2 < \infty$ $(i = 1, 2, \cdots, n)$ *and* $\|d_j(k)\|_2 < \infty$ $(j = 1, 2, \cdots, n)$, *then* $\|e_l(k)\|$ $(l = 1, 2, \cdots, n)$ *become bounded (i.e., the system is BIBO stable) when the matrix of the left side of inequality* (5.14) *is Ostrowski's M-matrix.*

Proof By using matrix expression (5.15), the system inequality (5.14) can be written as follows:

$$\begin{bmatrix} a_{11} & a_{12} & \dots & a_{1n} \\ a_{21} & a_{22} & \dots & a_{2n} \\ \vdots & \vdots & \ddots & \vdots \\ a_{n1} & a_{n2} & \dots & a_{nn} \end{bmatrix} \begin{bmatrix} x_1 \\ x_2 \\ \vdots \\ x_n \end{bmatrix} = \begin{bmatrix} \tilde{y}_1 \\ \tilde{y}_2 \\ \vdots \\ \tilde{y}_n \end{bmatrix} \leqq \begin{bmatrix} y_1 \\ y_2 \\ \vdots \\ y_n \end{bmatrix}, \tag{5.16}$$

where elements of vectors x_i and y_i $(i = 1, 2, \cdots, n)$ are non-negative, and all non-diagonal elements of matrix (a_{ij}) $(i, j = 1, 2, \cdots, n)$ on the left side of (5.16) are non-positive (i.e., $x_i \geq 0$, $y_i \geq 0$, and $a_{ij} \leq 0$ for $i \neq j$).

The equality part of (5.16) can be rewritten as

$$\begin{bmatrix} a_{11}^{(1)} & a_{12}^{(1)} & \dots & a_{1n}^{(1)} \\ 0 & a_{22}^{(2)} & \dots & a_{2n}^{(2)} \\ \vdots & \vdots & \ddots & \vdots \\ 0 & 0 & \dots & a_{nn}^{(n)} \end{bmatrix} \begin{bmatrix} x_1^{(1)} \\ x_2^{(1)} \\ \vdots \\ x_n^{(1)} \end{bmatrix} = \begin{bmatrix} \tilde{y}_1^{(1)} \\ \tilde{y}_2^{(2)} \\ \vdots \\ \tilde{y}_n^{(n)} \end{bmatrix}, \tag{5.17}$$

where $a_{ij}^{(1)} = a_{ij}$, $x_j^{(1)} = x_j$, $\tilde{y}_i^{(1)} = \tilde{y}_i$, and furthermore,

$$\begin{cases} a_{ij}^{(2)} = \dfrac{1}{a_{11}^{(1)}} \begin{vmatrix} a_{11}^{(1)} & a_{1j}^{(1)} \\ a_{i1}^{(1)} & a_{ij}^{(1)} \end{vmatrix} \\ a_{ij}^{(3)} = \dfrac{1}{a_{22}^{(2)}} \begin{vmatrix} a_{22}^{(2)} & a_{2j}^{(2)} \\ a_{i2}^{(2)} & a_{ij}^{(2)} \end{vmatrix} \\ \vdots \\ a_{ij}^{(n)} = \dfrac{1}{a_{n-1\,n-1}^{(n-1)}} \begin{vmatrix} a_{n-1\,n-1}^{(n-1)} & a_{n-1\,j}^{(n-1)} \\ a_{i\,n-1}^{(n-1)} & a_{i\,j}^{(n-1)} \end{vmatrix} \quad (i, j = 2, 3, \cdots, n). \end{cases}$$

[4]As a result, all diagonal elements become positive.

Then, the right side of (5.13) can be written as

$$\tilde{y}_1^{(1)} = \tilde{y}_1,\ \tilde{y}_2^{(2)} = \tilde{y}_2^{(1)} - \frac{a_{21}^{(1)}}{a_{11}^{(1)}}\tilde{y}_1^{(1)},\ \tilde{y}_3^{(3)} = \tilde{y}_3^{(2)} - \frac{a_{32}^{(2)}}{a_{22}^{(2)}}\tilde{y}_2^{(2)},$$

$$\cdots\cdots,\ \tilde{y}_n^{(n)} = \tilde{y}_n^{(n-1)} - \frac{a_{n\,n-1}^{(n-1)}}{a_{n-1\,n-1}^{(n-1)}}\tilde{y}_{n-1}^{(n-1)}$$

provided $a_{11}^{(1)} > 0, a_{22}^{(2)} > 0, \cdots, a_{n-1\,n-1}^{(n-1)} > 0$, where $\tilde{y}_j^{(j)}$ $(j = 1, 2, \cdots, n)$. Thus, one can see that these values are non-negative and bounded if each vector y_i is bounded (i.e., $y_i^{(1)} < \infty,\ i = 1, 2, \cdots, n$). In addition, if $a_{nn}^{(n)} > 0$ is satisfied, then $x_n^{(1)} < \infty$, $x_{n-1}^{(1)} < \infty$, $\cdots$, and $x_1^{(1)} < \infty$ are obtained in reverse order. We note here that these conditions can be rewritten as:

$$\begin{cases} a_{11}^{(1)} = \Delta_1 = a_{11} > 0 \\ a_{22}^{(2)} = \dfrac{\Delta_2}{\Delta_1} = \dfrac{\begin{vmatrix} a_{11} & a_{12} \\ a_{21} & a_{22} \end{vmatrix}}{a_{11}} > 0 \\ a_{33}^{(3)} = \dfrac{\Delta_3}{\Delta_2} = \dfrac{\begin{vmatrix} a_{11} & a_{12} & a_{13} \\ a_{21} & a_{22} & a_{23} \\ a_{31} & a_{32} & a_{33} \end{vmatrix}}{\begin{vmatrix} a_{11} & a_{12} \\ a_{21} & a_{22} \end{vmatrix}} > 0 \\ \vdots \\ a_{nn}^{(n)} = \dfrac{\Delta_n}{\Delta_{n-1}} = \dfrac{\begin{vmatrix} a_{11} & a_{12} & \cdots & a_{1n} \\ a_{21} & a_{22} & \cdots & a_{2n} \\ \vdots & \vdots & \ddots & \vdots \\ a_{n1} & a_{n2} & \cdots & a_{nn} \end{vmatrix}}{\begin{vmatrix} a_{11} & \cdots & a_{1\,n-1} \\ \vdots & \ddots & \vdots \\ a_{n-1\,1} & \cdots & a_{n-1\,n-1} \end{vmatrix}} > 0. \end{cases} \tag{5.18}$$

On the other hand, if the solution of (5.16) is calculated using some numerical method as $x_1 < \infty, x_2 < \infty, \cdots, x_n < \infty$, then $a_{11}^{(1)} > 0, a_{22}^{(2)} > 0), \cdots, a_{nn}^{(n)}$ will be obtained in consequence thereof. These conditions say that all principal minors of matrix $\mathcal{A}$ are positive, which means that the matrix becomes an M-matrix.

The conditions say that all principal minors of matrix $\boldsymbol{A}$ are positive.[5] That is, the matrix becomes an M-matrix. Thus, we can prove that

$$\|e_i^*\| < \infty \text{ and } \|e_i\| < \infty, \quad i = 1, 2, 3,$$

when the nominal control system with gains K, K_n, and K_f is asymptotically stable. Based on the concept of BIBO stability, the robust stability of multi-loop discrete control systems can be proved.

Remark As is clear from Fig. 5.26 (see Appendix B), the condition "all principal minors of matrix $\boldsymbol{A}$ are positive" can be rewritten as: "the solution of $\boldsymbol{Ax} = \boldsymbol{0}$ exists in the first quadrant ($x_i > 0$, $\forall i$)."

5.4 Input-Output Stability for Two-Control-Input Systems

In the case of two-input and two-output systems, Fig. 5.1 is redrawn as shown in Fig. 5.2. As is given in (5.2), cross terms of nonlinearities are not considered here. It is assumed that the nonlinear time-varying elements are written as $f_1(e_1, k)$, $f_2(e_2, k)$, and

$$\frac{|f_1(e_1, k)|}{|e_1(k)|} \le \rho_1 < \infty, \qquad \frac{|f_2(e_2, k)|}{|e_2(k)|} \le \rho_2 < \infty. \tag{5.19}$$

Then, the norm of the output of each nonlinear element is written in the following vector-matrix form:

$$\begin{bmatrix} \|v_1(k)\|_2 \\ \|v_2(k)\|_2 \end{bmatrix} \leqq \begin{bmatrix} \rho_1 & 0 \\ 0 & \rho_2 \end{bmatrix} \begin{bmatrix} \|e_1(k)\|_2 \\ \|e_2(k)\|_2 \end{bmatrix}. \tag{5.20}$$

As is obvious from Fig. 5.2,

$$\begin{bmatrix} \|u_1(k)\|_2 \\ \|u_2(k)\|_2 \end{bmatrix} \leqq \begin{bmatrix} \rho_1 & 0 \\ 0 & \rho_2 \end{bmatrix} \begin{bmatrix} \|e_1(k)\|_2 \\ \|e_2(k)\|_2 \end{bmatrix} + \begin{bmatrix} \|d_1(k)\|_2 \\ \|d_2(k)\|_2 \end{bmatrix}. \tag{5.21}$$

On the other hand, with respect to the controlled systems the following inequality holds for $z = \mathrm{e}^{j\omega h}$:

$$\begin{bmatrix} \|\tilde{y}_1(z)\|_2 \\ \|\tilde{y}_2(z)\|_2 \end{bmatrix} \leqq \begin{bmatrix} \sup_{|z|=1} |G_{11}(z)| & \sup_{|z|=1} G_{12}(z) \\ \sup_{|z|=1} |G_{21}(z)| & \sup_{|z|=1} G_{22}(z) \end{bmatrix} \begin{bmatrix} \|\tilde{u}_1(z)\|_2 \\ \|\tilde{u}_2(z)\|_2 \end{bmatrix}. \tag{5.22}$$

[5] The derivation of the result in general form was proved by the author; see [4].

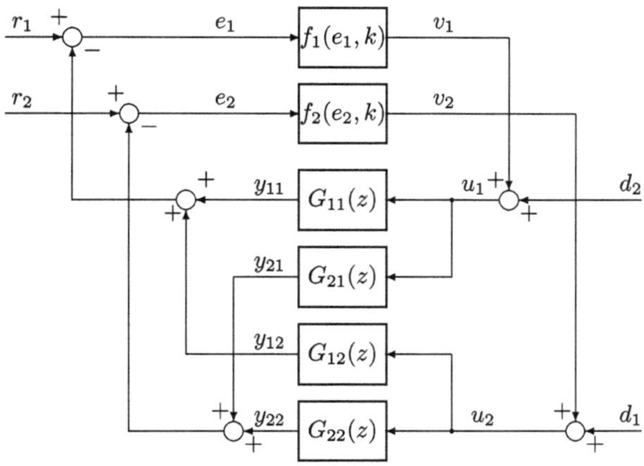

Fig. 5.2 Nonlinear time-varying two-control-input feedback system

Here, $|z| = 1$ corresponds to $\omega := -\pi/h \to \pi/h$, i.e.,

$$\sup_{|z|=1} |G_{ij}(z)| = \sup_{-\pi/h \leq \omega \leq \pi/h} |G_{ij}(e^{j\omega h})|, \quad i, j = 1, 2.$$

By applying Parseval's identity,

$$\begin{bmatrix} \|y_1(k)\|_2 \\ \|y_2(k)\|_2 \end{bmatrix} \leqq \begin{bmatrix} |G_{11}(e^{j\omega h})| & |G_{12}(e^{j\omega h})| \\ |G_{21}(e^{j\omega h})| & |G_{22}(e^{j\omega h})| \end{bmatrix} \begin{bmatrix} \|u_1(k)\|_2 \\ \|u_2(k)\|_2 \end{bmatrix}. \tag{5.23}$$

From Fig. 5.2,

$$\begin{bmatrix} \|e_1(k)\|_2 \\ \|e_2(k)\|_2 \end{bmatrix} \leqq \begin{bmatrix} \|r_1(k)\|_2 \\ \|r_2(k)\|_2 \end{bmatrix} + \begin{bmatrix} |G_{11}(e^{j\omega h})| & |G_{12}(e^{j\omega h})| \\ |G_{21}(e^{j\omega h})| & |G_{22}(e^{j\omega h})| \end{bmatrix} \begin{bmatrix} \|u_1(k)\|_2 \\ \|u_2(k)\|_2 \end{bmatrix}. \tag{5.24}$$

Thus,

$$\begin{bmatrix} \|e_1(k)\|_2 \\ \|e_2(k)\|_2 \end{bmatrix} \leqq \begin{bmatrix} \|r_1(k)\|_2 \\ \|r_2(k)\|_2 \end{bmatrix} + \begin{bmatrix} |G_{11}(e^{j\omega h})| & |G_{12}(e^{j\omega h})| \\ |G_{21}(e^{j\omega h})| & |G_{22}(e^{j\omega h})| \end{bmatrix}$$
$$\cdot \left\{ \begin{bmatrix} \rho_1 & 0 \\ 0 & \rho_2 \end{bmatrix} \begin{bmatrix} \|e_1(k)\|_2 \\ \|e_2(k)\|_2 \end{bmatrix} + \begin{bmatrix} \|d_1(k)\|_2 \\ \|d_2(k)\|_2 \end{bmatrix} \right\}. \tag{5.25}$$

Rearranging inequality (5.25), the following is obtained:

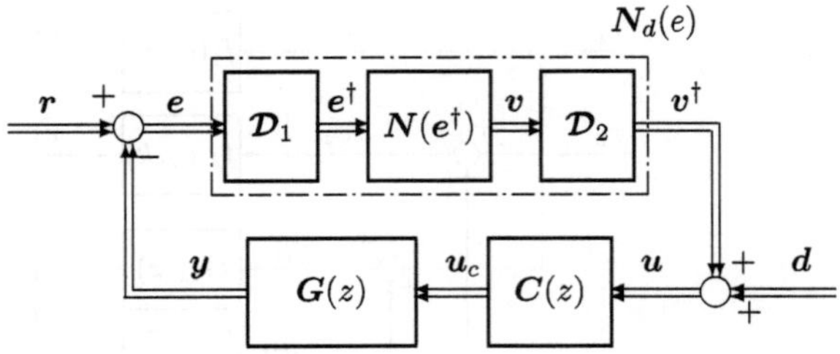

Fig. 5.3 Multi-loop discretized PID control system

$$\begin{bmatrix} 1-\rho_1|G_{11}(\mathrm{e}^{j\omega h})| & -\rho_2|G_{12}(\mathrm{e}^{j\omega h})| \\ -\rho_1|G_{21}(\mathrm{e}^{j\omega h})| & 1-\rho_2|G_{22}(\mathrm{e}^{j\omega h})| \end{bmatrix} \begin{bmatrix} \|e_1(k)\|_2 \\ \|e_2(k)\|_2 \end{bmatrix}$$

$$\leqq \begin{bmatrix} \|r_1(k)\|_2 \\ \|r_2(k)\|_2 \end{bmatrix} + \begin{bmatrix} |G_{11}(\mathrm{e}^{j\omega h})| & |G_{12}(\mathrm{e}^{j\omega h})| \\ |G_{21}(\mathrm{e}^{j\omega h})| & |G_{22}(\mathrm{e}^{j\omega h})| \end{bmatrix} \begin{bmatrix} \|d_1(k)\|_2 \\ \|d_2(k)\|_2 \end{bmatrix}. \tag{5.26}$$

Corollary 5.2 *If* $\|r_1\|_2 < \infty$, $\|r_2\|_2 < \infty$, $\|d_1\|_2 < \infty$, *and* $\|d_2\|_2 < \infty$, *then* $\|e_1\|_2$ *and* $\|e_2\|_2$ *become bounded when the following inequalities are satisfied*:

$$1-\rho_1 \cdot |G_{11}(\mathrm{e}^{j\omega h})| > 0, \tag{5.27}$$
$$(\text{ or } 1-\rho_2 \cdot |G_{22}(\mathrm{e}^{j\omega h})| > 0\ ),\quad 0 < \omega < \omega_c,$$

and

$$(1-\rho_1|G_{11}(\mathrm{e}^{j\omega h})|)(1-\rho_2|G_{22}(\mathrm{e}^{j\omega h})|) > 0. \tag{5.28}$$

Proof The above conditions are obvious from the proof of Theorem 5.1. □

5.5 Multi-Loop Discretized PID Control Systems

For a multi-loop structure, the discretized nonlinear control system is drawn as shown in Fig. 5.3. In this figure, the dynamical system $\boldsymbol{G}(z)$ is given in matrix form as follows:

$$\boldsymbol{G}(z) = \begin{bmatrix} G_{11}(z) & G_{12}(z) & \dots & G_{1n}(z) \\ G_{21}(z) & G_{22}(z) & \dots & G_{2n}(z) \\ \vdots & \vdots & \ddots & \vdots \\ G_{n1}(z) & G_{n2}(z) & \dots & G_{nn}(z) \end{bmatrix}. \tag{5.29}$$

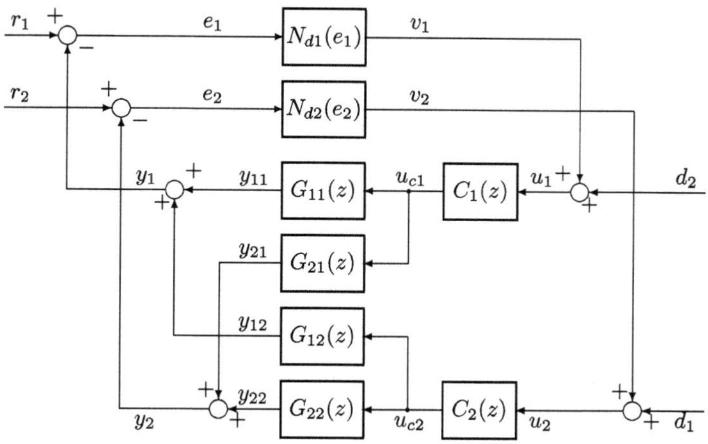

Fig. 5.4 Nonlinear two-input/two-output PID control system

Moreover, controller $\boldsymbol{C}(z)$ is a diagonal matrix that can be written as

$$\boldsymbol{C}(z) = \begin{bmatrix} C_1(z) & 0 & \dots & 0 \\ 0 & C_2(z) & \dots & 0 \\ \vdots & \vdots & \ddots & \vdots \\ 0 & 0 & \dots & C_n(z) \end{bmatrix}. \tag{5.30}$$

When considering two-input and two-output systems (i.e., $n = 2$), the discretized control system can be drawn as shown in Fig. 5.4.

In (5.30), each controller $C_\iota(z)$ is given by

$$\hat{u}_{ci}(z) = K_{pi}\hat{u}_i(z) + C_{Ii}\frac{\hat{u}(z)}{\delta} + C_{Di} \cdot \delta\hat{u}(z) \tag{5.31}$$

as shown in (4.4), where δ is a bilinear operator written as

$$\delta = \frac{2}{h} \cdot \frac{1 - z^{-1}}{1 + z^{-1}}.$$

Therefore, controller matrix $\boldsymbol{C}(z)$ can also be written as

$$\begin{aligned} \boldsymbol{C}(\delta) &= \text{diag}\{C_1(\delta), C_2(\delta), \cdots, C_n(\delta)\}, \\ C_i(\delta) &= K_{pi}\left(1 + \frac{1}{T_{Ii}\delta} + T_{Di}\delta\right). \end{aligned} \tag{5.32}$$

In the frequency domain,

$$C_i(j\Omega) = K_{pi}\left(1 + \frac{1}{jT_{Ii}\Omega} + jT_{Dj}\Omega\right) = K_{pi}\left[1 + j\left(T_{Di}\Omega - \frac{1}{T_{Ii}\Omega}\right)\right], \tag{5.33}$$

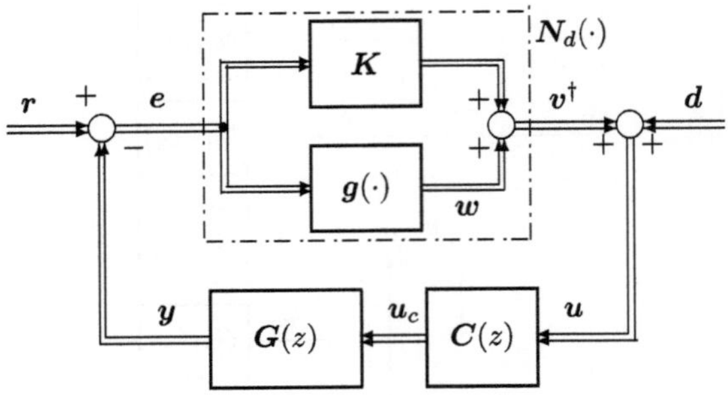

Fig. 5.5 Discretized nonlinear PID control system

where $\Omega = (2/h)\tan(\omega h/2)$. As described in Chap. 4, if using the direct difference method, the control algorithm is given by

$$u_{ci}(k) = K_{pi}u_i(k) + C_{Ii}\sum_{j=0}^{k} u(j) + C_{Di}\Delta u(k). \tag{5.34}$$

The multi-loop discretized control system shown in Fig. 5.3 is redrawn as shown in Fig. 5.5 in regard to the nominal gain of discretized nonlinearity. Here, the nominal gain matrix is given by

$$\boldsymbol{K} = \begin{bmatrix} K_1 & 0 & \dots & 0 \\ 0 & K_2 & \dots & 0 \\ \vdots & \vdots & \ddots & \vdots \\ 0 & 0 & \dots & K_n \end{bmatrix}, \tag{5.35}$$

and the nonlinear part of the system is written as

$$\begin{aligned} &\boldsymbol{g} : \mathbb{R}^n \to \mathbb{R}^n \\ &w_i = g_i(e_i) \quad i = 1, 2, \cdots, n. \end{aligned} \tag{5.36}$$

In addition, each variable is written in the following vector forms:

$$\begin{aligned} \boldsymbol{e} &= (e_1(k), e_2(k), \cdots, e_n(k))^T, \quad \boldsymbol{w} = (w_1(k), w_2(k), \cdots, w_n(k))^T \\ \boldsymbol{v}^\dagger &= (v_1^\dagger(k), v_2^\dagger(k), \cdots, v_n^\dagger(k))^T, \quad \boldsymbol{y} = (y_1(k), y_2(k), \cdots, y_n(k))^T. \end{aligned}$$

The exogenous inputs are also written as

$$\begin{aligned} \boldsymbol{r} &= (r_1(k), r_2(k), \cdots, r_n(k))^T, \\ \boldsymbol{d} &= (d_1(k), d_2(k), \cdots, d_n(k))^T. \end{aligned}$$

From the block diagram of Fig. 5.5,

$$\boldsymbol{v}^\dagger(k) = \boldsymbol{w}(k) + \boldsymbol{K}\boldsymbol{e}(k).$$

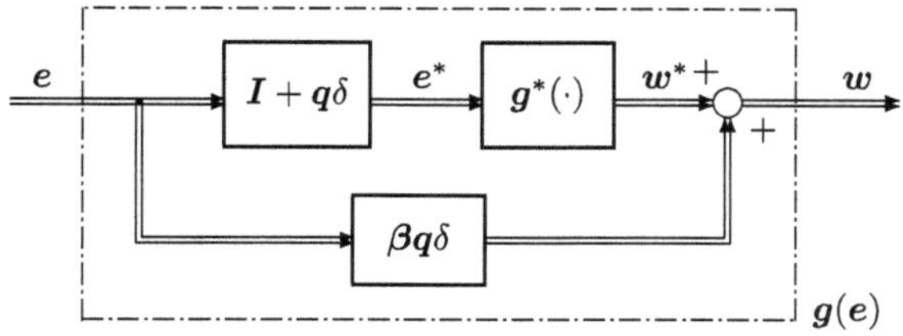

Fig. 5.6 Nonlinear multivariable subsystem

In regard to the z-transform,

$$\hat{\boldsymbol{v}}^{\dagger}(z) = \hat{\boldsymbol{w}}(z) + \boldsymbol{K}\hat{\boldsymbol{e}}(z).$$

Moreover,

$$\begin{aligned}\hat{\boldsymbol{y}}(z) &= \boldsymbol{G}(z)\boldsymbol{C}(z)\hat{\boldsymbol{u}}(z) = \boldsymbol{G}(z)\boldsymbol{C}(z)(\hat{\boldsymbol{v}}^{\dagger}(z) + \hat{\boldsymbol{d}}(z)) \\ &= \boldsymbol{G}(z)\boldsymbol{C}(z)[\hat{\boldsymbol{w}}(z) + \boldsymbol{K}\hat{\boldsymbol{e}}(z) + \hat{\boldsymbol{d}}(z)].\end{aligned}$$

Since $\hat{\boldsymbol{e}}(z) = \hat{\boldsymbol{r}}(z) - \hat{\boldsymbol{y}}(z)$,

$$\hat{\boldsymbol{e}}(z) = \hat{\boldsymbol{r}}(z) - \boldsymbol{G}(z)\boldsymbol{C}(z)\hat{\boldsymbol{w}}(z) - \boldsymbol{G}(z)\boldsymbol{C}(z)\boldsymbol{K}\hat{\boldsymbol{e}}(z) - \boldsymbol{G}(z)\boldsymbol{C}(z)\hat{\boldsymbol{d}}(z).$$

Here, as is shown in Fig. 5.6,

$$\hat{\boldsymbol{w}}(z) = \hat{\boldsymbol{w}}^{*}(z) + \boldsymbol{\beta}\boldsymbol{q}\delta\hat{\boldsymbol{e}}(z).$$

Therefore,

$$\hat{\boldsymbol{e}}(z) = \hat{\boldsymbol{r}}(z) - \boldsymbol{G}(z)\boldsymbol{C}(z)[\hat{\boldsymbol{w}}^{*}(z) + \boldsymbol{\beta}\boldsymbol{q}\delta\hat{\boldsymbol{e}}(z)] - \boldsymbol{G}(z)\boldsymbol{C}(z)\boldsymbol{K}\hat{\boldsymbol{e}}(z) - \boldsymbol{G}(z)\boldsymbol{C}(z)\hat{\boldsymbol{d}}(z),$$

and furthermore,

$$[\boldsymbol{I} + \boldsymbol{G}(z)\boldsymbol{C}(z)(\boldsymbol{K} + \boldsymbol{\beta}\boldsymbol{q}\delta)]\hat{\boldsymbol{e}}(z) = \hat{\boldsymbol{r}}(z) - \boldsymbol{G}(z)\boldsymbol{C}(z)\hat{\boldsymbol{w}}^{*}(z) - \boldsymbol{G}(z)\boldsymbol{C}(z)\hat{\boldsymbol{d}}(z),$$

where $\boldsymbol{I}$ is the identity matrix. Thus, $\boldsymbol{H}(\boldsymbol{\beta}, \boldsymbol{q}, z)$ can be obtained as follows:

$$\boldsymbol{H}(\boldsymbol{\beta}, \boldsymbol{q}, z) = [\boldsymbol{I} + \boldsymbol{G}(z)\boldsymbol{C}(z)(\boldsymbol{K} + \boldsymbol{\beta}\boldsymbol{q}\delta)]^{-1}\boldsymbol{G}(z)\boldsymbol{C}(z). \tag{5.37}$$

As was described in Chap. 2, the discretized control system shown in Fig. 5.5 is redrawn as shown in Fig. 5.7. In the figure, $\boldsymbol{q}$ is a diagonal matrix with non-negative elements that is written as

$$\boldsymbol{q} = \begin{bmatrix} q_1 & 0 & \dots & 0 \\ 0 & q_2 & \dots & 0 \\ \vdots & \vdots & \ddots & \vdots \\ 0 & 0 & \dots & q_n \end{bmatrix}. \tag{5.38}$$

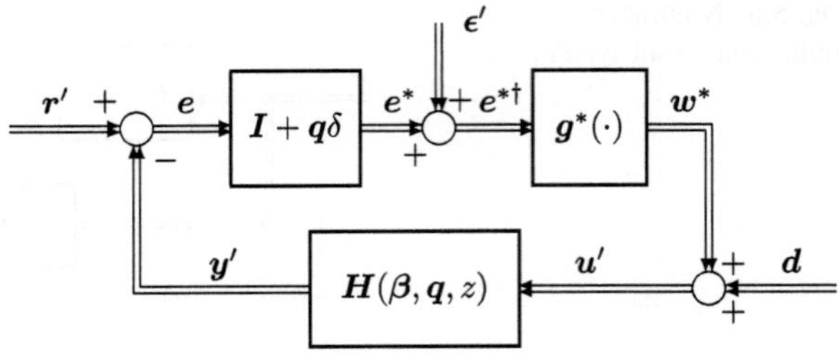

Fig. 5.7 Equivalent multi-loop system

Moreover, $\boldsymbol{\beta}$ is also a diagonal matrix that can be written as

$$\boldsymbol{\beta} = \begin{bmatrix} \beta_1 & 0 & \dots & 0 \\ 0 & \beta_2 & \dots & 0 \\ \vdots & \vdots & \ddots & \vdots \\ 0 & 0 & \dots & \beta_n \end{bmatrix}. \tag{5.39}$$

From the equivalent multi-loop system as shown in Fig. 5.7, the following relations are obtained:

$$\boldsymbol{w}^*(k) = \boldsymbol{g}^*[\boldsymbol{e}^{*\dagger}(k)], \tag{5.40}$$

$$\boldsymbol{e}^{*\dagger}(k) = \boldsymbol{\epsilon}'(k) + \boldsymbol{e}^*(k). \tag{5.41}$$

Here,

$$\boldsymbol{w}^*(k) = \boldsymbol{g}^*[\boldsymbol{e}^{*\dagger}(k)] = \begin{bmatrix} g_1^*(e^{*\dagger}(k)) & 0 & \dots & 0 \\ 0 & g_2^*(e^{*\dagger}(k)) & \dots & 0 \\ \vdots & \vdots & \ddots & \vdots \\ 0 & 0 & \dots & g_n^*(e^{*\dagger}(k)) \end{bmatrix}, \tag{5.42}$$

and, as shown in (3.25), it is assumed that the following sectors can be considered in regard to each nonlinear part:

$$\frac{|g_i^*[e^{*\dagger}(k)]|}{|e^{*\dagger}(k)|} \le \beta_i < \infty, \quad i = 1, 2, \cdots, n. \tag{5.43}$$

Clearly, from Fig. 5.7,

$$\hat{\boldsymbol{e}}^{*\dagger}(z) = \boldsymbol{\epsilon}'(z) + (\boldsymbol{I} + \boldsymbol{q}\delta)\hat{\boldsymbol{e}}(z)),$$

$$\hat{\boldsymbol{e}}(z) = \hat{\boldsymbol{r}}'(z) - \hat{\boldsymbol{y}}'(z).$$

On the other hand, the following relations hold:

$$\boldsymbol{y}'(z) = \boldsymbol{H}(\boldsymbol{\beta}, \boldsymbol{q}, z)\hat{\boldsymbol{u}}'(z),$$

$$\hat{\boldsymbol{u}}'(z) = \hat{\boldsymbol{w}}^*(z) + \boldsymbol{d}'(z).$$

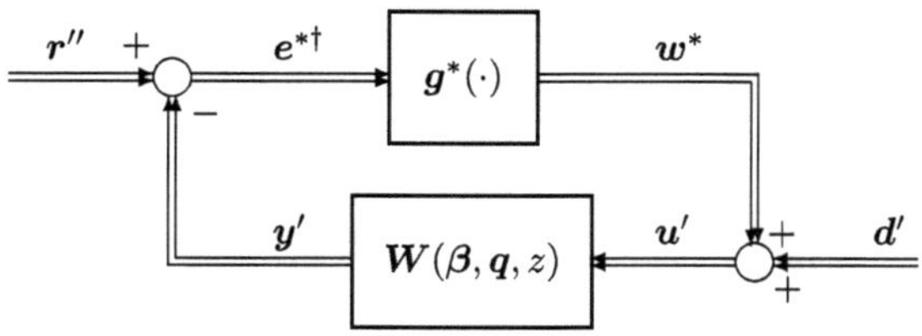

Fig. 5.8 Equivalent small-gain system

The block diagram of the discrete control system shown in Fig. 5.7 can be transformed into Fig. 5.8, where

$$\begin{aligned} \boldsymbol{W}(\boldsymbol{\beta},\boldsymbol{q},z) &= (\boldsymbol{I}+\boldsymbol{q}\delta)\boldsymbol{H}(\boldsymbol{\beta},\boldsymbol{q},z) \\ &= (\boldsymbol{I}+\boldsymbol{q}\delta)[\boldsymbol{I}+(\boldsymbol{K}+\boldsymbol{\beta}\boldsymbol{q}\delta)\boldsymbol{G}(z)\boldsymbol{C}(z)]^{-1}\boldsymbol{G}(z)\boldsymbol{C}(z). \end{aligned} \tag{5.44}$$

Thus,

$$\boldsymbol{e}^{*\dagger}(z) = \boldsymbol{r}''(z) - \boldsymbol{W}(\boldsymbol{\beta},\boldsymbol{q},z)(\boldsymbol{w}^*(z)+\boldsymbol{d}'(z)), \tag{5.45}$$

where

$$\boldsymbol{r}''(z) = \boldsymbol{\epsilon}(z) + (\boldsymbol{I}+\boldsymbol{q}\delta(z))\boldsymbol{r}'(z).$$

The discrete control system shown in Fig. 5.8 corresponds to Fig. 5.1. Therefore, inequalities (5.11) through (5.14) can be applied to this multi-loop system. That is, the following inequality is obtained with respect to $z = \mathrm{e}^{j\omega h}$:[6]

$$\begin{aligned} \begin{bmatrix} \|e_1^{*\dagger}(k)\|_2 \\ \|e_2^{*\dagger}(k)\|_2 \\ \vdots \\ \|e_n^{*\dagger}(k)\|_2 \end{bmatrix} \leqq \begin{bmatrix} \|r_1''(k)\|_2 \\ \|r_2''(k)\|_2 \\ \vdots \\ \|r_n''(k)\|_2 \end{bmatrix} + \begin{bmatrix} |W_{11}(\mathrm{e}^{j\omega h})| & |W_{12}(\mathrm{e}^{j\omega h})| & \dots & |W_{1n}(\mathrm{e}^{j\omega h})| \\ |W_{21}(\mathrm{e}^{j\omega h})| & |W_{22}(\mathrm{e}^{j\omega h})| & \dots & |W_{2n}(\mathrm{e}^{j\omega h})| \\ \vdots & \vdots & \ddots & \vdots \\ |W_{n1}(\mathrm{e}^{j\omega h})| & |W_{n2}(\mathrm{e}^{j\omega h})| & \dots & |W_{nn}(\mathrm{e}^{j\omega h})| \end{bmatrix} \\ \cdot \left\{ \begin{bmatrix} \beta_1 & 0 & \dots & 0 \\ 0 & \beta_2 & \dots & 0 \\ \vdots & \vdots & \ddots & \vdots \\ 0 & 0 & \dots & \beta_n \end{bmatrix} \begin{bmatrix} \|e_1^{*\dagger}(k)\|_2 \\ \|e_2^{*\dagger}(k)\|_2 \\ \vdots \\ \|e_n^{*\dagger}(k)\|_2 \end{bmatrix} + \begin{bmatrix} \|d_1'(k)\|_2 \\ \|d_2'(k)\|_2 \\ \vdots \\ \|d_n'(k)\|_2 \end{bmatrix} \right\}. \end{aligned} \tag{5.46}$$

Then,

[6] In the following inequalities, vectors of sector parameters $\boldsymbol{\beta}$ and arbitrary non-negative parameters $\boldsymbol{q}$ are omitted in $W_{ij}(\cdot,\cdot,\mathrm{e}^{j\omega h})$, $(i, j = 1, 2, \cdots, n)$.

$$\begin{bmatrix} 1-\beta_1|W_{11}(e^{j\omega h})| & -\beta_2|W_{12}(e^{j\omega h})| & \dots & -\beta_n|W_{1n}(e^{j\omega h})| \\ -\beta_1|W_{21}(e^{j\omega h})| & 1-\beta_2|W_{22}(e^{j\omega h})| & \dots & -\beta_n|W_{2n}(e^{j\omega h})| \\ \vdots & \vdots & \ddots & \vdots \\ -\beta_1|W_{n1}(e^{j\omega h})| & -\beta_2|W_{n2}(e^{j\omega h})| & \dots & 1-\beta_n|W_{nn}(e^{j\omega h})| \end{bmatrix} \begin{bmatrix} \|e_1^{*\dagger}(k)\|_2 \\ \|e_2^{*\dagger}(k)\|_2 \\ \vdots \\ \|e_n^{*\dagger}(k)\|_2 \end{bmatrix}$$

$$\leqq \begin{bmatrix} \|r_1''(k)\|_2 \\ \|r_2''(k)\|_2 \\ \vdots \\ \|r_n''(k)\|_2 \end{bmatrix} + \begin{bmatrix} |W_{11}(e^{j\omega h})| & |W_{12}(e^{j\omega h})| & \dots & |W_{1n}(e^{j\omega h})| \\ |W_{21}(e^{j\omega h})| & |W_{22}(e^{j\omega h})| & \dots & |W_{2n}(e^{j\omega h})| \\ \vdots & \vdots & \ddots & \vdots \\ |W_{n1}(e^{j\omega h})| & |W_{n2}(e^{j\omega h})| & \dots & |W_{nn}(e^{j\omega h})| \end{bmatrix} \begin{bmatrix} \|d_1'(k)\|_2 \\ \|d_2'(k)\|_2 \\ \vdots \\ \|d_n'(k)\|_2 \end{bmatrix}. \tag{5.47}$$

As was described in (5.14), the matrix of the left side of (5.47) can be written as

$$\boldsymbol{A} = \begin{bmatrix} a_{11} & a_{12} & \dots & a_{1n} \\ a_{21} & a_{22} & \dots & a_{2n} \\ \vdots & \vdots & \ddots & \vdots \\ a_{n1} & a_{n2} & \dots & a_{nn} \end{bmatrix}. \tag{5.48}$$

Obviously, all non-diagonal elements of (5.48) are non-positive. If all principal minors of (5.48) are positive, the matrix $\boldsymbol{A}$ becomes an Ostrowski M-matrix.

Based on the above premise, we obtain the following theorem for multi-loop discretized nonlinear systems.

Theorem 5.2 *If there exists a $q_i \geq 0$, $\forall i$ in which matrix* (5.48) *becomes an Ostrowski M-matrix, the discretized multi-loop control system with sector nonlinearities* (5.43) *is robustly stable in an ℓ_2 sense, when the linearized system with nominal gains K_i is asymptotically stable.*

Proof By using matrix expression (5.48), the system inequality (5.47) can be written as follows:

$$\begin{bmatrix} a_{11} & a_{12} & \dots & a_{1n} \\ a_{21} & a_{22} & \dots & a_{2n} \\ \vdots & \vdots & \ddots & \vdots \\ a_{n1} & a_{n2} & \dots & a_{nn} \end{bmatrix} \begin{bmatrix} x_1 \\ x_2 \\ \vdots \\ x_n \end{bmatrix} = \begin{bmatrix} \tilde{y}_1 \\ \tilde{y}_2 \\ \vdots \\ \tilde{y}_n \end{bmatrix} \leqq \begin{bmatrix} y_1 \\ y_2 \\ \vdots \\ y_n \end{bmatrix}, \tag{5.49}$$

where elements of vectors x_i and y_i $(i = 1, 2, \cdots, n)$ are non-negative, and all non-diagonal elements of matrix (a_{ij}) $(i, j = 1, 2, \cdots, n)$ on the left side of (5.49) are non-positive (i.e., $x_i \geq 0$, $y_i \geq 0$, and $a_{ij} \leq 0$ for $i \neq j$).

The equality part of (5.49) can be rewritten as

$$\begin{bmatrix} a_{11}^{(1)} & a_{12}^{(1)} & \cdots & a_{1n}^{(1)} \\ 0 & a_{22}^{(2)} & \cdots & a_{2n}^{(2)} \\ \vdots & \vdots & \ddots & \vdots \\ 0 & 0 & \cdots & a_{nn}^{(n)} \end{bmatrix} \begin{bmatrix} x_1^{(1)} \\ x_2^{(1)} \\ \vdots \\ x_n^{(1)} \end{bmatrix} = \begin{bmatrix} \tilde{y}_1^{(1)} \\ \tilde{y}_2^{(2)} \\ \vdots \\ \tilde{y}_n^{(n)} \end{bmatrix}, \tag{5.50}$$

where $a_{ij}^{(1)} = a_{ij}$, $x_j^{(1)} = x_j$, $\tilde{y}_i^{(1)} = \tilde{y}_i$, and furthermore,

$$\begin{cases} a_{ij}^{(2)} = \dfrac{1}{a_{11}^{(1)}} \begin{vmatrix} a_{11}^{(1)} & a_{1j}^{(1)} \\ a_{i1}^{(1)} & a_{ij}^{(1)} \end{vmatrix} \\ a_{ij}^{(3)} = \dfrac{1}{a_{22}^{(2)}} \begin{vmatrix} a_{22}^{(2)} & a_{2j}^{(2)} \\ a_{i2}^{(2)} & a_{ij}^{(2)} \end{vmatrix} \\ \vdots \\ a_{ij}^{(n)} = \dfrac{1}{a_{n-1\,n-1}^{(n-1)}} \begin{vmatrix} a_{n-1\,n-1}^{(n-1)} & a_{n-1\,j}^{(n-1)} \\ a_{i\,n-1}^{(n-1)} & a_{i\,j}^{(n-1)} \end{vmatrix} \quad (i, j = 2, 3, \cdots, n). \end{cases}$$

Then, the right side of (5.50) can be written as

$$\tilde{y}_1^{(1)} = \tilde{y}_1, \ \tilde{y}_2^{(2)} = \tilde{y}_2^{(1)} - \frac{a_{21}^{(1)}}{a_{11}^{(1)}} \tilde{y}_1^{(1)}, \ \tilde{y}_3^{(3)} = \tilde{y}_3^{(2)} - \frac{a_{32}^{(2)}}{a_{22}^{(2)}} \tilde{y}_2^{(2)},$$

$$\cdots\cdots, \ \tilde{y}_n^{(n)} = \tilde{y}_n^{(n-1)} - \frac{a_{n\,n-1}^{(n-1)}}{a_{n-1\,n-1}^{(n-1)}} \tilde{y}_{n-1}^{(n-1)}$$

provided $a_{11}^{(1)} > 0, a_{22}^{(2)} > 0, \cdots, a_{n-1\,n-1}^{(n-1)} > 0$, where $\tilde{y}_j^{(j)}$ $(j = 1, 2, \cdots, n)$. Thus, these values are non-negative and bounded if each vector y_i is bounded (i.e., $y_i^{(1)} < \infty$, $i = 1, 2, \cdots, n$). In addition, if $a_{nn}^{(n)} > 0$ is satisfied, then $x_n^{(1)} < \infty$, $x_{n-1}^{(1)} < \infty$, $\cdots$, and $x_1^{(1)} < \infty$ are obtained in reverse order.

On the other hand, if the solution of (5.49) is calculated using some numerical method as $x_1 < \infty, x_2 < \infty, \cdots, x_n < \infty$, then $a_{11}^{(1)} > 0, a_{22}^{(2)} > 0), \cdots, a_{nn}^{(n)}$ will be obtained in consequence thereof. These conditions say that all principal minors of matrix $\mathcal{A}$ are positive, meaning that the matrix becomes an M-matrix.

The conditions say that all principal minors of matrix $\boldsymbol{A}$ are positive, and this means that the matrix becomes an M-matrix. Thus, we can prove that

$$\|e_i^*\| < \infty \ \text{ and } \ \|e_i\| < \infty, \quad i = 1, 2, 3,$$

when the nominal control system with gains K, K_n, and K_f is asymptotically stable. Based on the concept of BIBO stability, the robust stability of multi-loop discrete control systems can be proved.

Example 5.1 Consider the following 2×2 controlled system:

$$\boldsymbol{G}(s) = \begin{bmatrix} G_{11}(s) & G_{12}(s) \\ G_{21}(s) & G_{22}(s) \end{bmatrix} = \begin{bmatrix} \dfrac{0.2}{(s+0.5)(s+1.0)} & \dfrac{0.04}{s+0.3} \\ \dfrac{0.04}{s+0.2} & \dfrac{0.2}{(s+0.4)(s+0.8)} \end{bmatrix}. \tag{5.51}$$

By using computerized transformation, we can obtain the following transfer functions with respect to z:

$$\begin{cases} G_{11}(z) = \dfrac{0.061z+0.037}{z^2-0.97z+0.22}, \quad G_{12}(z) = \dfrac{0.036}{z-0.82}, \\ G_{21}(z) = \dfrac{0.035}{z-0.74}, \quad G_{22}(z) = \dfrac{0.068z+0.046}{z^2-1.12z+0.30}. \end{cases} \tag{5.52}$$

It is assumed that the controller is given as the following two-channel controller:

$$\begin{cases} u_{c1}(k) = K_{p1}u_1(k) + C_{I1}\sum_{j=0}^{k} u_1(j) + C_{D1}\Delta u_1(k), \\ u_{c2}(k) = K_{p2}u_i(k) + C_{I2}\sum_{j=0}^{k} u_2(j) + C_{D2}\Delta u_2(k), \end{cases} \tag{5.53}$$

as shown in (4.11), where $\Delta u_i(k) = u_i(k) - u_i(k-1)$ $(i = 1, 2)$. Using the z-transform expression, the controller actions are written as

$$\begin{cases} \hat{u}_{c1}(z) = K_{p1}\hat{u}_1(z) + C_{I1} \cdot \dfrac{1}{1-z^{-1}}\hat{u}_1(z) + C_{D1}(1-z^{-1})\hat{u}_1(z) \\ \hat{u}_{c2}(z) = K_{p2}\hat{u}_2(z) + C_{I2} \cdot \dfrac{1}{1-z^{-1}}\hat{u}_2(z) + C_{D2}(1-z^{-1})\hat{u}_2(z). \end{cases} \tag{5.54}$$

The discrete controller $\boldsymbol{C}(z)$ is given by

$$\boldsymbol{C}(z) = \begin{bmatrix} C_1(z) & 0 \\ 0 & C_2(z) \end{bmatrix}, \tag{5.55}$$

where

$$C_i(z) = K_{pi} + C_{Ii} \cdot \frac{1}{1-z^{-1}} + C_{Di}(1-z^{-1}), \quad i = 1, 2.$$

The closed-loop characteristic $\boldsymbol{W}(\boldsymbol{\beta}, \boldsymbol{q}, z)$ as shown in Fig. 5.8 is derived as follows:

$$\boldsymbol{W}(\boldsymbol{\beta}, \boldsymbol{q}, z) = \begin{bmatrix} W_{11}(\boldsymbol{\beta}, \boldsymbol{q}, z) & W_{12}(\boldsymbol{\beta}, \boldsymbol{q}, z) \\ W_{21}(\boldsymbol{\beta}, \boldsymbol{q}, z) & W_{22}(\boldsymbol{\beta}, \boldsymbol{q}, z) \end{bmatrix} = \begin{bmatrix} 1+q_1\delta & 0 \\ 0 & 1+q_2\delta \end{bmatrix} \tag{5.56}$$

$$\cdot \left\{ \begin{bmatrix} 1 & 0 \\ 0 & 1 \end{bmatrix} + \begin{bmatrix} K_1 + \beta_1 q_1 \delta & 0 \\ 0 & K_2 + \beta_2 q_2 \delta \end{bmatrix} \begin{bmatrix} G_{11} & G_{12} \\ G_{21} & G_{22} \end{bmatrix} \begin{bmatrix} C_1 & 0 \\ 0 & C_2 \end{bmatrix} \right\}^{-1}$$
$$\cdot \begin{bmatrix} G_{11} & G_{12} \\ G_{21} & G_{22} \end{bmatrix} \begin{bmatrix} C_1 & 0 \\ 0 & C_2 \end{bmatrix}$$

$$\begin{bmatrix} W_{11}(\beta_1, \beta_2, q_1, q_2, z) & W_{12}(\beta_1, \beta_2, q_1, q_2, z) \\ W_{21}(\beta_1, \beta_2, q_1, q_2, z) & W_{22}(\beta_1, \beta_2, q_1, q_2, z) \end{bmatrix} = \begin{bmatrix} 1 + q_1 \delta & 0 \\ 0 & 1 + q_2 \delta \end{bmatrix}$$
$$\cdot \left\{ \begin{bmatrix} 1 & 0 \\ 0 & 1 \end{bmatrix} + \begin{bmatrix} (K_1 + \beta_1 q_1 \delta) G_{11} C_1 & (K_1 + \beta_1 q_1 \delta) G_{12} C_2) \\ (K_2 + \beta_2 q_2 \delta) G_{21} C_1 & (K_2 + \beta_2 q_2 \delta) G_{22} C_2) \end{bmatrix} \right\}^{-1}$$
$$\cdot \begin{bmatrix} G_{11} C_1 & G_{12} C_2 \\ G_{21} C_1 & G_{22} C_2 \end{bmatrix}. \tag{5.57}$$

Here, we use the following symbol in (5.57):

$$\Pi_{ij}(\beta_i, q_i, z) = (K_i + \beta_i q_i \delta) G_{ij} C_j, \tag{5.58}$$

in order to rewrite (5.57) as

$$\frac{1}{(1 + \Pi_{11})(1 + \Pi_{22}) - \Pi_{12}\Pi_{21}} \begin{bmatrix} 1 + q_1 \delta & 0 \\ 0 & 1 + q_2 \delta 1 \end{bmatrix} \begin{bmatrix} 1 + \Pi_{22} & -\Pi_{12} \\ -\Pi_{21} & 1 + \Pi_{11} \end{bmatrix}$$
$$\cdot \begin{bmatrix} G_{11} C_1 & G_{12} C_2 \\ G_{21} C_1 & G_{22} C_2 \end{bmatrix}$$

Thus, each element of $\boldsymbol{W}$ is given as follows:

$$\begin{cases} W_{11} = \dfrac{1}{(1 + \Pi_{11})(1 + \Pi_{22}) - \Pi_{12}\Pi_{21}} (1 + q_1 \delta)[(1 + \Pi_{22}) G_{11} C_1 - \Pi_{12} G_{21} C_1] \\ W_{12} = \dfrac{1}{(1 + \Pi_{11})(1 + \Pi_{22}) - \Pi_{12}\Pi_{21}} (1 + q_1 \delta)[(1 + \Pi_{22}) G_{12} C_2 - \Pi_{12} G_{22} C_2] \\ W_{21} = \dfrac{1}{(1 + \Pi_{11})(1 + \Pi_{22}) - \Pi_{12}\Pi_{21}} (1 + q_2 \delta)[(1 + \Pi_{11}) G_{21} C_1 - \Pi_{21} G_{11} C_1] \\ W_{22} = \dfrac{1}{(1 + \Pi_{11})(1 + \Pi_{22}) - \Pi_{12}\Pi_{21}} (1 + q_2 \delta)[(1 + \Pi_{11}) G_{22} C_2 - \Pi_{21} G_{12} C_2] \end{cases} \tag{5.59}$$

Here, we note that Π_{ij} $(i, j = 1, 2)$ are complex functions and $1 + q_i \delta = 1 + j q_i \Omega$ when $z = \mathrm{e}^{j\omega h}$.

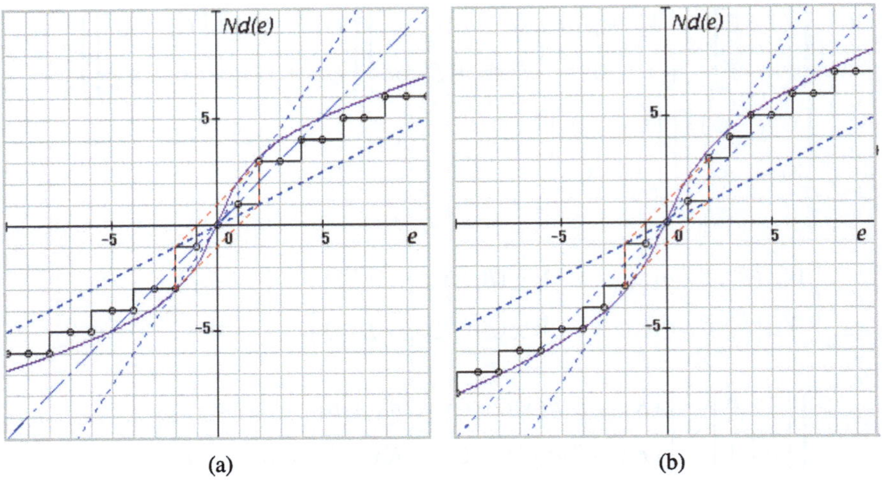

Fig. 5.9 Discretized nonlinear characteristics for Example 5.1

The discretized nonlinear characteristics are assumed to be as shown in Figs. 5.9(a) and (b). If using C-language expressions, they can be written for (a),

$$\begin{aligned} e_1^{\dagger} &= \gamma * \texttt{(double)(int)}(e_1/\gamma) \\ v_1 &= 0.3 * e_1^{\dagger} + 2.7 * \texttt{atan}(0.7 * e_1^{\dagger}) \\ v_1^{\dagger} &= \gamma * \texttt{(double)(int)}(v_1/\gamma), \end{aligned} \tag{5.60}$$

and for (b),

$$\begin{aligned} e_2^{\dagger} &= \gamma * \texttt{(double)(int)}(e_2/\gamma) \\ v_2 &= 0.4 * e_2^{\dagger} + 2.9 * \texttt{atan}(0.6 * e_2^{\dagger}) \\ v_2^{\dagger} &= \gamma * \texttt{(double)(int)}(v_2/\gamma). \end{aligned} \tag{5.61}$$

First, it is assumed that the PID parameters in (5.54) are given by

$$K_{p1} = 1.0, \quad C_{I1} = 0.9, \quad C_{D1} = 0.0,$$
$$K_{p2} = 1.0, \quad C_{I2} = 0.9, \quad C_{D2} = 0.0,$$

that is, only PI control is executed. In this case, the modified Hall diagram and Nyquist curves are as shown in Fig. 5.10. Moreover, the step responses of each variable are given as shown in Fig. 5.11. Here, the red and orange curves show crossed output responses. Figure 5.12 shows Δ_1 and Δ_2 vs. ω. It can be seen from these figures that the two-input and two-output control system is stabilized.

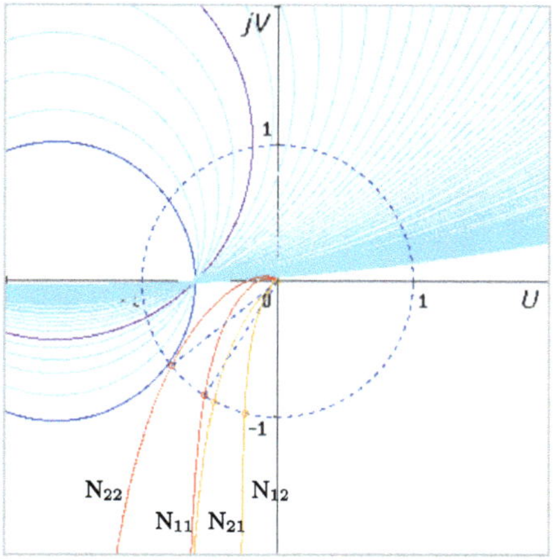

Fig. 5.10 Modified Hall diagram for Example 5.1 (PI control). N_{ij}: Nyquist curve for path $i \to j$

Next, the following PID parameters are applied:

$$K_{p1} = 1.0, \quad C_{I1} = 1.0, \quad C_{D1} = 0.2,$$

$$K_{p2} = 1.0, \quad C_{I2} = 0.8, \quad C_{D2} = 0.2.$$

The modified Hall diagram and Nyquist curves are as shown in Fig. 5.13, and the step responses of each variable are given as shown in Fig. 5.14. As is clear from the figure, the PID control system is well stabilized.

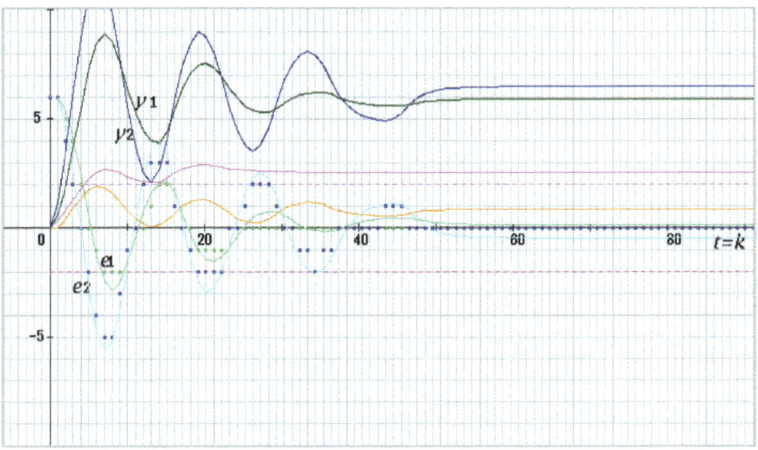

Fig. 5.11 Step responses for Example 5.1 (PI control)

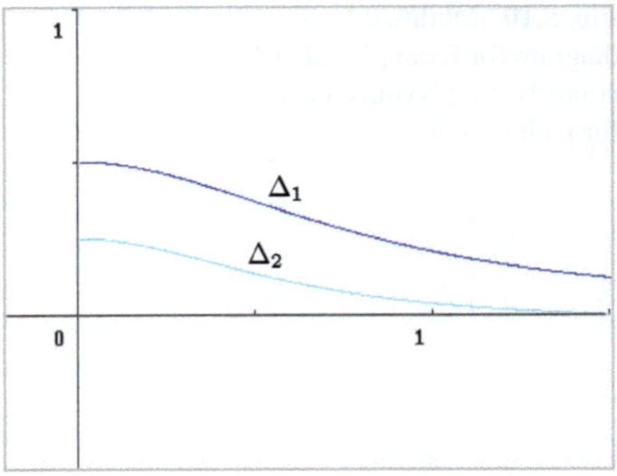

Fig. 5.12 Δ_1 and Δ_2 curves for Example 5.1 (PI control)

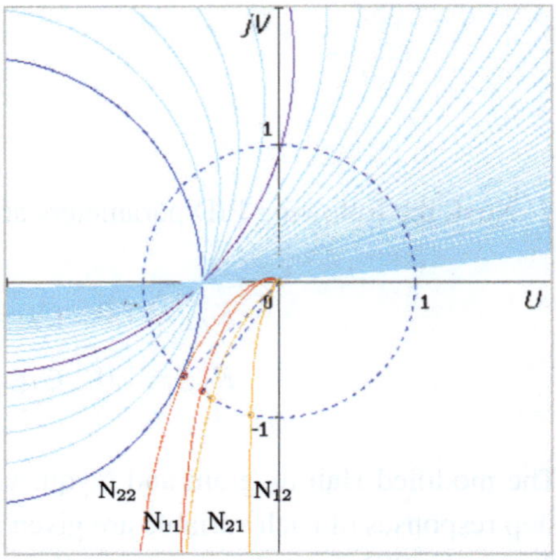

Fig. 5.13 Modified Hall diagram for Example 5.1 (PID control). N_{ij}: Nyquist curve for path $i \to j$

Example 5.2 Next, consider the case of a 2 × 2 controlled systems with transmission delay,

$$\boldsymbol{G}(s) = \begin{bmatrix} \dfrac{0.2}{(s+0.5)(s+1.0)} & \dfrac{0.04}{s+0.3} \\ \dfrac{0.04}{s+0.2} & \dfrac{0.2}{(s+0.4)(s+0.8)} \end{bmatrix}. \tag{5.62}$$

In this example, there are transmission delays $L_1 = 2.0$ and $L_2 = 3.0$. Therefore, the PID algorithm is given as

$$\begin{cases} u_{c1}(k) = K_{p1}u_1(k-2) + C_{I1}\displaystyle\sum_{j=0}^{k} u_1(j-2) + C_{D1}\Delta u_1(k-2), \\ u_{c2}(k) = K_{p2}u_i(k-3) + C_{I2}\displaystyle\sum_{j=0}^{k} u_2(j-3) + C_{D2}\Delta u_2(k-3). \end{cases} \tag{5.63}$$

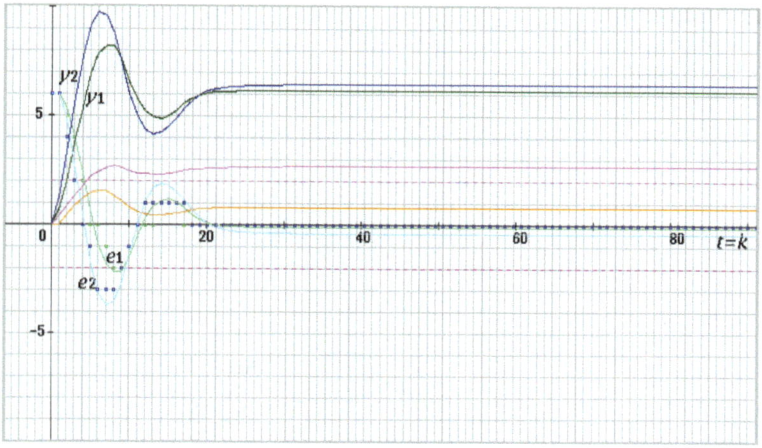

Fig. 5.14 Step responses for Example 5.1 (PID control)

Using the z-transform expression, the controller actions are written as

$$\begin{cases} \hat{u}_{c1}(z) = K_{p1}\hat{u}_1(z)z^{-2} + C_{I1} \cdot \dfrac{1}{1-z^{-1}}\hat{u}_1(z)z^{-2} + C_{D1}(1-z^{-1})\hat{u}_1(z)z^{-2}, \\ \hat{u}_{c2}(z) = K_{p2}\hat{u}_2(z)z^{-3} + C_{I2} \cdot \dfrac{1}{1-z^{-1}}\hat{u}_2(z)z^{-3} + C_{D2}(1-z^{-1})\hat{u}_2(z)z^{-3}. \end{cases} \tag{5.64}$$

Thus, the PID controller matrix is given by

$$\boldsymbol{C}(z) = \begin{bmatrix} C_1(z) & 0 \\ 0 & C_2(z) \end{bmatrix}, \tag{5.65}$$

where

$$C_1(z) = K_{pi}z^{-2} + C_{I1} \cdot \frac{z^{-2}}{1-z^{-1}} + C_{D1}z^{-2}(1-z^{-1}),$$

$$C_2(z) = K_{p2}z^{-3} + C_{I2} \cdot \frac{z^{-3}}{1-z^{-1}} + C_{D2}z^{-3}(1-z^{-1}).$$

First, the following PID parameters are considered (i.e., PI control):

$$K_{p1} = 1.0, \quad C_{I1} = 0.5, \quad C_{D1} = 0.0,$$

$$K_{p2} = 1.0, \quad C_{I2} = 0.3, \quad C_{D2} = 0.0.$$

The modified Hall diagram and Nyquist curves are given as shown in Fig. 5.15. The step responses of each variable are as shown in Fig. 5.16. Figure 5.17 shows the robust stability measures Δ_1 and Δ_2.

Next, the following PID control parameters are considered:

$$K_{p1} = 1.0, \quad C_{I1} = 0.5, \quad C_{D1} = 0.2,$$

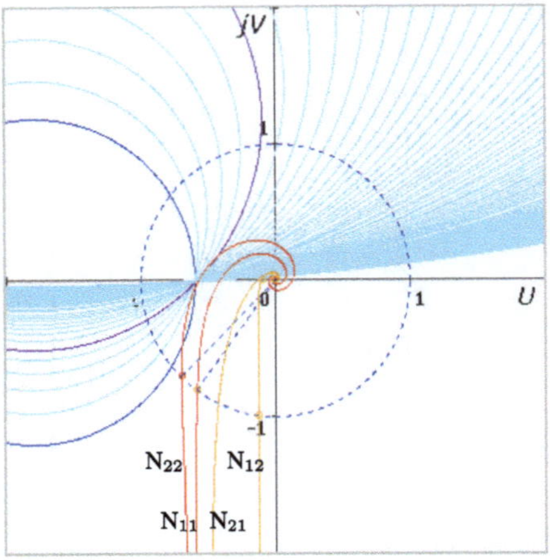

Fig. 5.15 Modified Hall diagram for Example 5.2 (PI control). N_{ij}: Nyquist curve for path $i \to j$

$$K_{p2} = 1.0, \quad C_{I2} = 0.3, \quad C_{D2} = 0.2.$$

The modified Hall diagram and Nyquist curves are shown in Fig. 5.18. The step responses of each variable are as shown in Fig. 5.19. The responses are well stabilized by the derivative action.

Figure 5.20 shows the robust stability measures Δ_1 and Δ_2. If considering a restricted frequency range for the control system, $\Delta_1 > 0$ and $\Delta_2 > 0$ will be satisfied in either of (5.17) and (5.20).

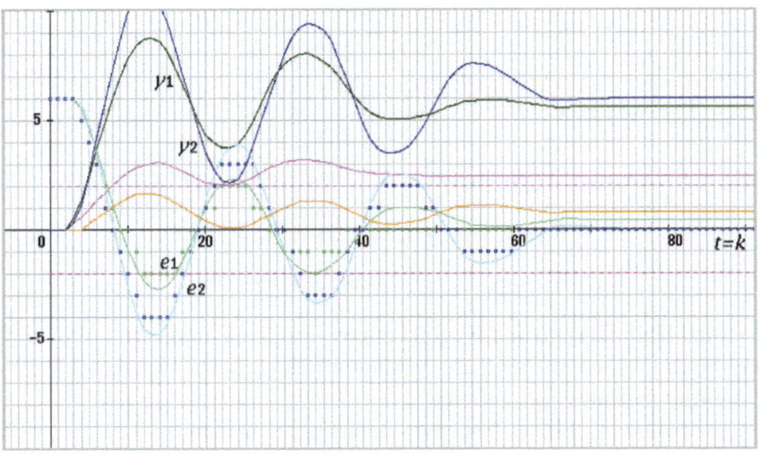

Fig. 5.16 Step responses for Example 5.2 (PI control)

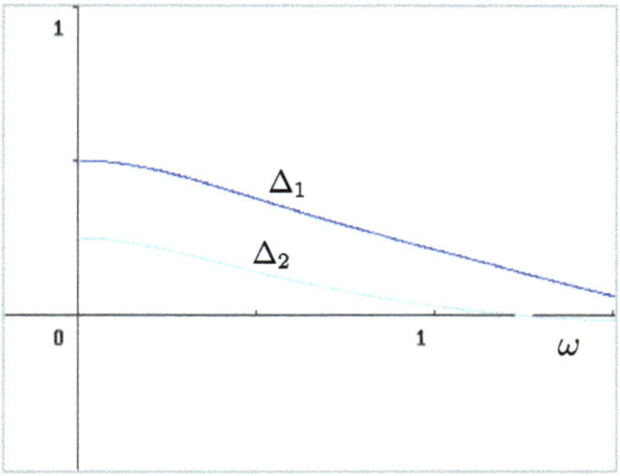

Fig. 5.17 Δ_1 and Δ_2 curves for Example 5.2 (PI control)

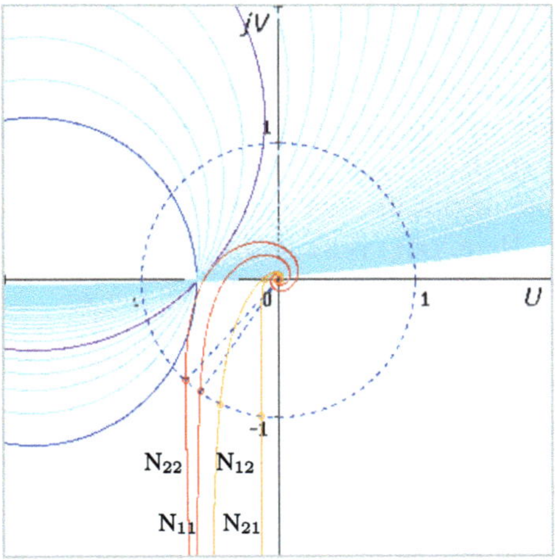

Fig. 5.18 Modified Hall diagram for Example 5.2 (PID control). N_{ij}: Nyquist curve for path $i \to j$

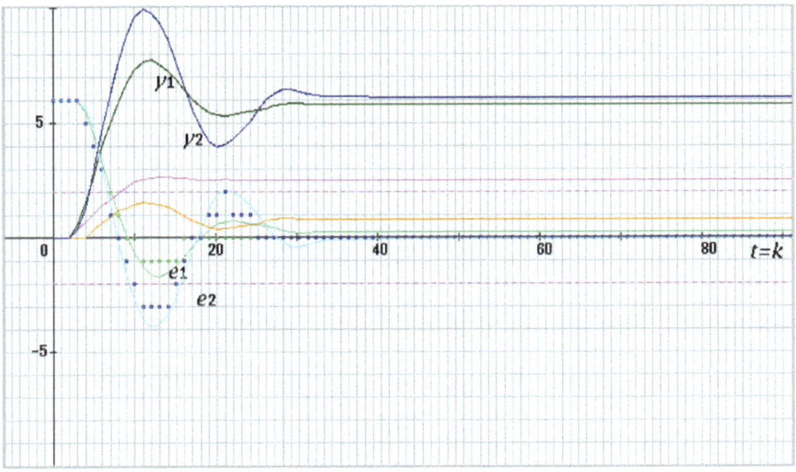

Fig. 5.19 Step responses for Example 5.2 (PID control)

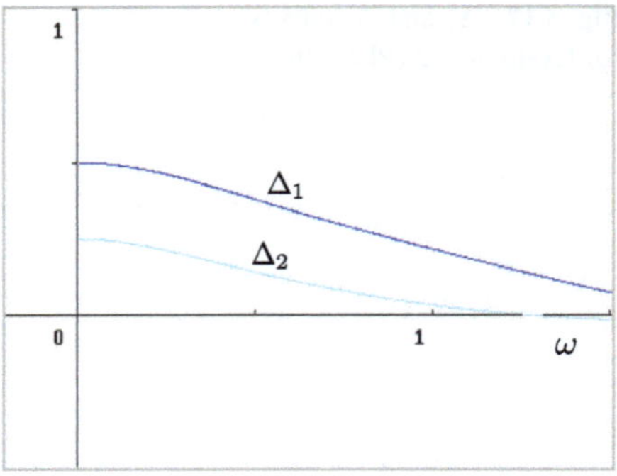

Fig. 5.20 Δ_1 and Δ_2 curves for Example 5.2 (PID control)

5.6 Model Reference Multi-Loop Control Systems

As was described in Chap. 4, model reference discretized control systems are considered here. Figure 5.21 shows the model reference control system extended to a multi-loop feedback structure. Figure 5.22 is an equivalent expression for this structure. In this figure, disturbance $\boldsymbol{d}'(k) = 0$ is assumed for simplicity.

Here, $\boldsymbol{G}(z)$ is a controlled system (plant) matrix which is written by

$$\boldsymbol{G}(z) = \begin{bmatrix} G_{11}(z) & G_{12}(z) & \dots & G_{1n}(z) \\ G_{21}(z) & G_{22}(z) & \dots & G_{2n}(z) \\ \vdots & \vdots & \ddots & \vdots \\ G_{n1}(z) & G_{n2}(z) & \dots & G_{nn}(z) \end{bmatrix}. \tag{5.66}$$

Furthermore, in this chapter, the following model system is considered:

$$\boldsymbol{G}_m(z) = \begin{bmatrix} G_{m1}(z) & 0 & \dots & 0 \\ 0 & G_{m2}(z) & \dots & 0 \\ \vdots & \vdots & \ddots & \vdots \\ 0 & 0 & \dots & G_{mn}(z) \end{bmatrix}. \tag{5.67}$$

Each model is given as

$$G_{mi}(z) = \frac{1}{1 + C_{i1}\delta + C_{i2}\delta^2}.$$

It is assumed that the feedback compensator is also a diagonal matrix,

$$\boldsymbol{F}(z) = \begin{bmatrix} F_1(z) & 0 & \dots & 0 \\ 0 & F_2(z) & \dots & 0 \\ \vdots & \vdots & \ddots & \vdots \\ 0 & 0 & \dots & F_n(z) \end{bmatrix}, \tag{5.68}$$

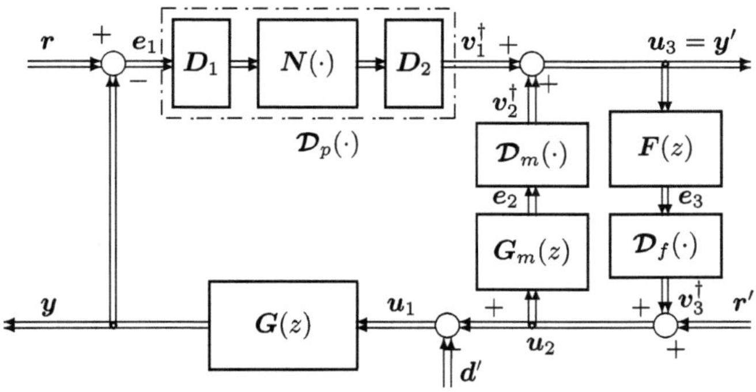

Fig. 5.21 Model reference multi-loop control system

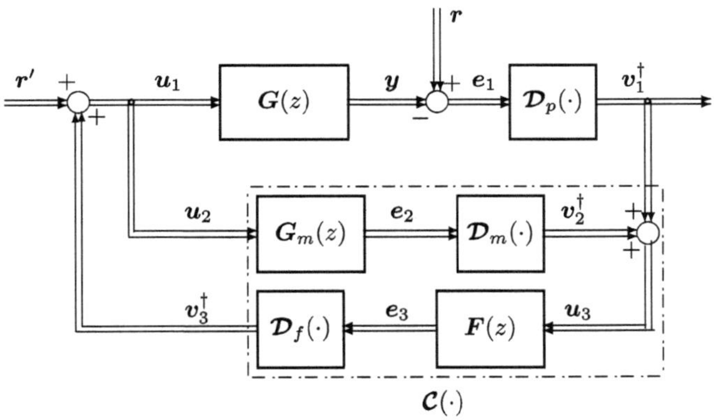

Fig. 5.22 Model reference multi-loop control system 2

where

$$F_i(z) = \frac{1 + C_{i1}\delta + C_{i2}\delta^2}{K_{mi}(1 + c_{i1}\delta + c_{i2}\delta^2)},$$

$$\delta(\mathrm{e}^{j\omega h}) = j\Omega(\omega) = j\frac{2}{h}\tan\left(\frac{\omega h}{2}\right), \quad j = \sqrt{-1}. \tag{5.69}$$

Using this expression, (5.67) and (5.68) can be written as follows:

$$G_{mi}(\mathrm{e}^{j\omega h}) = \tilde{P}_{mi}(j\Omega) = \frac{K_{mi}}{1 - C_{i2}\Omega^2 + jC_{i1}\Omega}, \tag{5.70}$$

and

$$F_i(\mathrm{e}^{j\omega h}) = \tilde{F}_i(j\Omega) = \frac{1 - C_{i2}\Omega^2 + jC_{i1}\Omega}{K_{mi}(1 - c_{i2}\Omega^2 + jc_{i1}\Omega)}. \tag{5.71}$$

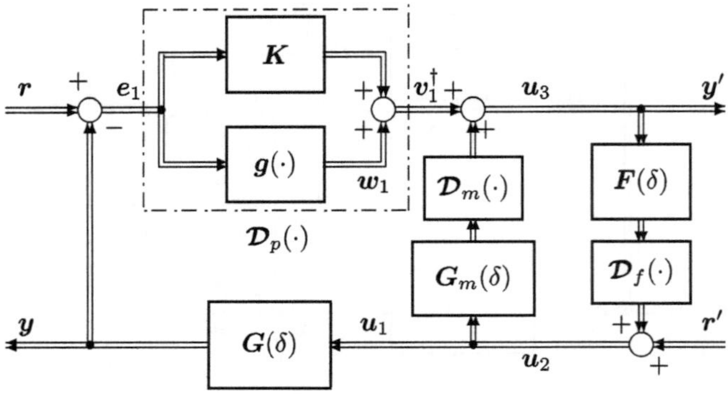

Fig. 5.23 Model reference multi-loop control system (equivalent expression)

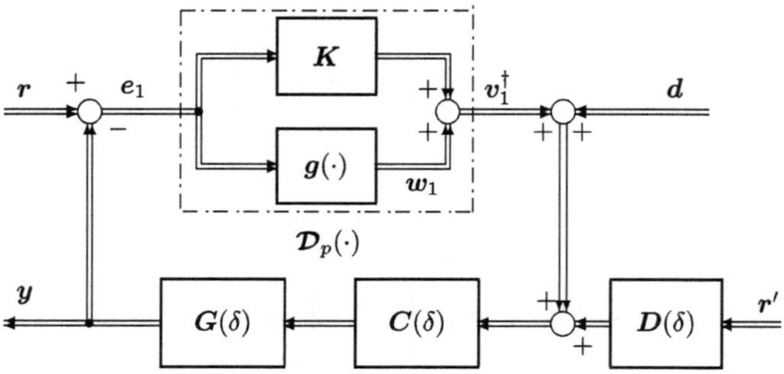

Fig. 5.24 Approximate multi-loop PID control system

When the controllers are in high resolution (i.e., $\gamma \to 0$), the model reference control system as shown in Fig. 5.23 can be transformed into Fig. 5.24. Here, $\boldsymbol{d}'$ is a disturbance signal generated by the discretization of controllers. Since $\boldsymbol{G}_m$ and $\boldsymbol{F}$ are diagonal matrices, $\boldsymbol{C}$ and $\boldsymbol{D}$ are also diagonals:

$$\boldsymbol{C}(\delta) = \text{diag}\{C_1, C_2, \cdots, C_n\} \tag{5.72}$$

$$\boldsymbol{D}(\delta) = \text{diag}\{D_1, D_2, \cdots, D_n\}, \tag{5.73}$$

where C_i and D_i are given by

$$C_i(\delta) = \frac{1 + C_{i1}\delta + C_{i2}\delta^2}{K_{mi}(c_{i1}\delta + c_{i2}\delta^2)}, \tag{5.74}$$

$$D_i(\delta) = \frac{K_{mi}(1 + c_{i1}\delta + c_{i2}\delta^2)}{1 + C_{i1}\delta + C_{i2}\delta^2}. \tag{5.75}$$

Here, $C_i(\delta)$ can be considered to be a controller when $c_{i2} \ll c_{i1}$ $(i = 1, 2, \cdots, n)$. If $c_{i2} \to 0$, the controller is approximately written as

$$C_i(\delta) = \frac{1}{\kappa_i}\delta^{-1} + \frac{C_{i1}}{\kappa_i} + \frac{C_{i2}}{\kappa_i}\delta, \tag{5.76}$$

where $\kappa_i = K_{mi}c_{i1}$. When the sampling period is $h \to 0$, δ becomes the Laplace transform variable s. Therefore, the scheme will correspond to a traditional continuous PID control.

In regard to the robust stability of a model reference control system as shown in Fig. 5.23, the following equations are obtained:

$$\begin{aligned}
\boldsymbol{v}_1^{\dagger}(k) &= \boldsymbol{w}_1(k) + \boldsymbol{K}\boldsymbol{e}_1(k),\\
\hat{\boldsymbol{u}}_2(z) &= \hat{\boldsymbol{r}}'(z) + \boldsymbol{F}(\delta)\hat{\boldsymbol{u}}_3(z),\\
\hat{\boldsymbol{u}}_3(z) &= \hat{\boldsymbol{v}}_1^{\dagger}(z) + \boldsymbol{G}_m(\delta)\hat{\boldsymbol{u}}_2(z).
\end{aligned}$$

Therefore,

$$\hat{\boldsymbol{u}}_2(z) = \hat{\boldsymbol{r}}'(z) + \boldsymbol{F}(\delta)(\hat{\boldsymbol{v}}_1^{\dagger}(z) + \boldsymbol{G}_m(\delta)\hat{\boldsymbol{u}}_2(z)),$$

$$(\boldsymbol{I} - \boldsymbol{F}(\delta)\boldsymbol{G}_m(\delta))\hat{\boldsymbol{u}}_2(z) = \hat{\boldsymbol{r}}'(z) + \boldsymbol{F}(\delta)\hat{\boldsymbol{v}}_1^{\dagger}(z),$$

and then,

$$\hat{\boldsymbol{u}}_2(z) = (\boldsymbol{I} - \boldsymbol{F}(\delta)\boldsymbol{G}_m(\delta))^{-1}\hat{\boldsymbol{r}}'(z) + (\boldsymbol{I} - \boldsymbol{F}(\delta)\boldsymbol{G}_m(\delta))^{-1}\boldsymbol{F}(\delta)\hat{\boldsymbol{v}}_1^{\dagger}(z).$$

Thus, the following expressions can be given:

$$\boldsymbol{C}(\delta) = (\boldsymbol{I} - \boldsymbol{F}(\delta)\boldsymbol{G}_m(\delta))^{-1}\boldsymbol{F}(\delta) \tag{5.77}$$

$$\boldsymbol{D}(\delta) = \boldsymbol{F}^{-1}(\delta). \tag{5.78}$$

With respect to the multi-loop PID control structure as shown in Fig. 5.24, we obtain the equation

$$\hat{\boldsymbol{y}}(z) = \boldsymbol{G}(\delta)[\boldsymbol{C}(\delta)(\hat{\boldsymbol{w}}_1(z) + \boldsymbol{K}\hat{\boldsymbol{e}}_1(z) + \hat{\boldsymbol{d}}'(z) + \boldsymbol{D}(\delta)\hat{\boldsymbol{r}}'(z)). \tag{5.79}$$

Since $\hat{\boldsymbol{e}}(z) = \hat{\boldsymbol{r}}(z) - \hat{\boldsymbol{y}}(z)$,[7]

$$\hat{\boldsymbol{e}}(z) = \hat{\boldsymbol{r}}(z) - \boldsymbol{G}(\delta)\boldsymbol{C}(\delta)[\hat{\boldsymbol{w}}(z) + \boldsymbol{K}\hat{\boldsymbol{e}}(z) + \hat{\boldsymbol{d}}(z)] - \boldsymbol{G}(\delta)\boldsymbol{C}(\delta)\boldsymbol{D}(\delta)\hat{\boldsymbol{r}}'(z). \tag{5.80}$$

On the other hand, from Fig. 5.6,

$$\begin{aligned}
\hat{\boldsymbol{e}}^*(z) &= (\boldsymbol{I} + \boldsymbol{q}\delta)\hat{\boldsymbol{e}}(z)\\
\hat{\boldsymbol{w}}(z) &= \hat{\boldsymbol{w}}^*(z) + \boldsymbol{\beta}\boldsymbol{q}\delta\hat{\boldsymbol{e}}(z).
\end{aligned}$$

[7]For simplicity, the subscript 1, e.g., in e_1, will be omitted hereafter.

Then,

$$\hat{\boldsymbol{e}}(z) = \hat{\boldsymbol{r}}(z) - \boldsymbol{G}(\delta)\boldsymbol{C}(\delta)[\hat{\boldsymbol{w}}^*(z) + \boldsymbol{\beta q}\delta\hat{\boldsymbol{e}}(z) + \boldsymbol{K}\hat{\boldsymbol{e}}(z) + \hat{\boldsymbol{d}}(z)] - \boldsymbol{G}(\delta)\boldsymbol{C}(\delta)\boldsymbol{D}(\delta)\hat{\boldsymbol{r}}'(z),$$

and

$$[\boldsymbol{I} + \boldsymbol{G}(\delta)\boldsymbol{C}(\delta)(\boldsymbol{K} + \boldsymbol{\beta q}\delta)]\hat{\boldsymbol{e}}(z) = \hat{\boldsymbol{r}}(z) - \boldsymbol{G}(\delta)\boldsymbol{C}(\delta)\hat{\boldsymbol{w}}^*(z) - \boldsymbol{G}(\delta)\boldsymbol{C}(\delta)[\hat{\boldsymbol{d}}(z) + \boldsymbol{D}(\delta)\hat{\boldsymbol{r}}'(z)]. \tag{5.81}$$

Thus, $\boldsymbol{H}(\boldsymbol{\beta}, \boldsymbol{q}, \delta)$ can be regarded as

$$\boldsymbol{H}(\beta, q, \delta) = [\boldsymbol{I} + \boldsymbol{G}(\delta)\boldsymbol{C}(\delta)(\boldsymbol{K} + \boldsymbol{\beta q}\delta)]^{-1}\boldsymbol{G}(\delta)\boldsymbol{C}(\delta).$$

From the system as shown in Fig. 5.7,

$$\boldsymbol{w}^*(k) = \boldsymbol{g}^*(\boldsymbol{e}^{*\dagger}(k)),$$
$$\boldsymbol{e}^{*\dagger}(k) = \boldsymbol{\epsilon}(k) + \boldsymbol{e}^*(k).$$

Furthermore,

$$\hat{\boldsymbol{e}}^{*\dagger}(z) = \hat{\boldsymbol{\epsilon}}'(z) + (\boldsymbol{I} + \boldsymbol{q}\delta)\hat{\boldsymbol{e}}(z),$$
$$\hat{\boldsymbol{e}}(z) = \hat{\boldsymbol{r}}'(z) - \hat{\boldsymbol{y}}'(z),$$

and

$$\hat{\boldsymbol{y}}'(z) = \boldsymbol{H}(\boldsymbol{\beta}, \boldsymbol{q}, \delta)(\hat{\boldsymbol{w}}^*(z) + \hat{\boldsymbol{d}}(z)).$$

The block diagram of the discrete control system shown in Fig. 5.7 can be transformed into Fig. 5.8, where

$$\boldsymbol{W}(\boldsymbol{\beta}, \boldsymbol{q}, \delta) = (\boldsymbol{I} + \boldsymbol{q}\delta)\boldsymbol{H}(\boldsymbol{\beta}, \boldsymbol{q}, \delta) = (\boldsymbol{I} + \boldsymbol{q}\delta)[\boldsymbol{I} + (\boldsymbol{K} + \boldsymbol{\beta q}\delta)\boldsymbol{G}(\delta)\boldsymbol{C}(\delta)]^{-1}\boldsymbol{G}(\delta)\boldsymbol{C}(\delta). \tag{5.82}$$

Consequently,

$$\hat{\boldsymbol{e}}^{*\dagger}(z) = \hat{\boldsymbol{r}}''(z) - \boldsymbol{W}(\boldsymbol{\beta}, \boldsymbol{q}, \delta)(\hat{\boldsymbol{w}}^*(z) + \hat{\boldsymbol{d}}(z)), \tag{5.83}$$

where

$$\hat{\boldsymbol{r}}''(z) = \hat{\boldsymbol{\epsilon}}(z) + (\boldsymbol{I} + \boldsymbol{q}\delta)\hat{\boldsymbol{r}}'(z).$$

Thus, the robust stability of the model reference multi-loop control system can be discriminated by inequalities (5.46) and (5.47), i.e.,

$$\begin{bmatrix} \|e_1^{*\dagger}(k)\|_2 \\ \|e_2^{*\dagger}(k)\|_2 \\ \vdots \\ \|e_n^{*\dagger}(k)\|_2 \end{bmatrix} \leqq \begin{bmatrix} \|r_1''(k)\|_2 \\ \|r_2''(k)\|_2 \\ \vdots \\ \|r_n''(k)\|_2 \end{bmatrix} + \begin{bmatrix} |W_{11}(e^{j\omega h})| & |W_{12}(e^{j\omega h})| & \dots & |W_{1n}(e^{j\omega h})| \\ |W_{21}(e^{j\omega h})| & |W_{22}(e^{j\omega h})| & \dots & |W_{2n}(e^{j\omega h})| \\ \vdots & \vdots & \ddots & \vdots \\ |W_{n1}(e^{j\omega h})| & |W_{n2}(e^{j\omega h})| & \dots & |W_{nn}(e^{j\omega h})| \end{bmatrix}$$

$$\cdot \left\{ \begin{bmatrix} \beta_1 & 0 & \dots & 0 \\ 0 & \beta_2 & \dots & 0 \\ \vdots & \vdots & \ddots & \vdots \\ 0 & 0 & \dots & \beta_n \end{bmatrix} \begin{bmatrix} \|e_1^{*\dagger}(k)\|_2 \\ \|e_2^{*\dagger}(k)\|_2 \\ \vdots \\ \|e_n^{*\dagger}(k)\|_2 \end{bmatrix} + \begin{bmatrix} \|d_1'(k)\|_2 \\ \|d_2'(k)\|_2 \\ \vdots \\ \|d_n'(k)\|_2 \end{bmatrix} \right\} \tag{5.84}$$

and

$$\begin{bmatrix} 1-\beta_1|W_{11}(e^{j\omega h})| & -\beta_2|W_{12}(e^{j\omega h})| & \dots & -\beta_n|W_{1n}(e^{j\omega h})| \\ -\beta_1|W_{21}(e^{j\omega h})| & 1-\beta_2|W_{22}(e^{j\omega h})| & \dots & -\beta_n|W_{2n}(e^{j\omega h})| \\ \vdots & \vdots & \ddots & \vdots \\ -\beta_1|W_{n1}(e^{j\omega h})| & -\beta_2|W_{n2}(e^{j\omega h})| & \dots & 1-\beta_n|W_{nn}(e^{j\omega h})| \end{bmatrix} \begin{bmatrix} \|e_1^{*\dagger}(k)\|_2 \\ \|e_2^{*\dagger}(k)\|_2 \\ \vdots \\ \|e_n^{*\dagger}(k)\|_2 \end{bmatrix}$$

$$\leqq \begin{bmatrix} \|r_1''(k)\|_2 \\ \|r_2''(k)\|_2 \\ \vdots \\ \|r_n''(k)\|_2 \end{bmatrix} + \begin{bmatrix} |W_{11}(e^{j\omega h})| & |W_{12}(e^{j\omega h})| & \dots & |W_{1n}(e^{j\omega h})| \\ |W_{21}(e^{j\omega h})| & |W_{22}(e^{j\omega h})| & \dots & |W_{2n}(e^{j\omega h})| \\ \vdots & \vdots & \ddots & \vdots \\ |W_{n1}(e^{j\omega h})| & |W_{n2}(e^{j\omega h})| & \dots & |W_{nn}(e^{j\omega h})| \end{bmatrix} \begin{bmatrix} \|d_1'(k)\|_2 \\ \|d_2'(k)\|_2 \\ \vdots \\ \|d_n'(k)\|_2 \end{bmatrix}. \tag{5.85}$$

5.7 Exercises

(1) From Fig. 5.4, determine the 2×2 loop transfer matrix which corresponds to $\boldsymbol{W}(\boldsymbol{\beta}, \boldsymbol{q}, z)$ in Fig. 5.8.

(2) Show that if all principal minors in (5.15) and (5.48) are positive, all diagonal elements become positive.

(3) Prove that if the solution of the left side of simultaneous equations (5.16) is calculated using some numerical method as $x_1 < \infty$, $x_2 < \infty$, $\cdots$, $x_n < \infty$ (in regard to $y_j \geq 0$, $J = 1, 2, \cdots, n$ and $a_{ij} \leq 0$, $i \neq j$), then the matrix (5.15) becomes an M-matrix.

(4) The colored area shown in Fig. 5.26 is given by the following (vector-matrix) inequality:

$$\begin{bmatrix} 0.8 & -0.3 & -0.3 \\ -0.1 & 0.8 & -0.2 \\ -0.1 & -0.3 & 0.6 \end{bmatrix} \begin{bmatrix} x_1 \\ x_2 \\ x_3 \end{bmatrix} \leqq \begin{bmatrix} 3.6 \\ 3.9 \\ 2.6 \end{bmatrix}.$$

Confirm that the matrix on the left side of the above inequality is an M-matrix, and determine that $x_1 = \tilde{x}_1 < \infty$, $x_2 = \tilde{x}_2 < \infty$, and $x_3 = \tilde{x}_3 < \infty$.

(5) Replace x_2 with x_3 ($x_2 \leftrightarrow x_3$) and write the vector-matrix inequality that corresponds to the above inequality in (4). Confirm that the matrix on the left side of this inequality is an M-matrix.

(6) For a general vector-matrix expression, $\boldsymbol{Ax} \leqq \boldsymbol{y}$, where

$$\boldsymbol{A} = \begin{bmatrix} a_{11} & a_{12} & \dots & a_{1n} \\ a_{21} & a_{22} & \dots & a_{2n} \\ \vdots & \vdots & \ddots & \vdots \\ a_{n1} & a_{n2} & \dots & a_{nn} \end{bmatrix}, \quad \boldsymbol{x} = \begin{bmatrix} x_1 \\ x_2 \\ \vdots \\ x_n \end{bmatrix}, \quad \boldsymbol{y} = \begin{bmatrix} y_1 \\ y_2 \\ \vdots \\ y_n \end{bmatrix},$$

show that the following operations are valid as described in [10] for simultaneous equations:

(a) interchanging two rows,
(b) multiplying each term in one row by a positive constant,
(c) adding a positive multiple of one row to another,

considering the influence on variables x_i and y_j $(i, j = 1, 2, \cdots, n)$.

(7) Show that the condition of the M-matrix is not influenced by the above operations (a), (b), and (c) in general. (Refer to Exercises (5) and (6) in Chap. 4.)

Appendix A: Definition of Ostrowski's M-Matrix

An M-matrix is a real square matrix $\boldsymbol{A}$ with the following properties [1, 2, 5–7]:

(1) $\boldsymbol{A} = \rho \boldsymbol{I} - \boldsymbol{P}$,
$\boldsymbol{P}$: a real square matrix with non-negative elements,
ρ: a positive number that is larger than the absolute value of all the eigenvalues of $\boldsymbol{P}$.

(2) In general, with respect to a real square matrix $\boldsymbol{A}$ with non-positive off-diagonal elements,

(i) there exists $\boldsymbol{x} > \boldsymbol{0}$ that satisfies $\boldsymbol{Ax} > \boldsymbol{0}$;
(ii) $\boldsymbol{A}$ is nonsingular and all the elements of $\boldsymbol{A}^{-1}$ are non-negative;
(iii) the principal minors of $\boldsymbol{A}$ are positive.

Appendix B: Graphical Interpretation of M-Matrices

(1) For the first-order case, $\Delta_1 = a_{11} > 0$ is a trivial problem. Obviously,

$$x_1 = \frac{\tilde{y}_1}{a_{11}} \leq \frac{y_1}{a_{11}}. \tag{5.86}$$

Therefore, if $0 < \tilde{y}_1 \leq y_1 < \infty$ and $x_1 \geq 0$, then $0 \leq x_1 < \infty$ is obtained.

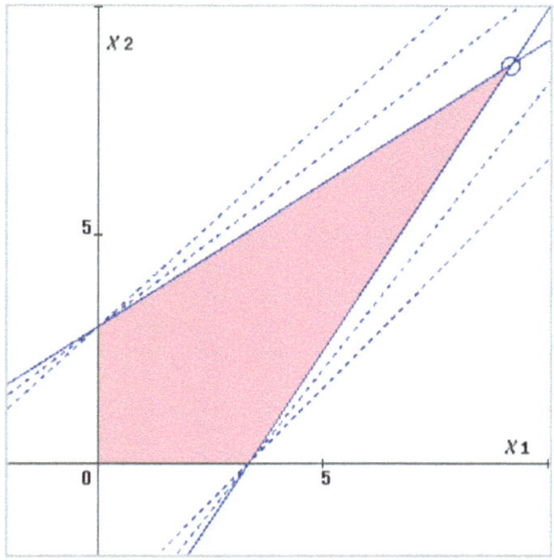

Fig. 5.25 A graphical interpretation of a two-dimensional M-matrix

(2) For the second-order case, the principal-minor condition

$$\Delta_2 = \begin{vmatrix} a_{11} & a_{12} \\ a_{21} & a_{22} \end{vmatrix} > 0 \tag{5.87}$$

is interpreted graphically as shown in Fig. 5.25. The colored area is enclosed by the following two line equations:

$$\begin{cases} \mathcal{L}_1 \ : \ a_{11}x_1 + a_{12}x_2 = \tilde{y}_1 \leq y_1 \\ \mathcal{L}_2 \ : \ a_{21}x_1 + a_{22}x_2 = \tilde{y}_2 \leq y_2 \end{cases} \tag{5.88}$$

and lines $x_1 = 0$ and $x_2 = 0$ (i.e., the area given by $x_1 \geq 0$ and $x_2 \geq 0$). Here, $a_{11} > 0$ and $a_{22} > 0$, with $a_{12} \leq 0$ and $a_{21} \leq 0$ from the premises of M-matrices. In the figure, the outside and inside lines, $\mathcal{L}_1$ and $\mathcal{L}_2$, become parallel when $\Delta_2 = 0$ and $\tilde{y}_1 \neq \tilde{y}_2$. As is obvious from (5.88),

$$x_1 = \frac{1}{a_{11}}(\tilde{y}_1 - a_{12}x_2) \ \text{ and } \ x_2 = \frac{1}{\Delta_2}(-a_{21}\tilde{y}_1 + a_{11}\tilde{y}_2). \tag{5.89}$$

Therefore, if $0 < \tilde{y}_1 \leq y_1$, $0 < \tilde{y}_2 < y_2$, and $\Delta_2 > 0$, then $0 \leq x_1 < \infty$ and $0 \leq x_2 < \infty$ are determined based on the premise of M-matrices (i.e., $a_{11} > 0$, $a_{22} > 0$, $a_{12} \leq 0$, and $a_{21} \leq 0$). That is, (5.88) has a solution (intersection) in the first quadrant.

On the other hand, from (5.88)

$$a_{11}x_1 = \tilde{y}_1 - a_{12}x_2 \ \text{ and } \ \Delta_2 x_2 = -a_{21}\tilde{y}_1 + a_{11}\tilde{y}_2. \tag{5.90}$$

Therefore, if the solution (x_1, x_2) calculated by the numerical method exists in the first quadrant (the strictly positive orthant, i.e., $x_1 > 0$ and $x_2 > 0$), and $0 < \tilde{y}_1 < \infty$, $0 < \tilde{y}_2 < \infty$, the principal-minor condition (5.87) must be satisfied. That is, the corresponding matrix becomes an Ostrowski M-matrix.

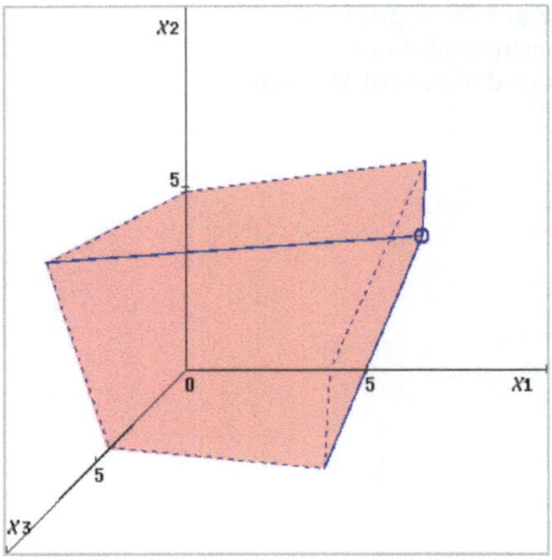

Fig. 5.26 A graphical interpretation of a three-dimensional M-matrix

(3) For the third-order case, the principal-minor condition

$$\Delta_3 = \begin{vmatrix} a_{11} & a_{12} & a_{13} \\ a_{21} & a_{22} & a_{23} \\ a_{31} & a_{32} & a_{33} \end{vmatrix} > 0 \tag{5.91}$$

is interpreted graphically as shown in Fig. 5.26. The colored area is determined (but not drawn in detail because of the three-dimensional expression) by the following three plane equations:

$$\begin{cases} \mathcal{P}_1 \ : \ a_{11}x_1 + a_{12}x_2 + a_{13}x_3 = \tilde{y}_1 \le y_1 \\ \mathcal{P}_2 \ : \ a_{21}x_1 + a_{22}x_2 + a_{23}x_3 = \tilde{y}_2 \le y_2 \\ \mathcal{P}_3 \ : \ a_{31}x_1 + a_{32}x_2 + a_{33}x_3 = \tilde{y}_3 \le y_3 \end{cases} \tag{5.92}$$

and planes $x_1 = 0$, $x_2 = 0$, and $x_3 = 0$ (i.e., the area given by $x_1 \ge 0$, $x_2 \ge 0$, and $x_3 \ge 0$). Here, $a_{11} > 0$, $a_{22} > 0$, and $a_{33} > 0$; $a_{12} \le 0$, $a_{13} \le 0$, $a_{21} \le 0$, $a_{23} \le 0$, $a_{31} \le 0$, and $a_{32} \le 0$ from the premises of M-matrices.

In the figure, three of the edges are given from the following two plane equations that meet in a line in the positive orthant (i.e., the area determined by $x_1 \ge 0$, $x_2 \ge 0$, and $x_3 \ge 0$):

$$\begin{cases} \mathcal{P}_1 \ : \ a_{11}x_1 + a_{12}x_2 + a_{13}x_3 = \tilde{y}_1 \\ \mathcal{P}_2 \ : \ a_{21}x_1 + a_{22}x_2 + a_{23}x_3 = \tilde{y}_2, \end{cases} \tag{5.93}$$

$$\begin{cases} \mathcal{P}_2 \ : \ a_{21}x_1 + a_{22}x_2 + a_{23}x_3 = \tilde{y}_2 \\ \mathcal{P}_3 \ : \ a_{31}x_1 + a_{32}x_2 + a_{33}x_3 = \tilde{y}_3, \end{cases} \tag{5.94}$$

$$\begin{cases} \mathcal{P}_1 \ : \ a_{11}x_1 + a_{12}x_2 + a_{13}x_3 = \tilde{y}_1 \\ \mathcal{P}_3 \ : \ a_{31}x_1 + a_{32}x_2 + a_{33}x_3 = \tilde{y}_3. \end{cases} \tag{5.95}$$

From (5.92),

$$x_1 = \frac{1}{a_{11}}(\tilde{y}_1 - a_{12}x_2 - a_{13}x_3),$$

$$x_2 = \frac{a_{11}}{\Delta_2}\left(\tilde{y}_2 - \frac{a_{11}a_{23} - a_{21}a_{13}}{a_{11}}x_3\right),$$

$$x_3 = \frac{\Delta_2}{\Delta_3} \cdot \tilde{y}_3.$$

Thus, if $0 < \tilde{y}_1 < \infty$, $0 < \tilde{y} < \infty$, $0 < \tilde{y}_3 < \infty$, $\Delta_2 > 0$, and $\Delta_3 > 0$, then $0 \le x_1 < \infty$, $0 \le x_2 < \infty$, and $0 \le x_3 < \infty$ are determined from the premise of M-matrices.

On the other hand, from (5.92)

$$a_{11}x_1 = \tilde{y}_1 - a_{12}x_2 - a_{13}x_3$$

$$\Delta_2 x_2 = a_{11}(\tilde{y}_2 - \frac{a_{11}a_{23} - a_{21}a_{13}}{a_{11}}x_3$$

$$\Delta_3 x_3 = \Delta_2 \tilde{y}_3.$$

Therefore, if the solution (x_1, x_2, x_3) calculated by the numerical method exists in the strictly positive orthant (i.e., $x_1 > 0$, $x_2 > 0$, and $x_3 > 0$), and $0 < \tilde{y}_1 < \infty$, $0 < \tilde{y}_2$, $0 < \tilde{y}_3$, the principal-minor conditions (5.87) and (5.91) must be satisfied. That is, the corresponding matrix becomes an Ostrowski's M-matrix.

References

1. Fan M (1960) Note on M-matrices. Q J Math, 43–49
2. Fiedler M, Pták V (1962) On matrices with non-positive off-diagonal elements and positive principal minors. Czechoslov Math J, 382–400
3. Harris CJ, Valenca JME (1983) The stability of input-output dynamical systems. Academic Press, New York
4. Okuyama Y (1967) On the L_2-stability of linear systems with time-varying parameters. Trans SICE 3:252–259 (in Japanese)
5. Okuyama Y (1974) Relative stability of linear dynamic systems with bounded uncertain coefficients. Trans SICE 10:527–532 (in Japanese)
6. Ostrowski AM (1937) Über die Determinanten mit überwiegender Hauptdiagonale. Comment Math Helv 22:69–96 (in German)
7. Ostrowski AM (1955) Note on bounds for some determinants. Duke Math J 22:95–102
8. Safonov MG (1980) Stability and robustness of multivariable feedback systems. MIT Press, Cambridge
9. Vidyasagar M (1993) Nonlinear systems analysis, 2nd edn. Prentice-Hall, New York, republished by SIAM, 2002
10. Wright DJ (1999) Introduction to linear algebra. McGraw-Hill, New York

Chapter 6
Interval Polynomials and Robust Performance

6.1 Introduction

In the previous chapters, control systems with discretized and nonlinear characteristics have been considered. However, the problems of these nonlinear control systems are generally difficult to analyze. Only the stability (input-output stability) problem was discussed in the frequency domain, and the method of design from the stability margin was presented. In this chapter, a nonlinear characteristic is treated as a set of linear characteristics, in other words, as interval parameters.[1] This concept is based on the validity of Aizerman's conjecture.

There are many works in the literature [1–3, 5, 7] in which the analysis and design of a control system with uncertainties and nonlinearities are discussed in the concept of an *interval system*. In those studies, it is assumed that the physical parameters of controlled systems are uncertain and are accompanied by nonlinearity. Therefore, the transfer function (and the state representation) is expressed by interval parameters.

6.2 Sector Nonlinearities and Interval Systems

First, consider a multi-loop interval system, as shown in Fig. 6.1. Here, $\boldsymbol{G}(z)$ is a multivariable dynamical system with interconnections (e.g., an interconnected plant with some controllers), written as

$$\boldsymbol{G}(z) = \begin{bmatrix} G_{11}(z) & G_{12}(z) & \dots & G_{1n}(z) \\ G_{21}(z) & G_{22}(z) & \dots & G_{2n}(z) \\ \vdots & \vdots & \ddots & \vdots \\ G_{n1}(z) & G_{n2}(z) & \dots & G_{nn}(z) \end{bmatrix} \tag{6.1}$$

[1] Basically, the concept of an interval parameter is not different from that of a sector for a nonlinear characteristic.

Y. Okuyama, *Discrete Control Systems*,
DOI 10.1007/978-1-4471-5667-3_6, © Springer-Verlag London 2014

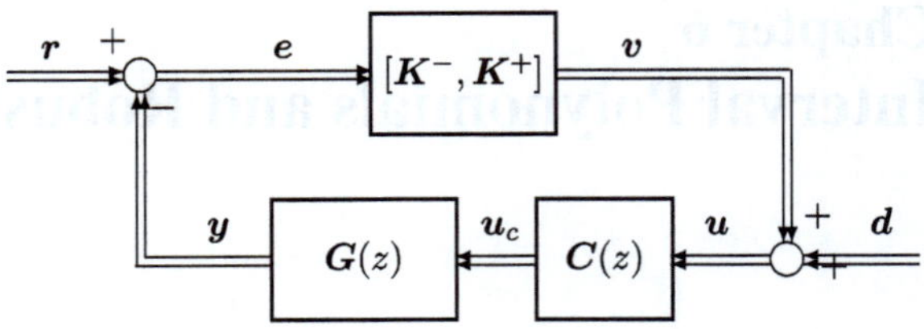

Fig. 6.1 Multi-loop nonlinear/interval feedback system

and

$$C(z) = \begin{bmatrix} C_1(z) & 0 & \dots & 0 \\ 0 & C_2(z) & \dots & 0 \\ \vdots & \vdots & \ddots & \vdots \\ 0 & 0 & \dots & C_n(z) \end{bmatrix}. \tag{6.2}$$

Moreover, interval gains are written as

$$[K^-, K^+] = \begin{bmatrix} [K_1^-, K_1^+] & 0 & \dots & 0 \\ 0 & [K_2^-, K_2^+] & \dots & 0 \\ \vdots & \vdots & \ddots & \vdots \\ 0 & 0 & \dots & [K_n^-, K_n^+] \end{bmatrix}. \tag{6.3}$$

Figures 6.2(a) and (b) show examples of sectors of discretized nonlinearity i.e., interval gains. The loop characteristics from $\boldsymbol{e}$ to $\boldsymbol{y}$ are given as

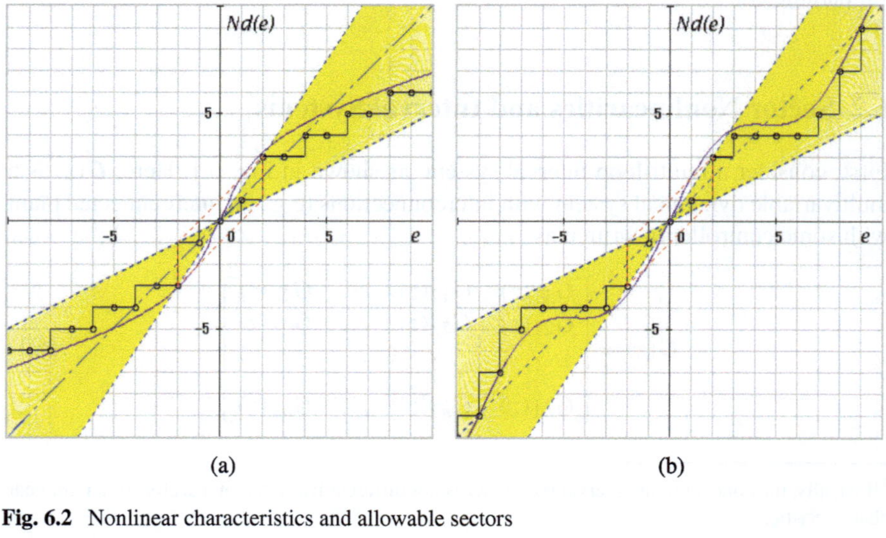

Fig. 6.2 Nonlinear characteristics and allowable sectors

$$G(z)C(z)[K^-, K^+]$$
$$= \begin{bmatrix} [K_1^-, K_1^+]C_1G_{11} & [K_2^-, K_2^+]C_2G_{12} & \dots & [K_n^-, K_n^+]C_nG_{1n} \\ [K_1^-, K_1^+]C_1G_{21} & [K_2^-, K_2^+]C_2G_{22} & \dots & [K_n^-, K_n^+]C_nG_{2n} \\ \vdots & \vdots & \ddots & \vdots \\ [K_1^-, K_1^+]C_1G_{n1} & [K_2^-, K_2^+]C_2G_{n2} & \dots & [K_n^-, K_n^+]C_nG_{nn} \end{bmatrix}, \quad (6.4)$$

whereas the loop characteristics from u to v are given as

$$[K^-, K^+]C(z)G(z)$$
$$= \begin{bmatrix} [K_1^-, K_1^+]C_1G_{11} & [K_1^-, K_1^+]C_1G_{12} & \dots & [K_1^-, K_1^+]C_1G_{1n} \\ [K_2^-, K_2^+]C_2G_{21} & [K_2^-, K_2^+]C_2G_{22} & \dots & [K_2^-, K_2^+]C_2G_{2n} \\ \vdots & \vdots & \ddots & \vdots \\ [K_n^-, K_n^+]C_nG_{n1} & [K_n^-, K_n^+]C_nG_{n2} & \dots & [K_n^-, K_n^+]C_nG_{nn} \end{bmatrix}. \quad (6.5)$$

Thus, the characteristic equation of this multivariable control system is given by

$$\det\{I + G(z)C(z)[K^-, K^+]\} = 0 \quad (6.6)$$

or

$$\det\{I + [K^-, K^+]C(z)G(z)\} = 0, \quad (6.7)$$

where I is the identity matrix. Note that Eqs. (6.6) and (6.7) are equivalent when $C(z)$ and $\tilde{K} \in [K^-, K^+]$ are diagonal matrices with nonzero diagonal elements.

Operations of Interval Parameters For interval parameters $[p_\ell^-, p_\ell^+]$ ($\ell = 1, 2, \cdots$), the following operations can be defined:

- Addition:
$$[p^-, p^+] := [p_1^-, p_1^+] + [p_2^-, p_2^+], \quad (6.8)$$
where $p^- = p_1^- + p_2^-$ and $p^+ = p_1^+ + p_2^+$.
- Subtraction:
$$[p^-, p^+] := [p_1^-, p_1^+] - [p_2^-, p_2^+], \quad (6.9)$$
where $p^- = p_1^- - p_2^+$ and $p^+ = p_1^+ - p_2^-$.
- Multiplication:
$$[p^-, p^+] := [p_1^-, p_1^+] \cdot [p_2^-, p_2^+], \quad (6.10)$$
where $p^- = p_1^- \cdot p_2^-$ and $p^+ = p_1^+ \cdot p_2^+$.
- Division:
$$[p^-, p^+] := \frac{[p_1^-, p_1^+]}{[p_2^-, p_2^+]}, \quad (6.11)$$
where $p^- = p_1^- / p_2^+$ and $p^+ = p_1^+ / p_2^-$.

Example 6.1 Consider a two-loop interval system having the following controlled system (plant) and controller:

$$\boldsymbol{G}(z) = \begin{bmatrix} G_{11}(z) & G_{12}(z) \\ G_{21}(z) & G_{22}(z) \end{bmatrix}, \quad \boldsymbol{C}(z) = \begin{bmatrix} C_1(z) & 0 \\ 0 & C_2(z) \end{bmatrix}. \tag{6.12}$$

Then, the characteristic equation of the interval system is given by

$$\begin{vmatrix} 1 + G_{11}(z)C_1(z)[K_1^-, K_1^+] & G_{12}(z)C_2(z)[K_2^-, K_2^+] \\ G_{21}(z)C_1(z)[K_1^-, K_1^+] & 1 + G_{22}(z)C_2(z)[K_2^-, K_2^+] \end{vmatrix} = 0 \tag{6.13}$$

or

$$\begin{vmatrix} 1 + [K_1^-, K_1^+]C_1(z)G_{11}(z) & [K_1^-, K_1^+]C_1(z)G_{12}(z) \\ [K_2^-, K_2^+]C_2(z)G_{21}(z) & 1 + [K_2^-, K_2^+]C_2(z)G_{22}(z) \end{vmatrix} = 0. \tag{6.14}$$

Therefore, each of the characteristic equations (6.13) and (6.14) can be written as:

$$\begin{aligned} &1 + [K_1^-, K_1^+]C_1(z)G_{11}(z) + [K_2^-, K_2^+]C_2(z)G_{22}(z) \\ &+[K_1^-, K_1^+] \cdot [K_2^-, K_2^+]C_1(z)C_2(z)G_{11}(z)G_{22}(z) \\ &-[K_1^-, K_1^+] \cdot [K_2^-, K_2^+]C_1(z)C_2(z)G_{12}(z)G_{21}(z) = 0. \end{aligned} \tag{6.15}$$

Here, $[K_1^-, K_1^+] \cdot [K_2^-, K_2^+]$ can be given as

$$[K^-, K^+] = [K_1^- K_2^-, K_1^+ K_2^+] = [K_1^-, K_1^+] \cdot [K_2^-, K_2^+] \tag{6.16}$$

from multiplication operation (6.10).

If the transfer function is expressed by the numerator and denominator polynomials, $G_{ij}(z)$ $(i, j = 1, 2)$ can be written as:

$$\begin{cases} G_{11}(z) = \dfrac{N_{11}(z)}{D_{11}(z)}, \quad G_{12} = \dfrac{N_{12}(z)}{D_{12}(z)}, \quad G_{21}(z) = \dfrac{N_{21}(z)}{D_{21}(z)}, \quad G_{22} = \dfrac{N_{22}(z)}{D_{22}(z)}, \\ C_1(z) = \dfrac{N_{c1}(z)}{D_{c1}(z)}, \quad C_2(z) = \dfrac{N_{c2}(z)}{D_{c2}(z)}. \end{cases}$$

The characteristic equation is given by

$$\begin{aligned} \tilde{F}(z) &= D_{c1}(z)D_{c_2}(z)D_{11}(z)D_{22}(z)D_{12}(z)D_{21}(z) \\ &+[K_1^-, K_1^+]N_{c1}(z)N_{11}(z)D_{c2}D_{22}(z)D_{12}(z)D_{21}(z) \\ &+[K_2^-, K_2^+]N_{c2}(z)N_{22}(z)D_{c1}D_{11}(z)D_{12}(z)D_{21}(z) \\ &+[K^-, K^+]N_{c1}(z)N_{c2}(z)N_{11}(z)N_{22}(z)D_{12}(z)D_{21}(z) \\ &-[K^-, K^+]N_{c1}(z)N_{c2}(z)N_{12}(z)N_{21}(z)D_{11}(z)D_{22}(z) = 0. \end{aligned} \tag{6.17}$$

Equation (6.17) is a characteristic equation with interval coefficients. This type of polynomials, $\tilde{F}(z)$, is called an *interval polynomial*.

Example 6.2 Consider the following interconnected plant, which was described in Example 5.2:

$$\boldsymbol{G}(s)=\begin{bmatrix}\dfrac{0.2}{(s+0.5)(s+1.0)} & \dfrac{0.02}{s+0.6}\\ \dfrac{0.02}{s+0.4} & \dfrac{0.2}{(s+0.4)(s+0.8)}\end{bmatrix}. \tag{6.18}$$

By using computerized transformation, the following transfer functions with respect to z are obtained:

$$\begin{cases} G_{11}(z)=\dfrac{0.061z+0.037}{z^2-0.97z+0.22}, \quad G_{12}(z)=\dfrac{0.036}{z-0.82},\\ G_{21}(z)=\dfrac{0.035}{z-0.74}, \quad G_{22}(z)=\dfrac{0.068z+0.046}{z^2-1.12z+0.30}. \end{cases} \tag{6.19}$$

The discrete controller is assumed to be of proportional-integral (PI) type and is given as

$$\begin{cases} C_1(z)=K_{p1}+C_{I1}\cdot\dfrac{1}{1-z^{-1}},\\ C_2(z)=K_{p2}+C_{I2}\cdot\dfrac{1}{1-z^{-1}}. \end{cases} \tag{6.20}$$

In this example, the PI parameters are chosen as $K_{p1}=C_{I1}=1.0$ and $K_{p2}=C_{I2}=1.0$. Therefore,

$$C_1(z)=C_2(z)=\frac{2z-1}{z-1}.$$

Since

$$\begin{aligned} &N_{11}(z)=0.061z+0.037, \quad N_{12}(z)=0.036\\ &N_{21}(z)=0.035, \quad N_{22}(z)=0.068z+0.046\\ &D_{11}(z)=z^2-0.97z+0.22, \quad D_{12}(z)=z-0.82\\ &D_{21}(z)=z-0.74, \quad D_{22}(z)=z^2-1.12z+0.30,\\ &N_{c1}(z)=N_{c2}(z)=2z-1, \quad D_{c1}(z)=D_{c2}(z)=z-1, \end{aligned}$$

as shown in (6.17), the following characteristic equation with interval parameters can be defined:

$$\begin{aligned} &\tilde{F}(z)=(z-1)^2(z^2-0.97z+0.22)(z^2-1.12z+0.30)(z-0.82)(z-0.74)\\ &+[K_1^-,K_1^+](2z-1)(0.061z+0.037)(z-1)(z^2-1.12z+0.30)(z-0.82)(z-0.74)\\ &+[K_2^-,K_2^+](2z-1)(0.068z+0.046)(z-1)(z^2-0.97z+0.22)(z-0.82)(z-0.74)\\ &+[K^-,K^+](2z-1)^2(0.061z+0.037)(0.068z+0.046)(z-0.82)(z-0.74)\\ &+[K^-,K^+](2z-1)^2\times 0.036\times 0.035\times (z^2-0.97z+0.22)(z^2-1.12z+0.30)=0. \end{aligned}$$

Thus, we obtain the following eighth-order characteristic equation:

$$\begin{aligned}
\tilde{F}(z) &= z^8 - 1.65z^7 - 0.83z^6 + 2.99z^5 - 1.27z^4 - 0.98z^3 + 1.06z^2 - 0.35z + 0.04 \\
&+[K_1^-, K_1^+](0.12z^7 - 0.44z^6 + 0.57z^5 - 0.26z^4 - 0.085z^3 + 0.14z^2 - 0.052z + 0.0067) \\
&+[K_2^-, K_2^+](0.14z^7 - 0.46z^6 + 0.53z^5 - 0.17z^4 - 0.16z^3 + 0.16z^2 - 0.052z + 0.0061) \\
&+[K^-, K^+](0.022z^6 - 0.037z^5 + 0.012z^4 + 0.004z^3 + 0.0028z^2 - 0.0046z + 0.0011) \\
&= 0. \qquad (6.21)
\end{aligned}$$

Here, $K^- = K_1^- K_2^-$ and $K^+ = K_1^+ K_2^+$. Note that each coefficient of (6.21) is rounded to the nearest decimal number.

6.3 Characteristic Polynomials with Interval Parameters

In general, the transfer function and the characteristic polynomial of a discrete control system with uncertainty (and/or nonlinearity) is expressed by the following interval polynomial:

$$\tilde{F}(z) = \tilde{a}_0 z^n + \tilde{a}_1 z^{n-1} + \cdots + \tilde{a}_{n-1} z + \tilde{a}_n, \qquad (6.22)$$
$$\tilde{a}_k \in [a_k^-, a_k^+], \quad k = 0, 1, 2, \cdots, n.$$

In the following, (6.22) may be written simply as

$$\tilde{F}(z) = [a_0^-, a_0^+]z^n + [a_1^-, a_1^+]z^{n-1} + \cdots + [a_{n-1}^-, a_{n-1]}^+]z + [a_n^-, a_n^+] = 0. \quad (6.23)$$

Since the interval coefficients are not always independent of each other, the interval polynomial is written in the form (see, e.g., [1, 2])

$$\tilde{F}(z) = a_0(\tilde{\boldsymbol{p}})z^n + \cdots + a_{n-1}(\tilde{\boldsymbol{p}})z + a_n(\tilde{\boldsymbol{p}}), \qquad (6.24)$$
$$\tilde{\boldsymbol{p}} = (\tilde{p}_1, \tilde{p}_2, \cdots, \tilde{p}_m)^T,$$

where $\tilde{p}_\ell$ ($\ell = 1, 2, \cdots, m$) are uncertain parameters. The coefficients of this type of polynomial can be expressed as

$$a_k(\tilde{\boldsymbol{p}}) = \sum_{\ell=0}^{m} c_{\ell k} \tilde{p}_\ell, \qquad (6.25)$$
$$\tilde{p}_\ell \in [p_\ell^-, p_\ell^+], \quad k = 0, 1, 2, \cdots, n, \quad m \le n,$$

where coefficients $c_{\ell k}$ are real constants.

These polynomials can also be written affinely in the following general form:

$$\begin{aligned}\tilde{F}(z) &= \sum_{k=0}^{n}\left(\sum_{\ell=0}^{m} c_{\ell k}\tilde{p}_{\ell}\right) z^{n-k},\\ &= \sum_{\ell=0}^{m} \tilde{p}_{\ell} F_{\ell}(z) = \sum_{\ell=0}^{m} [p_{\ell}^{-}, p_{\ell}^{+}] F_{\ell}(z),\end{aligned} \tag{6.26}$$

where polynomial $F_{\ell}(z)$ is defined as

$$F_{\ell}(z) = \sum_{k=0}^{n} c_{\ell k} z^{n-k}. \tag{6.27}$$

Note that if the argument of $F_{\ell}(z)$ is a constant when variable z traces on a contour $\partial\Gamma$ in a z-plane (or in an s-plane), four polynomials can be defined with respect to the sufficiency of the stability. In particular, if the argument is invariant in real or imaginary numbers, the Kharitonov rectangle can be determined with respect to the necessary and sufficient condition.[2]

Interval polynomial (6.21) is based on the expression of (6.26). Of course, the expression contains (6.22) and (6.23), where

$$\begin{aligned}&[p_{\ell}^{-}, p_{\ell}^{+}] = [a_{\ell}^{-}, a_{\ell}^{+}]\\ &F_{\ell}(z) = z^{n-\ell}, \quad c_{\ell\ell} = 1, c_{\ell k} = 0, \ \ell \neq k.\end{aligned}$$

Segment Polynomials The discrimination of the stability based on expression (6.26) with many uncertain parameters is a considerably complicated problem. When variable z traces on a circular contour, e.g., $\partial\Gamma$ as shown in Fig. 6.4, the mapping of $\tilde{F}(z)$ becomes a set of rotated polytopes (parallelotopes) [4, 14]. The robust stability of the above interval system cannot be discriminated by its vertex polynomials (i.e., the weak Kharitonov theorem is not satisfied). Thus, as was described in [10], the concept of a set of segment polynomials is applied here.

First, consider the following (simple) interval polynomial (i.e., a polynomial with only one interval set coefficient):

$$\tilde{F}(z) = \sum_{\ell=0}^{m} \tilde{p}_{\ell} F_{\ell}(z), \tag{6.28}$$

$$\begin{aligned}&\tilde{p}_h \in [p_h^{-}, p_h^{+}], \quad \tilde{p}_{\ell} = p_{\ell}, \quad \ell \neq h,\\ &h, \ell = 0, 1, 2, \cdots, m.\end{aligned}$$

This type of polynomial, (6.28), is called a *segment polynomial.*

The segment polynomial can be written in the following form:

$$\tilde{F}(z) = \lambda F^{+}(z) + (1-\lambda) F^{-}(z), \tag{6.29}$$

[2]See Appendix A.

where λ is a real number, $\lambda \in [0,\ 1]$. Therefore, at both ends of (6.28), the polynomials are expressed as follows:

$$F^{+}(z) = \sum_{\ell=0}^{m} \tilde{p}_{\ell} F_{\ell}(z), \tag{6.30}$$

$$\tilde{p}_h = p_h^{+},\quad \tilde{p}_{\ell} = p_{\ell},\quad \ell \neq h,$$

$$F^{-}(z) = \sum_{\ell=0}^{m} \tilde{p}_{\ell} F_{\ell}(z), \tag{6.31}$$

$$\tilde{p}_h = p_h^{-},\quad \tilde{p}_{\ell} = p_{\ell},\quad \ell \neq h.$$

As for the above segment polynomial, when considering the algebraic equation $\tilde{F}(z) = 0$, segments of the characteristic root locus can be drawn on the z-plane. On the other hand, when considering mapping $\tilde{F}(z)$ for contour $z \in \partial\Gamma$ as shown in Fig. 6.4, a set of line segments will be drawn on a complex F-plane.

6.4 Sectorial $\mathcal{D}$-Stability

The sectorial $\mathcal{D}$-stability for a continuous-time system can be discriminated by the following radial lines on the complex s-plane:

$$s = (\gamma + j)\omega,\quad \gamma = \tan\phi,\quad j = \sqrt{-1}, \tag{6.32}$$

where ϕ is an inclination angle of the imaginary axis (the angular frequency ω-axis), and parameter γ is chosen as:[3]

$$\begin{cases} \gamma < 0 & \text{when } \omega : 0 \to +\infty \\ \gamma > 0 & \text{when } \omega : -\infty \to 0. \end{cases}$$

Since the complex roots of the algebraic equation with real coefficients are conjugates, it is sufficient to consider either of these cases. Thus, only the case of $\gamma < 0$ will be considered in this chapter.

As was described in the previous chapters, the contour that corresponds to (6.32) should be considered in the z-plane as follows:

$$z = \mathrm{e}^{(\gamma+j)\omega h}, \tag{6.33}$$

where h is a sampling period. The discrimination of the $\mathcal{D}$-stability by finite calculations cannot be achieved when a transcendental function (6.33) is used. Therefore,

[3]Note that the symbol γ is used differently from the resolution value in Chap. 2.

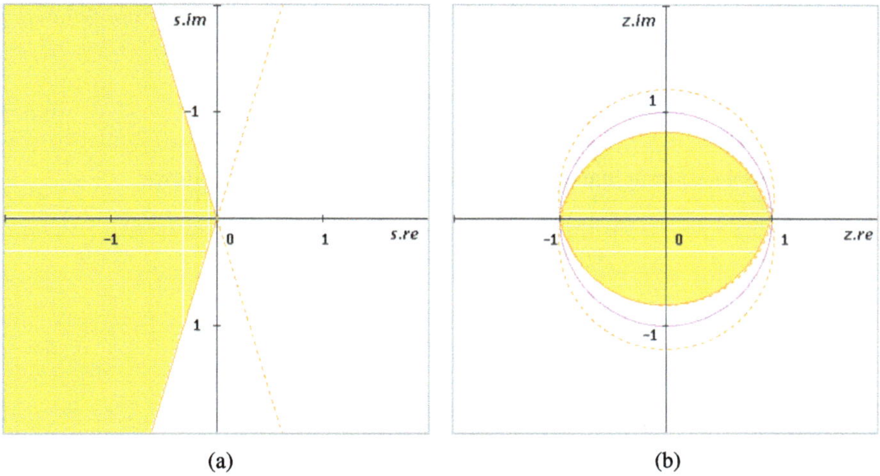

Fig. 6.3 Sectorial and pseudo-sectorial areas in the s- and z-planes

Fig. 6.4 Pseudo-sectorial area for discrete control system

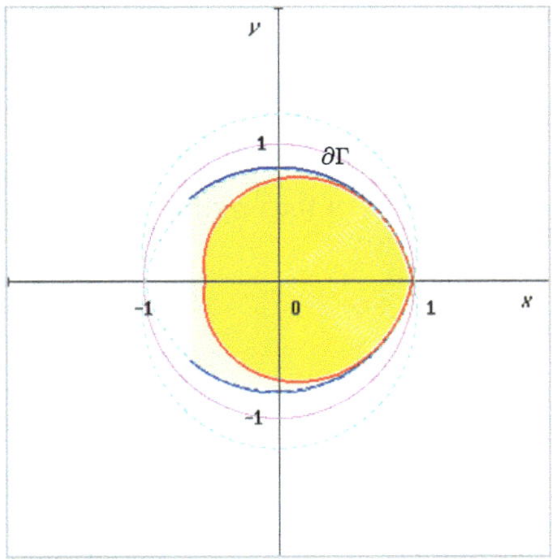

by applying a bilinear approximation (i.e., a rational function), the following contour and pseudo-sectorial area will be considered hereafter:

$$z = x + jy = \frac{1 + (\gamma + j)\theta}{1 - (\gamma + j)\theta}, \tag{6.34}$$

where $\theta = \dfrac{\omega h}{2}$. Figures 6.3(a) and (b) show a sectorial (stable) area in the s-plane and a pseudo-sectorial area in the z-plane. On the other hand, Fig. 6.4 shows shifted circles (blue) and pseudo-sectorial loci (red) in the z-plane that correspond to sectorial lines in the s-plane.

Lemma 6.1 *The locus of complex function* (6.34) *of* θ *is traced on the following shifted (off-axis) circle in the z-plane*:

$$x^2 + (y - \gamma)^2 = \gamma^2 + 1. \tag{6.35}$$

Proof The real and imaginary parts of (6.34) are expressed as follows:

$$x(\theta) = \frac{1 - (1+\gamma^2)\theta^2}{1 - 2\gamma\theta + (1+\gamma^2)\theta^2},$$
$$y(\theta) = \frac{2\theta}{1 - 2\gamma\theta + (1+\gamma^2)\theta^2}.$$

Thus, (6.35) is easily obtained. The proof is omitted. □

Equation (6.34) is also expressed in the form

$$z = \varrho \cdot e^{j\phi} + j\gamma, \tag{6.36}$$

where $\varrho = \sqrt{1+\gamma^2}$ is the radius of the shifted circle and

$$\phi = \tan^{-1}\left(\frac{2(1+\gamma^2)\theta - \gamma(1 + (1+\gamma^2)\theta^2)}{1 - (1+\gamma^2)\theta^2}\right)$$

is an argument of the original circle. Moreover, (6.36) can be rewritten as

$$z = \varrho \cdot \frac{1 + jv}{1 - jv} + j\gamma = \frac{A + Bv}{1 - jv}, \tag{6.37}$$

where

$$v = \tan\left(\frac{\phi + \tan^{-1}\gamma}{2}\right), \quad A = \varrho + j\gamma, \quad B = \gamma + j\varrho.$$

In these transformations, the following relation between nondimensional and distorted frequencies holds as shown in Fig. 6.5:

$$v : 0 \to +\infty, \quad \text{when } u : 0 \to +\infty \ (\omega : 0 \to +\infty)$$
$$v : -\infty \to 0, \quad \text{when } u : -\infty \to 0 \ (\omega : -\infty \to 0).$$

By applying the above transformation (6.37) to (6.28), the following numerator polynomial for the distorted frequency v can be obtained:

$$(1 - jv)^n \tilde{F}(z) = \sum_{i=1}^{m} \tilde{p}_i \left(\sum_{\ell=0}^{n} c_{i,\ell}(A + Bv)^{n-\ell}(1 - jv)^{\ell} \right). \tag{6.38}$$

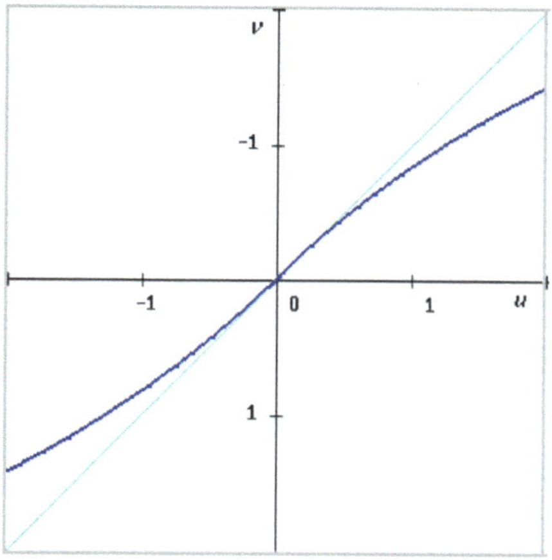

Fig. 6.5 Distorted frequency v vs. ω

Since (6.38) is a polynomial with complex coefficients, it can be written in the form

$$\tilde{\Phi}(jv) = (1 - jv)^n \tilde{F}(z) = \tilde{P}(v) + j\tilde{Q}(v), \tag{6.39}$$

where

$$\tilde{P}(v) = \tilde{a}_{0,0}v^n + \cdots + \tilde{a}_{0,n-1}v + \tilde{a}_{0,n}, \tag{6.40}$$

$$\tilde{Q}(v) = \tilde{b}_{0,0}v^n + \cdots + \tilde{b}_{0,n-1}v + \tilde{b}_{0,n}. \tag{6.41}$$

The coefficients in (6.40) and (6.41) can be calculated from the expansion of (6.73) in Appendix B.

6.5 Four Corner Polynomials

Using expression (6.29), Eq. (6.39) is rewritten as

$$\begin{aligned}\tilde{\Phi}(jv) &= (1 - jv)^n(\lambda F^+(z) + (1-\lambda)F^-(z)) \\ &= (\lambda P^+(v) + (1-\lambda)P^-(v)) \\ &\quad + j(\lambda Q^+(v) + (1-\lambda)Q^-(v)).\end{aligned} \tag{6.42}$$

The real and imaginary parts of (6.42) can be given as follows:

$$\tilde{P}(v) = \lambda P^+(v) + (1-\lambda)P^-(v),$$
$$\tilde{Q}(v) = \lambda Q^+(v) + (1-\lambda)Q^-(v).$$

Here, the extreme polynomials are expressed as:

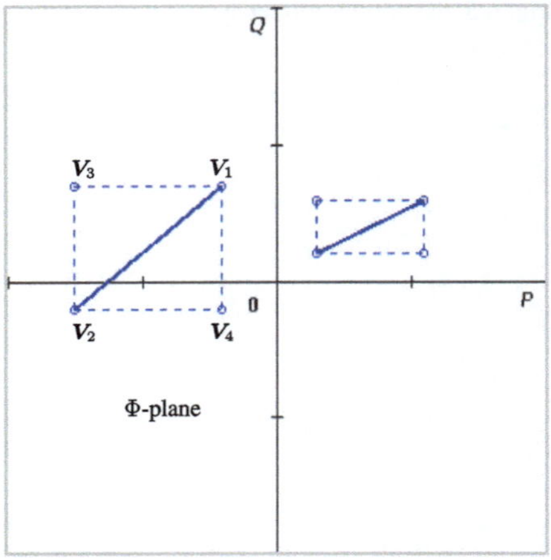

Fig. 6.6 Four corner points and rectangles

$$P^+(v) = a_{0,0}^+ v^n + \cdots + a_{0,n-1}^+ v + a_{0,n}^+,$$
$$Q^+(v) = b_{0,0}^+ v^n + \cdots + b_{0,n-1}^+ v + b_{0,n}^+, \tag{6.43}$$

$$P^-(v) = a_{0,0}^- v^n + \cdots + a_{0,n-1}^- v + a_{0,n}^-,$$
$$Q^-(v) = b_{0,0}^- v^n + \cdots + b_{0,n-1}^- v + b_{0,n}^-. \tag{6.44}$$

Thus, the following four corner points (vertices) can be given, and rectangles along with line segments (edges) can be drawn in the Φ-plane as shown in Fig. 6.6:

$$V_1 = (P^+,\ Q^+),\quad V_2 = (P^-,\ Q^-),$$
$$V_3 = (P^-,\ Q^+),\quad V_4 = (P^+,\ Q^-),$$

where the latter two points are additional ones. (In these expressions, note that polynomials P and Q and coefficients $a_{0,\ell}$ and $b_{0,\ell}$ with superscript $+$ do not always denote larger values than those with superscript $-$.)

Then, the following four pairs of polynomials can be defined in regard to $i = 1, 2, 3, 4$:

$$P^{(i)}(v) = a_{0,0}^{(i)} v^n + \cdots + a_{0,n-1}^{(i)} v + a_{0,n}^{(i)}, \tag{6.45}$$

$$Q^{(i)}(v) = b_{0,0}^{(i)} v^n + \cdots + b_{0,n-1}^{(i)} v + b_{0,n}^{(i)}, \tag{6.46}$$

where

$$P^{(1)}(v) = P^+(v),\quad Q^{(1)}(v) = Q^+(v),$$
$$P^{(2)}(v) = P^-(v),\quad Q^{(2)}(v) = Q^-(v),$$
$$P^{(3)}(v) = P^-(v),\quad Q^{(3)}(v) = Q^+(v),$$
$$P^{(4)}(v) = P^+(v),\quad Q^{(4)}(v) = Q^-(v).$$

For an expression of polynomials with complex coefficients, the following can be given:

$$\Phi^{(i)}(jv) = P^{(i)}(v) + jQ^{(i)}(v), \quad (i = 1, 2, 3, 4). \tag{6.47}$$

Note that, as for the edges in the F-plane, two additional polynomials with constant coefficients cannot generally be determined.

6.6 Division Algorithm

A division algorithm (Sturm's theorem) is applied in order to realize finite calculations [11, 12]. Here, the following notation is used in regard to the real and imaginary parts of (6.47):

$$f_0^{(i)}(v) = P^{(i)}(v), \quad f_1^{(i)}(v) = Q^{(i)}(v), \quad (i = 1, 2, 3, 4).$$

The division algorithm gives the following results [6, 13, 15]:

$$\begin{aligned} f_{2k-2}^{(i)}(v) &= q_{2k-1}^{(i)}(v) f_{2k-1}^{(i)}(v) - f_{2k}^{(i)}(v), \\ f_{2k-1}^{(i)}(v) &= q_{2k}^{(i)}(v) f_{2k}^{(i)}(v) - f_{2k+1}^{(i)}(v), \\ k &= 1, 2, \cdots, n. \end{aligned} \tag{6.48}$$

If $f_0^{(i)}(v)$ and $f_1^{(i)}(v)$ are n-th-order functions of v, then $f_2^{(i)}(v), f_3^{(i)}(v), \cdots, f_{2n}^{(i)}$ are expressed as follows:

$$\begin{aligned} f_{2h}^{(i)}(v) &= a_{h,h}^{(i)} v^{n-h} + \cdots + a_{h,n-1}^{(i)} v + a_{h,n}^{(i)} \\ f_{2h+1}^{(i)}(v) &= b_{h,h}^{(i)} v^{n-h} + \cdots + b_{h,n-1}^{(i)} v + b_{h,n}^{(i)} \\ &\cdots \\ f_{2n}^{(i)} &= a_{n,n}^{(i)}, \quad h = 1, 2, \cdots n-1. \end{aligned} \tag{6.49}$$

Here, each coefficient can be given by the following sequential operations:

$$\begin{aligned} a_{q,p}^{(i)} &= b_{q-1,p}^{(i)} \left(\frac{a_{q-1,q-1}^{(i)}}{b_{q-1,q-1}^{(i)}} \right) - a_{q-1,p}^{(i)}, \\ b_{q,p}^{(i)} &= a_{q,p+1}^{(i)} \left(\frac{b_{q-1,q-1}^{(i)}}{a_{q,q}^{(i)}} \right) - b_{q-1,p}^{(i)}, \\ &(q = 1, 2, \cdots, n, \quad p = q, q+1, \cdots, n-1) \\ &\cdots \\ a_{n,n}^{(i)} &= b_{n-1,n}^{(i)} \left(\frac{a_{n-1,n-1}^{(i)}}{b_{n-1,n-1}^{(i)}} \right) - a_{n-1,n}^{(i)}. \end{aligned} \tag{6.50}$$

Thus, as for $\tilde{\Phi}(jv)$, the change in the argument, $2\gamma\pi$, for polynomial $\tilde{F}(z)$ becomes $(2\mu - n)\pi$ by adding the change in the argument, $-n\pi$, for $(1 - jv)^n$. When $P^{(i)}/Q^{(i)}$ (or $-Q^{(i)}/P^{(i)}$) is considered, the number of sign changes that cross zero for $v : -\infty \to +\infty$ is $n - 2\mu$.

If the number of sign changes that cross the zero of $f_0^{(i)}(v)/f_1^{(i)}(v)$ for $v : v_1 \to v_2$ is expressed as $N^{(i)}(v_1, v_2)$ and the number of sign changes of sequence $f_0^{(i)}(v), f_1^{(i)}(v), \cdots, f_{2n}^{(i)}$ is expressed as $V^{(i)}(v)$, the following relationship is obtained:

$$N^{(i)}(v_1, v_2) = V^{(i)}(v_1) - V^{(i)}(v_2). \tag{6.51}$$

Since the condition is $N^{(i)}(-\infty, +\infty) = n - 2\mu$,

$$V^{(i)}(-\infty) - V^{(i)}(+\infty) = n - 2\mu \tag{6.52}$$

is obtainable. The condition for (6.52) corresponds to observing whether or not the following ratios are negative (the details were described in [10]):

$$\lim_{v\to+\infty} \frac{f_1^{(i)}(v)}{|v| f_2^{(i)}(v)}, \quad \cdots, \quad \lim_{v\to+\infty} \frac{f_{2n-1}^{(i)}(v)}{|v| f_{2n}^{(i)}(v)}. \tag{6.53}$$

Suppose that the number of negative ratios is N and the number of positive ratios is P. Thus, $P - N = n - 2\mu$ can be obtained from (6.52). Since $P + N = n$, $N = \mu$ is given.

For a segment polynomial with complex coefficients (6.47), the following lemma and theorem can be given using the result of (6.53).

Lemma 6.2 *When the coefficient ratios*

$$\frac{b_{0,0}^{(i)}}{a_{1,1}^{(i)}}, \quad \frac{b_{1,1}^{(i)}}{a_{2,2}^{(i)}}, \quad \cdots, \quad \frac{b_{n-1,n-1}^{(i)}}{a_{n,n}^{(i)}} \tag{6.54}$$

are calculated for an extreme polynomial $\Phi^{(i)}(v)$ $(i = 1$ *or* $2)$, *the number of ratios* (6.54) *that will be negative,* μ, *is equal to the number of characteristic roots for the polynomial in the specified contour* (6.34). *Here,* $a_{q,p}^{(i)}$ *and* $b_{q-1,q-1}^{(i)}$ $(q = 1, 2, \cdots, n)$ *are calculated by using the sequential operations* (6.50).

Proof This lemma is a necessary and sufficient condition in regard to the existing area of characteristic roots for the extreme polynomial. The proof is easily shown using (6.53). □

Based on the above premise, the following condition for the $\mathcal{D}$-stability of discrete-time interval systems is given with regard to the segment polynomial (6.28):

Theorem 6.1 *When the coefficient ratios*

$$\frac{b_{0,0}^{(i)}}{a_{1,1}^{(i)}},\ \frac{b_{1,1}^{(i)}}{a_{2,2}^{(i)}},\ \cdots,\ \frac{b_{n-1,n-1}^{(i)}}{a_{n,n}^{(i)}} \tag{6.55}$$

are calculated for each of the four corner polynomials $\Phi^{(i)}(v)$ $(i = 1, 2, 3, 4)$, *the following condition gives a sufficiency of the* $\mathcal{D}$*-stability of the discrete-time interval systems in question. If the number of ratios* (6.55) *that will be negative,* μ, *is equal to the system order,* n, *all the characteristic roots of the segment polynomial* (6.28) *exist in the specified contour* (6.34). *That is, if the number of ratios* (6.55) *that will be negative is not changed for the four corner polynomials,* $\mu = n$, *the robust* $\mathcal{D}$*-stability of the discrete-time control system is satisfied.*

Proof This theorem is a sufficient condition in regard to the existing area of characteristic roots for a segment polynomial. The proof is obvious from the zero exclusion of the Kharitonov-like rectangle that is composed of the four corner points (6.45) and (6.46). That is, none of the edges of the rectangle pass through the origin. As a natural consequence, the line segment in the Φ-plane (and also in the F-plane) does not pass through the origin [4]. □

6.7 Multiple Edges and Rectangles

Theorem 6.1 can also be applied to a discrete-time interval system with multiple uncertainties as shown in (6.26). When a complex variable z is fixed (frozen), a polytope (a parallelotope) is drawn on the Φ-plane as shown in Fig. 6.7. For an interval polynomial expressed by (6.26), the number of vertices is 2^m, and the number of edges becomes $m \cdot 2^{m-1}$. Obviously, the number of additional vertices is given by $2 \times m \cdot 2^{m-1}$. Thus, the number of total vertices that should be checked for interval polynomial (6.26) is given by

$$N = 2^m + 2m \cdot 2^{m-1} = (m+1) \cdot 2^m. \tag{6.56}$$

In other words, $(m+1) \cdot 2^{m-2}$ sets of the following four vertices (polynomials) should be checked:

$$\begin{aligned} &V_1^{(j)} = (P^{(j)+},\ Q^{(j)+}), \quad V_2^{(j)} = (P^{(j)-},\ Q^{(j)-}), \\ &V_3^{(j)} = (P^{(j)-},\ Q^{(j)+}), \quad V_4^{(j)} = (P^{(j)+},\ Q^{(j)-}), \\ &j = 1, 2, \cdots, (m+1) \cdot 2^{m-2}. \end{aligned} \tag{6.57}$$

By applying this concept, we have the following theorem for an interval system with multiple uncertainties.

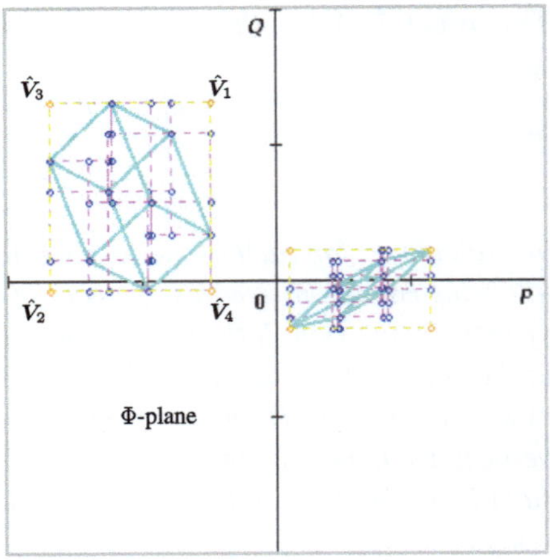

Fig. 6.7 Multiple edges and rectangles

Theorem 6.2 *When the coefficient ratios*

$$\frac{b_{0,0}^{(i)}}{a_{1,1}^{(i)}}, \frac{b_{1,1}^{(i)}}{a_{2,2}^{(i)}}, \cdots, \frac{b_{n-1,n-1}^{(i)}}{a_{n,n}^{(i)}} \tag{6.58}$$

are calculated for each of the $(m+1) \cdot 2^m$ *corner polynomials, the following condition gives a sufficiency of the* $\mathcal{D}$*-stability of discrete-time interval systems with the multiple uncertainties in question. If the number of ratios* (6.58) *that will be negative,* μ*, is equal to the system order,* n*, all the characteristic roots of the interval polynomial* (6.26) *exist in the specified contour* (6.34)*. That is, if the number of ratios* (6.58) *that will be negative is not changed for any of the vertex polynomials,* $\mu = n$*, the robust* $\mathcal{D}$*-stability of the discrete-time control system with multiple uncertainties is satisfied.*

Proof This theorem is a sufficient condition in regard to the existing area of characteristic roots for the interval polynomial. The proof is obvious from the result in Theorem 6.1, in which none of the edges of the rectangles pass through the origin. Consequently, as mentioned in the proof of Theorem 6.1, none of the edges of the parallelotope (a set of line segments) in the Φ-plane and in the F-plane pass through the origin. □

In Theorem 6.2, $(m+1) \cdot 2^{m-2}$ sets of four vertex calculations are necessary for the discrimination of the roots area. However, if the maximum and minimum of the real and imaginary parts of the vertex polynomials are determined, the sufficient condition of the robust $\mathcal{D}$-stability is reduced to the following corollary.

Corollary 6.1 *If the maximum and minimum of the vertex polynomials are determined as*

$$\hat{P}^+ = \max_j P^{(j)+} \quad and \quad \hat{P}^- = \min_j P^{(j)-},$$

Fig. 6.8 Outline of root areas

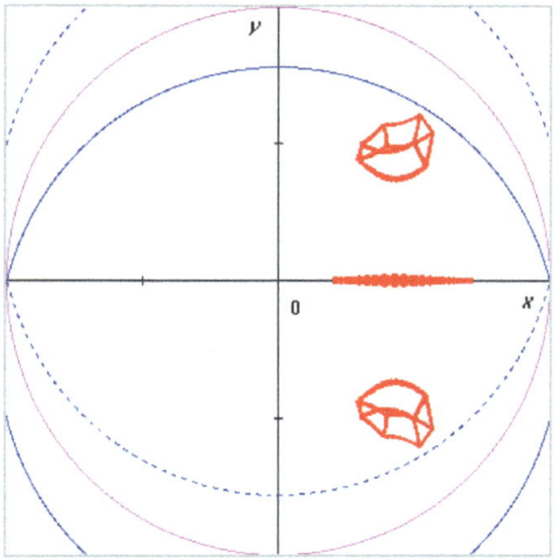

the four vertices

$$\hat{V}_1 = (\hat{P}^+,\ \hat{Q}^+),\quad \hat{V}_2 = (\hat{P}^-,\ \hat{Q}^-),$$
$$\hat{V}_3 = (\hat{P}^-,\ \hat{Q}^+),\quad \hat{V}_4 = (\hat{P}^+,\ \hat{Q}^-) \tag{6.59}$$

constitute a rectangle that covers all the $(m+1)\cdot 2^m$ *vertices, as shown in Fig.* 6.7.

The most simple case of this corollary will correspond to the Kharitonov and Kharitonov-like rectangles. Note that the above discrimination method contains the Routh-Hurwitz criterion for continuous-time systems (see Appendix B).

Example 6.3 Consider the following interval polynomial with three uncertain parameters:

$$\tilde{F}(z) = z^3 + \tilde{a}_1 z^2 + \tilde{a}_2 z + \tilde{a}_3, \tag{6.60}$$
$$\tilde{a}_1 \in [-1.82, -1.76], \tilde{a}_2 \in [1.19, 1.25], \tilde{a}_3 \in [-0.31, -0.25].$$

Since $m = 3$, the number of total vertices to be checked for (6.60) is obtained as $N = 32$ from (6.56). (Of course, the number of sets of four vertices is 8 from (6.57).) Based on Theorem 6.2, the coefficient ratios (6.58) were calculated for each of the 32 corner polynomials. The result shows that all the characteristic roots of interval polynomial (6.60) exist in the specified contour (6.34) when $\gamma = \pm 0.25$. That is, $\mu = 3$ was obtained for any of the sequential calculations. Figure 6.8 shows the outline of the areas of the true roots. In this case, a series of polytopes (parallelotopes) with 8 sets of four vertices for Φ becomes as shown in Fig. 6.9(b) when $\gamma = -0.25$. The $\mathcal{D}$-stability is guaranteed in the specified area as shown in Fig. 6.4. For reference, a series of parallelotopes for $\tilde{F}$ is also shown in Fig. 6.9(a).

On the other hand, when $\gamma = \pm 0.3$, the $\mathcal{D}$-stability is not guaranteed. The calculation result of 32 μ's is given as follows:

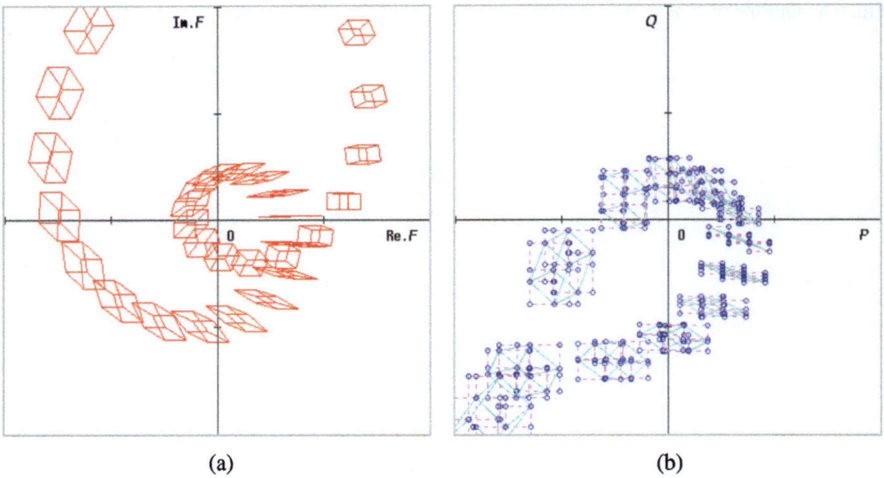

Fig. 6.9 Series of polytopes and rectangles for $\tilde{F}$- and Φ-functions ($\gamma = -0.25$)

$$3,\ 3,\ 3,\ 3,\ 3,\ 3,\ 3,\ 2,\ 3,\ 3,\ 3,\ 3,\ 3,\ 3,\ 3,\ 3,$$
$$3,\ 3,\ 3,\ 3,\ 3,\ 3,\ 3,\ 3,\ 3,\ 3,\ 3,\ 3,\ 3,\ 3,\ 3,\ 3.$$

Furthermore, when $\gamma = \pm 0.35$, the calculation result becomes

$$3,\ 3,\ 3,\ 3,\ 3,\ 3,\ 3,\ 2,\ 3,\ 3,\ 3,\ 3,\ 3,\ 3,\ 3,\ 2,$$
$$3,\ 3,\ 3,\ 3,\ 3,\ 3,\ 3,\ 3,\ 3,\ 3,\ 3,\ 3,\ 3,\ 3,\ 2,\ 3.$$

In this case, the series of polytopes (parallelotopes) with 8 sets of four vertices for Φ is as shown in Fig. 6.10(b). A series of parallelotopes for $\tilde{F}$ is shown in Fig. 6.10(a).

Example 6.4 Consider an interval polynomial expressed in the following general form:

$$\tilde{F}(z) = F_0(z) + \tilde{p}_1 F_1(z) + \tilde{p}_2 F_2(z) + \tilde{p}_3 F_3(z), \tag{6.61}$$
$$\tilde{p}_1 \in [0.95, 1.05],\ \tilde{p}_2 \in [-0.55, -0.45],\ \tilde{p}_3 \in [0.1, 0.2],$$

where

$$F_0(z) = 0.25z^2 + 0.125z, \quad F_1(z) = z^3 - z^2,$$
$$F_2(z) = z^2 - z, \quad F_3(z) = z - 1.$$

Theorem 6.2 can also be applied to this type of interval polynomial in Example 6.1. The result shows that all the roots of interval polynomial (6.61) exist in the contour (6.34) when $\gamma = \pm 0.25$. That is, $\mu = 3$ is obtained for any of the sequential

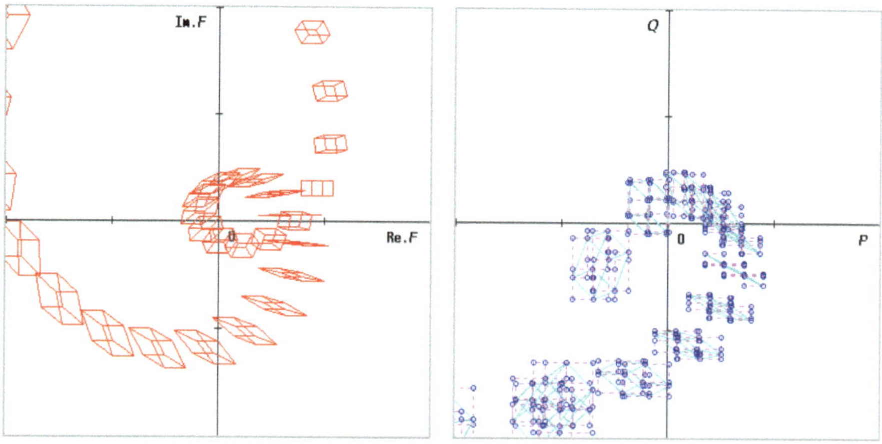

(a) (b)

Fig. 6.10 Series of polytopes and rectangles for $\tilde{F}$ and Φ functions ($\gamma = -0.35$)

Fig. 6.11 Outline of root areas

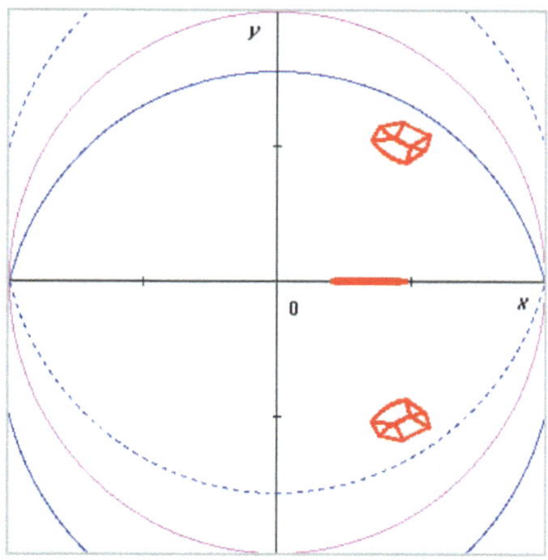

calculations. Figure 6.11 shows the outline of the areas of the true roots. In this example, a series of polytopes with 8 sets of four vertices for Φ becomes as shown in Fig. 6.12(b) when $\gamma = -0.25$. The $\mathcal{D}$-stability is guaranteed in the specified area as shown in Fig. 6.4. A series of parallelotopes for $\tilde{F}$ is also shown in Fig. 6.12(a). However, when $\gamma = 0.3$, the $\mathcal{D}$-stability is not guaranteed. The calculation result of 32 μ's is given as

$$3,\ 3,\ 3,\ 3,\ 3,\ 3,\ 3,\ 3,\ 3,\ 3,\ 3,\ 3,\ 3,\ 3,\ 3,\ 3,$$
$$3,\ 3,\ 3,\ 3,\ 3,\ 3,\ 3,\ 3,\ 3,\ 2,\ 3,\ 3,\ 3,\ 3,\ 3,\ 3.$$

Note that Corollary 6.1 cannot be applied to these examples, because the four vertices (6.59) are not expressed in the four polynomials.

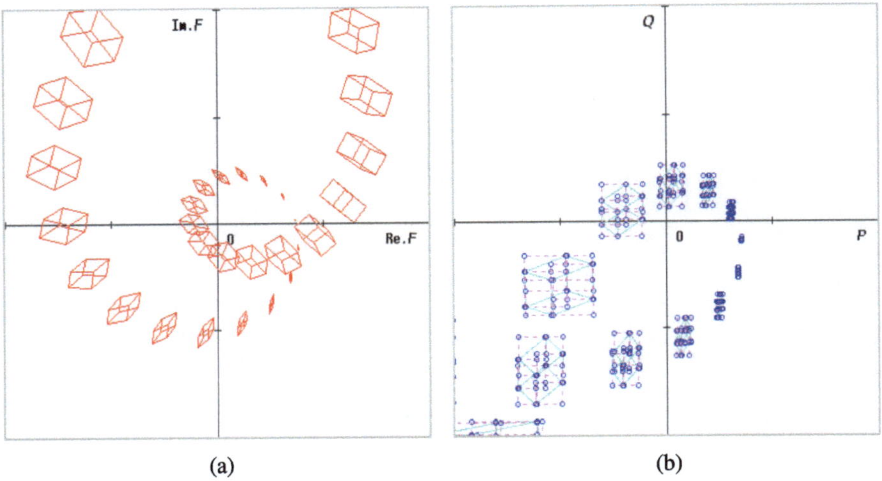

Fig. 6.12 Series of polytopes and rectangles for $\tilde{F}$- and Φ-functions ($\gamma = -0.3$)

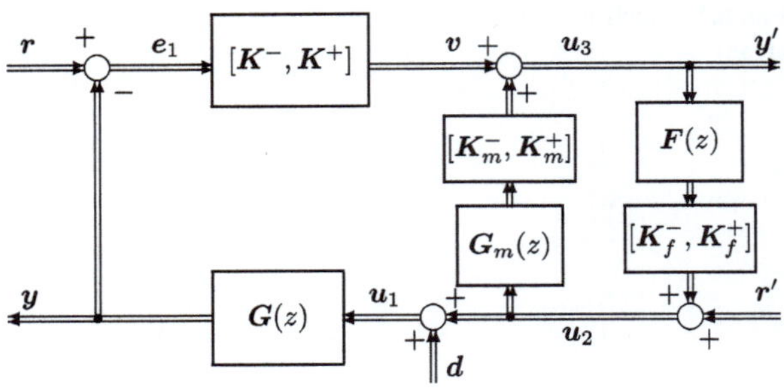

Fig. 6.13 Model reference multi-loop interval system

The concept of the discrimination method was extended to segment polynomials of a discrete-time control system and to the circular $\mathcal{D}$-stability, which corresponds to the sectorial $\mathcal{D}$-stability for a continuous-time control system. By applying a division algorithm to a set of the four corners of segment polynomials, a sufficient condition for the roots area, which is enclosed by a specified (circular) contour in a unit circle on a z-plane, was given. As a result, the discrimination of the roots area by finite calculations became possible. Although this result is only a sufficient condition, the discrimination method proposed here will be useful in designing robust control systems.

6.8 Robust Control System Design

Figure 6.13 is a model reference feedback system in which the discretized nonlinear elements in Fig. 5.16 are replaced by interval gains. As was shown in Chap. 5, when the model system and the feedback compensator can be considered linear (or

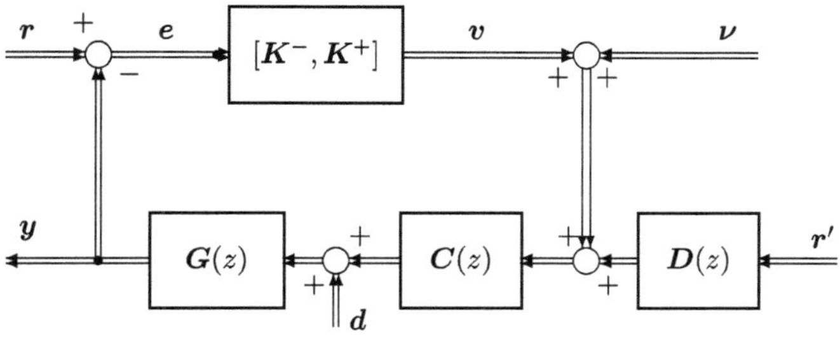

Fig. 6.14 Multi-loop PID control interval system

high-resolution) systems, the model reference feedback system [9] becomes a multi-loop proportional-integral-derivative (PID) control system as shown in Fig. 6.14. Obviously, the stability of the interval system is the same as that shown in Fig. 6.1. The characteristic equation of the multi-loop control system is given by

$$\det\{\boldsymbol{I} + \boldsymbol{G}(z)\boldsymbol{C}(z)[\boldsymbol{K}^-, \boldsymbol{K}^+]\} = 0 \tag{6.62}$$

and equivalently

$$\det\{\boldsymbol{I} + [\boldsymbol{K}^-, \boldsymbol{K}^+]\boldsymbol{C}(z)\boldsymbol{G}(z)\} = 0. \tag{6.63}$$

Therefore, the robust stabilization of model reference control systems with interval gains results in the roots area discrimination problem for (6.62) (or (6.63)) as described in this chapter.

The design of control systems containing many physical elements is not easy, because there are uncertainties, i.e., nonlinearities, in the connected devices. Therefore, a control system designer would like the connection of those elements (the control system structure) to be easy to see. Thus, the traditional block diagram (or signal flow graph [8]) becomes important in the analysis and design of control systems. Therefore, in this book, we do not treat some of the mathematically organized methods, such as quadratic forms and optimal and H_∞ control design techniques.

6.9 Exercises

(1) Show the relationship between characteristic equation $\tilde{F}(z) = 0$ and $\tilde{F}(\delta) = 0$, where

$$\tilde{F}(z) = z^3 - [1.3, 1.2]z^2 + [0.7, 0.8]z - [0.2, 0.1]$$

and

$$\delta = \frac{2}{h} \cdot \frac{z-1}{z+1}.$$

(2) Show the relationship between pseudo-sectorial loci in the z-plane and sectorial lines in the s-plane.

(3) With respect to the 2×2 feedback system shown in Fig. 5.4, determine the (interval) characteristic equation when $N_{d1}(e_1)/e_1$ and $N_{d2}(e_2)/e_2$ are replaced with interval gains $[K_1^-, K_1^+]$ and $[K_2^-, K_2^+]$, respectively.

(4) Show that the characteristic equations (6.6) and (6.7) are equivalent when $\boldsymbol{C}(z)$ and $\tilde{\boldsymbol{K}} \in [\boldsymbol{K}^-, \boldsymbol{K}^+]$ are diagonal matrices with nonzero diagonal elements.

(5) Prove Lemma 6.2 using (6.53).

(6) Show that if and only if four (corner) polynomials

$$\begin{cases} F^{(1)}(s) = a_0^+ s^3 + a_1^+ s^2 + a_2^- s + a_3^- \\ F^{(2)}(s) = a_0^- s^3 + a_1^+ s^2 + a_2^+ s + a_3^- \\ F^{(3)}(s) = a_0^+ s^3 + a_1^- s^2 + a_2^- s + a_3^+ \\ F^{(4)}(s) = a_0^- s^3 + a_1^- s^2 + a_2^+ s + a_3^+ \end{cases}$$

are stable, the interval polynomial

$$\tilde{F}(s) = [a_0^-, a_0^+]s^3 + [a_1^-, a_1^+]s^2 + [a_2^-, a_2 k]s + [a_3^-, a_3^+]$$

is stable.

(7) Show that the Routh series and Hurwitz determinant can be derived directly from

$$F(s) = a_0 s^n + a_1 s^{n-1} + \cdots + a_{n-1} s + a_n$$

by applying the division algorithm to

$$\begin{cases} f_0(s) = a_0 s^n + a_2 s^{n-2} + \cdots \\ f_1(s) = a_1 s^{n-1} + a_3 s^{n-3} + \cdots. \end{cases}$$

Appendix A: Kharitonov Rectangles

With respect to continuous-time systems, consider the following characteristic polynomials [7]:

$$\begin{aligned} \tilde{F}(s) &= \tilde{a}_0 s^n + \tilde{a}_1 s^{n-1} + \cdots + \tilde{a}_{n-1} s + \tilde{a}_n, \\ \tilde{a}_k &\in [a_k^-, a_k^+], \quad k = 0, 1, 2, \cdots, n. \end{aligned} \tag{6.64}$$

Considering the imaginary axis, i.e., $s = j\omega$, Eq. (6.64) can be given as

$$\tilde{F}(j\omega) = [P^-(\omega), P^+(\omega)] + j[Q^-(\omega), Q^+(\omega)]. \tag{6.65}$$

When n is odd,

$$P^-(\omega) = a_n^- - a_{n-2}^+ \omega^2 + \cdots + (-1)^{(n-1)/2} a_1^+ \omega^{n-1}$$

$$P^+(\omega) = a_n^+ - a_{n-2}^- \omega^2 + \cdots + (-1)^{(n-1)/2} a_1^- \omega^{n-1}$$

$$Q^-(\omega) = a_{n-1}^- \omega - a_{n-3}^+ \omega^3 + \cdots - (-1)^{(n-1)/2} a_0^+ \omega^n$$
$$Q^+(\omega) = a_{n-1}^+ \omega - a_{n-3}^- \omega^3 + \cdots - (-1)^{(n-1)/2} a_0^- \omega^n.$$

On the other hand, when n is even,

$$P^-(\omega) = a_n^- - a_{n-2}^+ \omega^2 + \cdots - (-1)^{n/2} a_0^+ \omega^n$$
$$P^+(\omega) = a_n^+ - a_{n-2}^- \omega^2 + \cdots - (-1)^{n/2} a_0^- \omega^n$$
$$Q^-(\omega) = a_{n-1}^- \omega - a_{n-3}^+ \omega^3 + \cdots + (-1)^{n/2} a_1^+ \omega^{n-1}$$
$$Q^+(\omega) = a_{n-1}^+ \omega - a_{n-3}^- \omega^3 + \cdots + (-1)^{n/2} a_1^- \omega^{n-1}.$$

For a third-order system, (6.65) can be written directly as follows:

$$\tilde{F}(s) = [a_0^-, a_0^+]s^3 + [a_1^-, a_1^+]s^2 + [a_2^-, a_2^+]s + [a_3^-, a_3^+].$$

Therefore,

$$\tilde{F}(j\omega) = [a_3^-, a_3^+] - [a_1^-, a_1^+]\omega^2 + j\{[a_2^-, a_2^+]\omega - [a_0^-, a_0^+]\omega^3\}. \tag{6.66}$$

When considering $\omega > 0$, the following vertices are obtained:

$$P^-(\omega) = a_3^- - a_1^+ \omega^2$$
$$P^+(\omega) = a_3^+ - a_1^- \omega^2$$
$$Q^-(\omega) = a_2^- \omega - a_0^+ \omega^3$$
$$Q^+(\omega) = a_2^+ \omega - a_0^- \omega^3.$$

Of course, when $\omega < 0$, the following vertices are obtained:

$$P^-(\omega) = a_3^- - a_1^+ \omega^2$$
$$P^+(\omega) = a_3^+ - a_1^- \omega^2$$
$$Q^-(\omega) = a_2^+ \omega - a_0^- \omega^3$$
$$Q^+(\omega) = a_2^- \omega - a_0^+ \omega^3$$

although they are conjugated with each other.

Example 6.5 Consider the following (monic) interval polynomial (the highest coefficient $a_0 = 1$):

$$\tilde{F}(s) = s^3 + [1.6, 2.4]s^2 + [1.6, 2.4]s + [0.8, 1.2].$$

Obviously, when $\omega > 0$,

$$P^-(\omega) = 0.8 - 2.4\omega^2$$

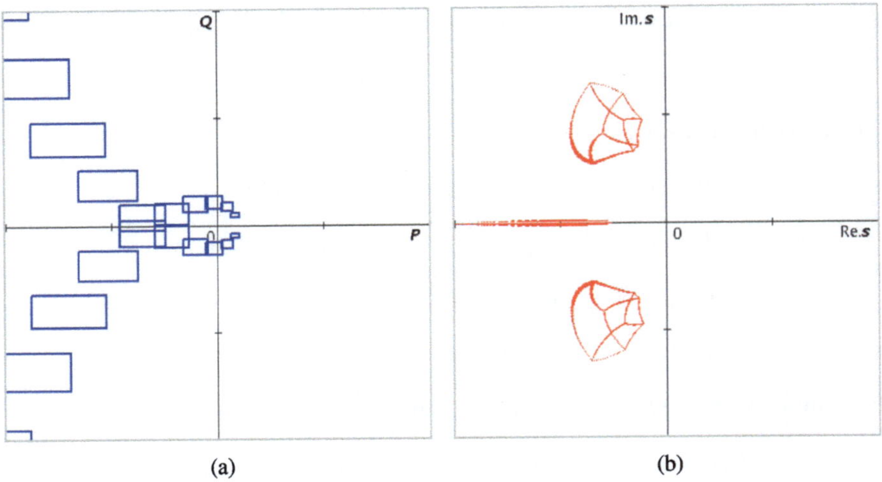

Fig. 6.15 Kharitonov rectangles and roots area

$$P^+(\omega) = 1.2 - 1.6\omega^2$$
$$Q^-(\omega) = 1.6\omega - \omega^3$$
$$Q^+(\omega) = 2.4\omega - \omega^3.$$

And, when $\omega < 0$,

$$P^-(\omega) = 0.8 - 2.4\omega^2$$
$$P^+(\omega) = 1.2 - 1.6\omega^2$$
$$Q^-(\omega) = 2.4\omega - \omega^3$$
$$Q^+(\omega) = 1.6\omega - \omega^3.$$

Figure 6.15(a) shows the trace of Kharitonov rectangles when $\omega: \ -\omega_c \to \omega_c$. From the figure, one can see that the stability of the interval system is determined by only the four vertices of the rectangular array. The robust stability condition is necessary and sufficient for interval (uncertain) systems. In this example, a view of the root areas is shown in Fig. 6.15(b).

Appendix B: Roots of Polynomials and Sturm's Theorem

The basic concept of Sturm's theorem is given as follows [6, 13]. Let $f(x)$ be a "real" polynomial.[4] Denote it by $f_0(x)$ and its derivative $f'(x)$ by $f_1(x)$. Applying

[4]First, it is assumed that variable x and the value of f (i.e., coefficients a_i of the polynomial) are real.

Euclid's division algorithm, the following sequence is obtained:

$$\begin{aligned} f_0(x) &= q_1(x) f_1(x) - f_2(x), \\ f_1(x) &= q_2(x) f_2(x) - f_3(x), \\ &\vdots \\ f_{n-2}(x) &= q_{n-1}(x) f_{n-1}(x) - f_n(x), \\ f_{n-1}(x) &= q_n(x) f_n, \end{aligned} \tag{6.67}$$

where $f_i(x)$ is of degree lower than that of $f_{i-1}(x)$ for $1 \leq i \leq n$. The signs of the remainders are negated from those in the algorithm. Note that the divisor f_k that yields a zero remainder is the greatest common divisor of $f(x)$ and $f'(x)$. The sequence $f_0, f_1, \cdots, f_n$ is called a *Sturm sequence* for the polynomial f.

Sturm's Theorem *The number of distinct (simple) zeros of a polynomial $f(x)$ in $x \in (x_1, x_2)$ is equal to the excess of the number of changes of sign in the sequence $f_0(x_1), \cdots, f_{n-1}(x_1), f_n(x_1)$ over the number of changes of sign in the sequence $f_0(x_2), \cdots, f_{n-1}(x_2), f_n(x_2)$.*

In other expressions, if the number of sign changes that cross the zero of $f_0(x)/f_1(x)$ for $x : x_1 \to x_2$ is defined as $N(x_1, x_2)$ and the number of sign changes of sequence $f_0(x_2), \cdots, f_{n-1}(x_2), f_n(x_2)$ is defined as $V(x)$, we obtain the relationship

$$N(x_1, x_2) = V(x_1) - V(x_2). \tag{6.68}$$

Here, $f(x)$ can be multiplied by a positive constant or a factor involving x provided that the factor remains positive in $x \in (x_1, x_2)$. Moreover, there is no common divisor between $f_0(x)$ and $f_1(x)$, because the zeros of $f(x)$ are distinct.

Example 6.6 (Real Roots) Consider the following polynomial to apply Sturm's theorem [16]:

$$f(x) = x^3 - 1.5x^2 - 0.16x - 0.24.$$

The first step polynomials f_0 and f_1 are given as:

$$\begin{aligned} f_0(x) &= x^3 - 1.5x^2 - 0.16x - 0.24, \\ f_1(x) &= 3x^2 + 3x - 0.16. \end{aligned}$$

Therefore, the results of the division algorithm are obtained as follows:

$$\begin{aligned} f_2(x) &= 0.607x + 0.213, \\ f_3(x) &= 0.844. \end{aligned}$$

Table 6.1 lists the signs of each Sturm sequence, and the number of sign changes is given in Table 6.2. From these tables, we obtain, for example,

Table 6.1 Sign changes of sequence f_i $(i = 1, 2, 3, 4)$

	$-\infty$	-2	-1	0	1	2	∞
f_0	$-$	$-$	$+$	$-$	$+$	$+$	$+$
f_1	$+$	$+$	$-$	$-$	$+$	$+$	$+$
f_2	$-$	$-$	$-$	$+$	$+$	$+$	$+$
f_3	$+$	$+$	$+$	$+$	$+$	$+$	$+$

Table 6.2 The number of sign changes $N(x_1, x_2)$

	$-\infty$	-2	-1	0	1	2	∞
f_0/f_1	1	1	1	0	0	0	0
f_1/f_2	1	1	0	1	0	0	0
f_2/f_3	1	1	1	0	0	0	0
$V(x_i)$	3	3	2	1	0	0	0

$$N(-2, 2) = V(-2) - V(2) = 3,$$
$$N(-1, 2) = V(-1) - V(1) = 2,$$
$$N(0, 1) = V(0) - V(1) = 1.$$

This result is clear from Fig. 6.16.

Complex Roots Problem The Sturm sequence can be extended to a complex roots problem. Consider the following circle as a closed contour in the complex plane:

$$z = \varrho e^{j\phi} + x_0 + j y_0, \quad (\phi : -\pi \to \pi), \tag{6.69}$$

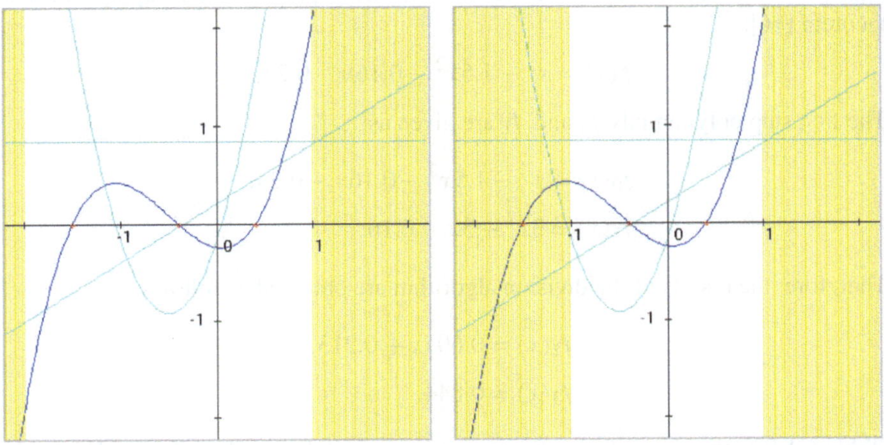

Fig. 6.16 Cubic curve and Sturm sequences ($f(x) = x^3 + 1.5x^2 - 0.16x - 0.24$)

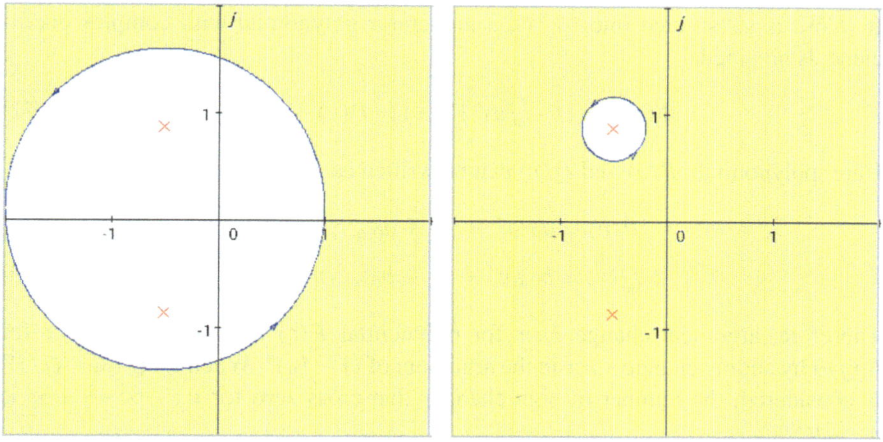

(a) (b)

Fig. 6.17 Circular contours ($f(z) = z^2 + z + 1$, $x_0 = -0.5$, (**a**) $y_0 = 0.1$, $\varrho = 1.5$, (**b**) $y_0 = 0.866$, $\varrho = 0.3$)

where ϱ, (x_0, y_0), and ϕ are the radius, the center, and the angle of rotation of a specified circle, respectively. The problem of a sectorial area in the complex plane can be considered in a large circle $\partial\Gamma$, as shown in Fig. 6.3(a).[5]

Figures 6.17(a), (b) show examples of the circular contour in a complex plane. The circular contour (6.36) can also be written as the following rational function:

$$z = \varrho \cdot \frac{1 + jv}{1 - jv} + x_0 + jy_0 = \frac{A + Bv}{1 - jv}, \tag{6.70}$$

where

$$v = \tan\left(\frac{\phi}{2}\right)$$

and A and B are complex constants written as $A = \varrho + x_0 + jy_0$ and $B = y_0 + j(\varrho - x_0)$. Clearly, the relationship between variables ϕ and v is

$$\begin{aligned}
&\phi = -\pi; \quad v = -\infty, \\
&\phi = 0; \quad\quad v = 0, \\
&\phi = +\pi; \quad v = +\infty.
\end{aligned}$$

Consider the following characteristic polynomial in general:

$$F(z) = a_0 z^n + a_1 z^{n-1} + \cdots + a_{n-1} z + a_n. \tag{6.71}$$

[5]The discrimination of the number of roots in a specified area is equivalent to the Routh-Hurwitz criterion.

If (6.69) is substituted into (6.71), a numerator polynomial with complex coefficients is obtained:

$$\Phi(jv) = (1 - jv)^n F(jv) = P(v) + jQ(v). \tag{6.72}$$

Here, polynomials $P(v)$ and $Q(v)$ can be written as

$$\begin{aligned} P(v) &= a_{0,0}v^n + \cdots + a_{0,n-1}v + a_{0,n}, \\ Q(v) &= b_{0,0}v^n + \cdots + b_{0,n-1}v + b_{0,n}. \end{aligned} \tag{6.73}$$

Therefore, argument change $2\mu\pi$ for polynomial $F(z)$ becomes $(2\mu - n)\pi$ for $\Phi(jv)$ by adding change $-n\pi$ in the argument of $(1-jv)^n$. When P/Q (or $-Q/P$) is considered, the number of sign changes that cross zero for $v : -\infty \to +\infty$ is $n - 2\mu$.

The coefficients in (6.73) are calculated by

$$\Phi(jv) = \sum_{k=0}^{n} a_k (A + Bv)^{n-k} (1 - jv)^k \tag{6.74}$$

from (6.69). This equation is expanded as

$$\sum_{m=0}^{n}\sum_{k=0}^{n}\sum_{l=k-m}^{n-m} a_k \binom{n-k}{n-m-l}\binom{k}{l}(-j)^l A^{m-k+l} B^{n-m-l} v^{n-m},$$

and thus we can obtain

$$a_{0,m} + jb_{0,m} = \sum_{k=0}^{n}\sum_{l=k-m}^{n-m} a_k \binom{n-k}{n-m-l}\binom{k}{l}(-j)^l A^{m-k+l} B^{n-m-l} v^{n-m},$$

where

$$\binom{k}{l} = {}_kC_l$$

denotes a combination symbol.

By setting $f_0(v) = P(v)$ and $f_1(v) = Q(v)$, the following division algorithm can be executed:

$$\begin{aligned} f_{2\kappa-2}(v) &= q_{2\kappa-1}(v) f_{2\kappa-1}(v) - f_{2\kappa}(v), \\ \kappa &= 1, 2, \cdots, n. \end{aligned} \tag{6.75}$$

If $f_0(v)$ and $f_1(v)$ are of the n-th order for v, then $f_2(v), f_3(v), \cdots, f_{2n}$ are expressed as:

$$\begin{aligned} f_2(v) &= a_{1,1}v^{n-1} + \cdots + a_{1,n} \\ f_3(v) &= b_{1,1}v^{n-1} + \cdots + b_{1,n} \end{aligned}$$

$$\cdots \tag{6.76}$$

$$f_{2n-2}(v) = a_{n-1,n-1}v + a_{n-1,n}$$

$$f_{2n-1}(v) = b_{n-1,n-1}v + b_{n-1,n}$$

$$f_{2n} = a_{n,n}.$$

Here, each coefficient can be given by the following sequential operations:

$$a_{1,p} = b_{0,p}\left(\frac{a_{0,0}}{b_{0,0}}\right) - a_{0,p},$$

$$b_{1,p} = a_{1,p+1}\left(\frac{b_{0,0}}{a_{1,1}}\right) - b_{0,p},$$

$$(p = 1, 2, \cdots, n) \tag{6.77}$$

$$\cdots$$

$$a_{q,p} = b_{q-1,p}\left(\frac{a_{q-1,q-1}}{b_{q-1,q-1}}\right) - a_{q-1,p},$$

$$b_{q,p} = a_{q,p+1}\left(\frac{b_{q-1,q-1}}{a_{q,q}}\right) - b_{q-1,p},$$

$$(p = q, \cdots, n)$$

$$\cdots$$

$$a_{n,n} = b_{n-1,n}\left(\frac{a_{n-1,n-1}}{b_{n-1,n-1}}\right) - a_{n-1,n},$$

$$(a_{q,n+1} = 0).$$

If the number of sign changes which cross the zero of $f_0(v)/f_1(v)$ for $v : v_1 \to v_2$ is expressed as $N(v_1, v_2)$ and the number of sign changes of sequence $f_0(v), f_1(v), \cdots, f_{2n}$ is expressed as $V(v)$, the following relationship is obtained:

$$N(v_1, v_2) = V(v_1) - V(v_2). \tag{6.78}$$

Since the condition is $N(-\infty, +\infty) = n - 2\mu$,

$$V(-\infty) - V(+\infty) = n - 2\mu \tag{6.79}$$

is obtainable.

On the other hand, the following expressions are derived from (6.76):

$$\lim_{v\to-\infty} \frac{f_0(v)}{f_1(v)} = \frac{a_{0,0}}{b_{0,0}},$$

$$\lim_{v\to-\infty} \frac{f_1(v)}{|v| f_2(v)} = -\frac{b_{0,0}}{a_{1,1}},$$

$$\vdots$$

$$\lim_{v\to-\infty}\frac{f_{2n-1}(v)}{|v|f_{2n}(v)}=-\frac{b_{n-1,n-1}}{a_{n,n}},$$

$$\begin{aligned}
&\lim_{v\to+\infty}\frac{f_0(v)}{f_1(v)}=\frac{a_{0,0}}{b_{0,0}},\\
&\lim_{v\to+\infty}\frac{f_1(v)}{|v|f_2(v)}=\frac{b_{0,0}}{a_{1,1}},\\
&\qquad\vdots\\
&\lim_{v\to+\infty}\frac{f_{2n-1}(v)}{|v|f_{2n}(v)}=\frac{b_{n-1,n-1}}{a_{n,n}},
\end{aligned}\tag{6.80}$$

The condition of (6.78) corresponds to observing whether the following ratios (ratios to the polynomial of different orders) are negative or not:

$$\begin{aligned}
&\lim_{v\to+\infty}\frac{f_1(v)}{|v|f_2(v)}=\frac{b_{0,0}}{a_{1,1}},\\
&\lim_{v\to+\infty}\frac{f_3(v)}{|v|f_4(v)}=\frac{b_{1,1}}{a_{2,2}},\\
&\qquad\vdots\\
&\lim_{v\to+\infty}\frac{f_{2n-1}(v)}{|v|f_{2n}(v)}=\frac{b_{n-1,n-1}}{a_{n,n}},
\end{aligned}\tag{6.81}$$

because ratios to the polynomial of same order (i.e., $\frac{a_{0,0}}{b_{0,0}}, \frac{a_{1,1}}{b_{1,1}}, \ldots$) are canceled based on (6.79). Suppose that the number of negative ratios is N and the number of positive ratios is P. Thus, $P-N=n-2\mu$ can be obtained from (6.79). Since $P+N=n$, $N=\mu$ is given.

Example 6.7 (Quadratic Equation) As a simple example, consider the following quadratic polynomial:

$$F(z)=z^2+z+1. \tag{6.82}$$

First, a specified circle is assumed to be $x_0=-0.5$, $y_0=0.1$, and $\varrho=1.5$, as shown in Fig. 6.17(a). Then, numerator polynomials $P(v)$ and $Q(v)$ written as

$$\Phi(jv)=(1-jv)^2F(jv)=P(v)+jQ(v)$$

are given as

$$\begin{aligned}
P(v)&=f_0(v)=-2.99v^2+2.99,\\
Q(v)&=f_1(v)=0.3v^2+3.02v-0.3.
\end{aligned}$$

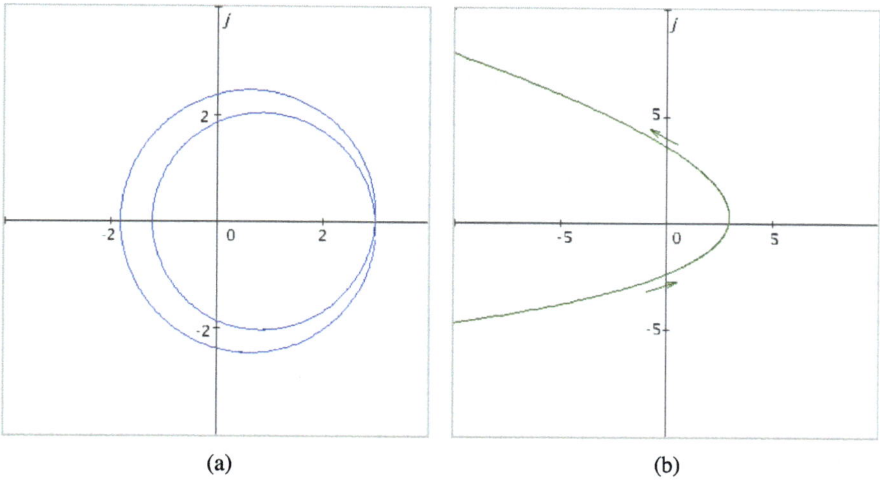

Fig. 6.18 Mapping $f(jv)$ and $\Phi(jv)$ for the specified circle

By setting $f_0(v) = P(v)$, $f_1(v) = Q(v)$, the following results are obtained from (6.76):

$$\begin{aligned} f_0(v) &= -2.99v^2 + 2.99, \\ f_1(v) &= 0.30v^2 + 3.02v - 0.30, \\ f_2(v) &= -30.1v - 5.98, \\ f_3(v) &= -2.96v - 0.30, \\ f_4 &= 2.93. \end{aligned} \tag{6.83}$$

Since the sequence of fractions, (6.81), is given as

$$\frac{b_{0,0}}{a_{1,1}} = \frac{0.30}{-30.1} < 0, \qquad \frac{b_{1,1}}{a_{2,2}} = \frac{-2.96}{2.93} < 0,$$

the number of roots in the specified circle is $\mu = 2$.

The discriminating procedure given above can be interpreted as follows. The specified contour encircles two roots of the quadratic equation. As a result, the mapping curve is as shown in Fig. 6.18(a). Thus, the argument change of $\Phi(jv)$ becomes $4\pi - 2\pi = 2\pi$ by adding a change of -2π to the argument of $(1 - jv)^2$. Figure 6.18(b) shows the Φ curve for this case.

On the other hand, when a circle is chosen as $x_0 = -0.5$, $y_0 = 0.8$, and $\varrho = 0.4$, as shown in Fig. 6.17(b), numerator polynomials $P(v)$ and $Q(v)$ are given by

$$\begin{aligned} P(v) &= f_0(v) = -0.27v^2 + 0.27, \\ Q(v) &= f_1(v) = 0.64v^2 + 0.10v + 0.64. \end{aligned}$$

The Sturm sequence is given as

$$\begin{aligned} f_0(v) &= -0.27v^2 + 0.27, \\ f_1(v) &= 0.64v^2 + 0.10v - 0.30, \\ f_2(v) &= -0.04v - 0.54, \\ f_3(v) &= 8.09v - 0.64, \\ f_4 &= 0.54. \end{aligned} \tag{6.84}$$

Since the sequence of fractions, (6.81), is given as

$$\frac{b_{0,0}}{a_{1,1}} = \frac{0.64}{-0.04} < 0, \quad \frac{b_{1,1}}{a_{2,2}} = \frac{8.09}{0.54} > 0,$$

the number of roots in the specified circle is $\mu = 1$.

Example 6.8 (Cubic Equation) Consider the following cubic polynomial:

$$F(z) = z^3 + 2z^2 + 2z + 1. \tag{6.85}$$

A specified circle is chosen as $x_0 = -0.5$, $y_0 = 0.1$, and $\varrho = 1.5$. Then, numerator polynomials $P(v)$ and $Q(v)$ written as

$$\Phi(jv) = (1 - jv)^3 F(jv) = P(v) + jQ(v)$$

are given as

$$\begin{aligned} P(v) &= -0.599v^3 - 9.03v^2 - 0.303v + 5.95, \\ Q(v) &= -2.96v^3 + 0.603v^2 + 9.06v + 0.899. \end{aligned}$$

By setting $f_0(v) = P(v)$, $f_1(v) = Q(v)$, the following results are obtained from (6.76):

$$\begin{aligned} f_0(v) &= -0.599v^3 - 9.03v^2 - 0.303v + 5.95, \\ f_1(v) &= -2.96v^3 + 0.603v^2 + 9.06v + 0.899, \\ f_2(v) &= 9.152v^2 + 2.1364v - 5,7681, \qquad (6.86) \\ f_3(v) &= -1.294v^2 - 7.1945v - 0.899, \\ f_4(v) &= 48.749v + 12.127, \qquad (6.87) \\ f_5(v) &= 6.8726v + 0.899, \\ f_6 &= -5.7497. \end{aligned}$$

Since the sequence of fractions, (6.81), is given as

$$\frac{b_{0,0}}{a_{1,1}} = \frac{-2.96}{9.152} < 0, \quad \frac{b_{1,1}}{a_{2,2}} = \frac{-1.294}{48.749} < 0, \quad \frac{b_{2,2}}{a_{3,3}} = \frac{6.8726}{-5.7497} < 0,$$

the number of roots in the specified circle is $\mu = 3$.

Routh-Hurwitz Criterion Finally, the relationship between the above results based on Sturm's theorem and the classical Routh-Hurwitz criterion is presented. In general, a characteristic equation for continuous-time systems is written as

$$F(s) = a_0 s^n + a_1 s^{n-1} + \cdots + a_{n-1} s + a_n = 0, \tag{6.88}$$

where s is the Laplace transform variable and a_i $(i = 1, 2, \cdots, n)$ are real coefficients. The stability of a control system having characteristic equation (6.88) is discriminated by a contour on the imaginary axis and a large half-contour in the right half-plane (RHP) or left half-plane (LHP) and its mapping in the F-plane.

On the imaginary axis $s = j\omega$ $(\omega : -\infty \to \infty)$,

$$F(j\omega) = P(\omega) + jQ(\omega). \tag{6.89}$$

The real and imaginary parts of (6.89) are given as follows.

(1) When $n = 1, 5, 9, \cdots$,

$$P^{(1)}(\omega) = a_1\omega^{n-1} - a_3\omega^{n-3} + a_5\omega^{n-5} - \cdots,$$
$$Q^{(1)}(\omega) = a_0\omega^{n} - a_2\omega^{n-2} + a_4\omega^{n-4} - \cdots.$$

(2) When $n = 2, 6, 10, \cdots$,

$$P^{(2)}(\omega) = -a_0\omega^{n} + a_2\omega^{n-2} - a_4\omega^{n-4} + \cdots,$$
$$Q^{(2)}(\omega) = a_1\omega^{n-1} - a_3\omega^{n-3} + a_5\omega^{n-5} - \cdots.$$

(3) When $n = 3, 7, 11, \cdots$,

$$P^{(3)}(\omega) = -a_1\omega^{n-1} + a_3\omega^{n-3} - a_5\omega^{n-5} + \cdots,$$
$$Q^{(3)}(\omega) = -a_0\omega^{n} + a_2\omega^{n-2} - a_4\omega^{n-4} + \cdots.$$

(4) When $n = 4, 8, 12, \cdots$,

$$P^{(4)}(\omega) = a_0\omega^{n} - a_2\omega^{n-2} + a_4\omega^{n-4} - \cdots,$$
$$Q^{(4)}(\omega) = -a_1\omega^{n-1} + a_3\omega^{n-3} - a_5\omega^{n-5} + \cdots.$$

If the following polynomials are considered:

$$\begin{aligned} f_0(\omega) &= a_0\omega^{n} - a_2\omega^{n-2} + a_4\omega^{n-4} - \cdots \\ f_1(\omega) &= a_1\omega^{n-1} - a_3\omega^{n-3} + a_5\omega^{n-5} - \cdots, \end{aligned} \tag{6.90}$$

the four cases of polynomials obviously become

$$P^{(1)}(\omega) = f_1(\omega), \quad Q^{(1)}(\omega) = f_0(\omega)$$
$$P^{(2)}(\omega) = -f_0(\omega), \quad Q^{(2)}(\omega) = f_1(\omega)$$

$$P^{(3)}(\omega) = -f_1(\omega), \quad Q^{(3)}(\omega) = -f_0(\omega)$$
$$P^{(4)}(\omega) = f_0(\omega), \quad Q^{(4)}(\omega) = -f_1(\omega).$$

Based on the above premise, the following division algorithm is executed:

$$f_{k-1}(\omega) = q_k(\omega) f_k(\omega) - f_{k+1}(\omega), \quad k = 1, 2, \cdots, n. \tag{6.91}$$

Then, $f_2(\omega), f_3(\omega), \cdots, f_n$ are obtained as follows:

$$\begin{aligned} f_0(\omega) &= a_{0,0}\omega^n + a_{0,2}\omega^{n-2} + a_{0,4}\omega^{n-4} + \cdots \\ f_1(\omega) &= a_{1,1}\omega^{n-1} + a_{1,3}\omega^{n-3} + a_{1,5}\omega^{n-5} + \cdots \\ f_2(\omega) &= a_{2,2}\omega^{n-2} + a_{2,4}\omega^{n-4} + \cdots \\ f_3(\omega) &= a_{3,3}\omega^{n-3} + a_{3,5}\omega^{n-5} + \cdots \\ &\cdots \\ f_n &= a_{n,n}. \end{aligned} \tag{6.92}$$

For an easier understanding, polynomials $f_0(\omega)$ and $f_1(\omega)$ are also written in (6.92). Here, $a_{0,0} = a_0$, $a_{0,2} = -a_2$, $a_{0,4} = a_4$, $\cdots$ and $a_{1,1} = a_1$, $a_{1,3} = -a_3$, $a_{1,5} = a_5$, $\cdots$. Moreover, each coefficient can be given by the following sequential operations:

$$\begin{aligned} & a_{2,p} = a_{1,p+1}\left(\frac{a_{0,0}}{a_{1,1}}\right) - a_{0,p} \\ & (p = 2, 4, \cdots) \\ & \quad \cdots \\ & a_{q,p} = a_{q-1,p+1}\left(\frac{a_{q-2,q-2}}{a_{q-1,q-1}}\right) - a_{q-2,p} \\ & (q = 3, 4, \cdots, \quad p = q, q+2, \cdots) \\ & \quad \cdots \\ & a_{n,n} = a_{n-1,n+1}\left(\frac{a_{n-2,n-2}}{a_{n-1,n-1}}\right) - a_{n-2,n}. \end{aligned} \tag{6.93}$$

In regard to the four cases, the following Sturm sequences are obtained.

(1) In this case, $f_0(\omega)/f_1(\omega) = Q^{(1)}(\omega)/P^{(1)}(\omega)$ should be calculated (when considering a large contour in the RHP). Since $a_{0,0} = a_0$, $a_{0,2} = -a_2$, $a_{0,4} = a_4, \cdots$, $a_{1,1} = a_1$, $a_{1,3} = -a_3$, $a_{1,5} = a_5, \cdots$, the following coefficients (Routh's series) are obtained:

$$a_{2,2} = -a_3 \left(\frac{a_0}{a_1}\right) + a_2, \quad a_{2,4} = a_5 \left(\frac{a_0}{a_1}\right) - a_4, \quad \cdots$$
$$a_{3,3} = a_{2,4} \left(\frac{a_1}{a_{2,2}}\right) + a_3, \quad a_{3,5} = a_{2,6} \left(\frac{a_1}{a_{2,2}}\right) - a_5, \quad \cdots$$
$$a_{4,4} = a_{3,5} \left(\frac{a_{2,2}}{a_{3,3}}\right) - a_{2,4}$$
$$\cdots .$$

(2) In this case, $f_0(\omega)/f_1(\omega) = -P^{(2)}(\omega)/Q^{(2)}(\omega)$ is assumed to be calculated (when considering a large contour in the RHP). Since $a_{0,0} = -a_0$, $a_{0,2} = a_2$, $a_{0,4} = -a_4, \cdots$, $a_{1,1} = a_1$, $a_{1,3} = -a_3$, $a_{1,5} = a_5, \cdots$, the following series are obtained:

$$a_{2,2} = -a_3 \left(\frac{a_0}{a_1}\right) + a_2, \quad a_{2,4} = a_5 \left(\frac{a_0}{a_1}\right) - a_4, \quad \cdots$$
$$a_{3,3} = a_{2,4} \left(\frac{a_1}{a_{2,2}}\right) + a_3, \quad a_{3,5} = a_{2,6} \left(\frac{a_1}{a_{2,2}}\right) - a_5, \quad \cdots$$
$$a_{4,4} = a_{3,5} \left(\frac{a_{2,2}}{a_{3,3}}\right) - a_{2,4}$$
$$\cdots .$$

In either case (1) or (2), the top of Routh's series can be written as

$$\begin{aligned}
a_{1,1} &= a_1 \\
a_{2,2} &= \begin{vmatrix} a_1 & a_3 \\ a_0 & a_2 \end{vmatrix} \Big/ a_1 \\
a_{3,3} &= \begin{vmatrix} a_1 & a_3 & a_5 \\ a_0 & a_2 & a_4 \\ 0 & a_1 & a_3 \end{vmatrix} \Big/ \begin{vmatrix} a_1 & a_3 \\ a_0 & a_2 \end{vmatrix} \\
a_{4,4} &= \begin{vmatrix} a_1 & a_3 & a_5 & a_7 \\ a_0 & a_2 & a_4 & a_6 \\ 0 & a_1 & a_3 & a_5 \\ 0 & a_0 & a_2 & a_4 \end{vmatrix} \Big/ \begin{vmatrix} a_1 & a_3 & a_5 \\ a_0 & a_2 & a_4 \\ 0 & a_1 & a_3 \end{vmatrix} \\
&\cdots \\
a_{n,n} &= a_n, \quad (\text{because } a_m = 0 \text{ for } m > n).
\end{aligned} \tag{6.94}$$

Each term of (6.94) corresponds to a Hurwitz determinant and its principal minors.

However, in regard to the remaining two cases, the following Sturm's sequences are calculated.

(3) In this case, if $f_0(\omega)/f_1(\omega) = Q^{(3)}(\omega)/P^{(3)}(\omega)$ is calculated as above for (1), the following series are obtained:

$$
\begin{aligned}
a_{2,2} &= a_3\left(\frac{a_0}{a_1}\right) - a_2, \quad a_{2,4} = -a_5\left(\frac{a_0}{a_1}\right) + a_4, \quad \cdots \\
a_{3,3} &= -a_{2,4}\left(\frac{a_1}{a_{2,2}}\right) - a_3, \quad a_{3,5} = -a_{2,6}\left(\frac{a_1}{a_{2,2}}\right) + a_5, \quad \cdots \\
a_{4,4} &= a_{3,5}\left(\frac{a_{2,2}}{a_{3,3}}\right) - a_{2,4} \\
&\cdots.
\end{aligned}
$$

(4) In this case, if $f_0(\omega)/f_1(\omega) = -P^{(4)}(\omega)/Q^{(4)}(\omega)$ is calculated as above for (2), the following series are obtained:

$$
\begin{aligned}
a_{2,2} &= a_3\left(\frac{a_0}{a_1}\right) - a_2, \quad a_{2,4} = -a_5\left(\frac{a_0}{a_1}\right) + a_4, \quad \cdots \\
a_{3,3} &= -a_{2,4}\left(\frac{a_1}{a_{2,2}}\right) - a_3, \quad a_{3,5} = -a_{2,6}\left(\frac{a_1}{a_{2,2}}\right) + a_5, \quad \cdots \\
a_{4,4} &= a_{3,5}\left(\frac{a_{2,2}}{a_{3,3}}\right) - a_{2,4} \\
&\cdots.
\end{aligned}
$$

In either case (3) or (4), the top of Routh's series can be written as

$$
\begin{aligned}
a_{1,1} &= -a_1 \\
a_{2,2} &= -\begin{vmatrix} a_1 & a_3 \\ a_0 & a_2 \end{vmatrix} \Big/ a_1 \\
a_{3,3} &= -\begin{vmatrix} a_1 & a_3 & a_5 \\ a_0 & a_2 & a_4 \\ 0 & a_1 & a_3 \end{vmatrix} \Big/ \begin{vmatrix} a_1 & a_3 \\ a_0 & a_2 \end{vmatrix} \\
a_{4,4} &= -\begin{vmatrix} a_1 & a_3 & a_5 & a_7 \\ a_0 & a_2 & a_4 & a_6 \\ 0 & a_1 & a_3 & a_5 \\ 0 & a_0 & a_2 & a_4 \end{vmatrix} \Big/ \begin{vmatrix} a_1 & a_3 & a_5 \\ a_0 & a_2 & a_4 \\ 0 & a_1 & a_3 \end{vmatrix} \\
&\cdots \\
a_{n,n} &= -a_n.
\end{aligned}
\tag{6.95}
$$

Each term of (6.95) corresponds to a Hurwitz determinant and its principal minors. Although it is different from (6.94) in sign, the number of sign changes is invariant (i.e., zero).

References

1. Ackermann J (1993) Robust control systems with uncertain physical parameters. Springer, Berlin
2. Barmish BR (1994) New tools for robustness of linear systems. Macmillan, New York
3. Barmish BR, Kang H (1993) A survey of extreme points results of robust control systems. Automatica, 13–15
4. Bartlett AC, Hollot CV, Huang L (1988) Root location of an entire polytope of polynomials: it suffices to check the edges. Math Control Signals Syst 1:61–71
5. Bhattacharyya SP, Chapellat H, Keel LH (1996) Robust control, the parametric approach. Prentice-Hall, New York
6. Cohn PM (2003) Basic algebra: group, rings and fields. Springer, Berlin
7. Kharitonov VL (1979) Asymptotic stability of an equilibrium position of a family of systems of linear differential equations. Differ Equ 14:1483–1485
8. Okuyama Y (1988) Sensitivity and robust stability of control systems with uncertainties: a graphical approach. Trans IEEJ 108-C:371–378 (in Japanese)
9. Okuyama Y, Takemori F (2000) Evaluation of robust performance for interval systems based on characteristic roots area. In: Kucera V, Sebek M (eds) Robust control design 2000. Pergamon Press, Oxford, pp 101–106
10. Okuyama Y, Takemori F (2002) A discrimination method of roots area for polytopic polynomials. In: Proc of the 15th IFAC world congress, T-Th-A21, Barcelona, Spain
11. Okuyama Y, Takemori F, Chen H (1998) On the Sturm theorem for interval polynomials. In: Proc of the American control conference, Philadelphia, PA, USA, pp 1883–1885
12. Okuyama Y, Takemori F, Chen H (1999) Discriminance of characteristic roots area for interval systems. In: Proc of the 14th IFAC world congress, vol G. Pergamon Press, Oxford, pp 13–18
13. Rahman QI, Schmeisser G (2002) Analytic theory of polynomials. Clarendon Press, Oxford
14. Soh YC, Evans RJ, Petersen IR, Betz RE (1997) Robust pole placement. Automatica 27:711–715
15. Takagi T (1948) Lecture notes in algebra (Daisugaku Kougi). Kyoritsu Shuppan, Tokyo (in Japanese)
16. Turnbull HW (1947) Theory of equations. Oliver and Boyd, Edinburgh

Chapter 7
Relation to Discrete Event Systems

7.1 Introduction

In the final chapter, the relationship between feedback control systems and finite state (more simply, finite value) control systems is described. The event-driven types of discrete systems are particularly important. There are many of those types of systems, e.g., manufacturing systems, industrial robots, computer networks, etc. However, the analysis and design of these discrete control systems have not been established, because the systems have severe nonlinear characteristics and do not respond continuously in time. In this chapter, the dynamics and stability of finite state and discrete event systems will be clarified.

Some authors have attempted to perform stability analyses of finite state and discrete event systems [6, 8]. However, many of them only discuss and define the stability (e.g., asymptotic, exponential, and Lyapunov stability) for specified discrete event systems. Thus, at the end of this chapter, a general description of the stability problem of finite state and discrete event systems will be proposed in connection with that of continuous feedback systems.

7.2 Finite State and Event-Driven Systems

As was described in Chap. 1, finite state systems can be written as [2, 3]:[1]

$$\boldsymbol{x}(k+1) = \boldsymbol{f}(\boldsymbol{x}(k), \boldsymbol{u}(k)) \tag{7.1}$$

$$\boldsymbol{y}(k) = \boldsymbol{g}(\boldsymbol{x}(k+1)) \tag{7.2}$$

$$\boldsymbol{x}(k) \in \mathbb{Z}^n, \quad \boldsymbol{u}(k) \in \mathbb{Z}^m, \quad \boldsymbol{y}(k) \in \mathbb{Z}^n, \quad \boldsymbol{f} : \mathbb{Z}^n \times \mathbb{Z}^m \to \mathbb{Z}^n, \quad \boldsymbol{g} : \mathbb{Z}^n \to \mathbb{Z}^n,$$

where $\mathbb{Z}$ is the set of (finite) integer numbers. In this expression, time-driven types

[1] In this chapter, variables u and y are generalized in multivariables $\boldsymbol{u}$ and $\boldsymbol{y}$.

Y. Okuyama, *Discrete Control Systems*,
DOI 10.1007/978-1-4471-5667-3_7, © Springer-Verlag London 2014

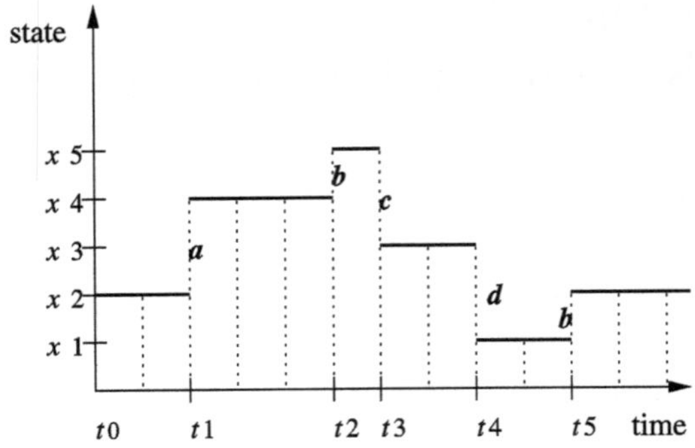

Fig. 7.1 State trajectory of a discrete event system

of discrete systems are considered in principle, although input sequence $\boldsymbol{u}(k)$ can correspond to an event-sequence vector. It is assumed that the state (or each value) of a control system continually changes as time changes. Even if the state is finite, those types of systems are regarded as being "time-driven." That is, $k \in \mathbb{Z}_+ (= \mathbb{N})$ is considered to be an independent variable that corresponds to an elapsed time t.

On the other hand, in "event-driven" systems, it is only the occurrence of an asynchronously generated discrete "event" that forces instantaneous state transitions [1]. A typical state trajectory for such a system is shown in Fig. 7.1. The state transitions that are called events may be labeled with alphabetical elements in the graph. These labels usually indicate the physical phenomena that caused the change in state. For example, in a manufacturing system, events of interest are "part accepted", "machine finished processing", "machine deadlocked"; in a communication protocol, typical events are "packet received", "packet sent", "time out" [8]. As shown in the figure, these events do not always occur at equal intervals of time.

In previous chapters, sampling at equal intervals was assumed in the feedback control systems. Moreover, the analysis and design of controlled systems (plants) was devoted to (Laplace or z) transformed variables. In other words, the characteristic of the frequency responses was considered in the analysis and design of some kinds of (discrete) control systems. However, a frequency response analysis cannot be applied to systems of unequal time intervals as shown in Fig. 7.1, even if the system is time-driven.

7.3 State and Event Trajectories

In this section, the concepts of "state trajectories" and "event trajectories" will be defined. The many areas in which discrete event systems arise and the different aspects of behavior relevant to each area have led to a variety of discrete event system models. For example, a common simplifying assumption is to ignore the times of occurrence of the events and consider only the order in which they occur. This leads to "logical" discrete event system models. In these models, a system trajectory is specified simply by listing (in order) the events that occur along the original sample path. In a logical model, the trajectory shown in Fig. 7.1 is reduce to the string of events $\{abcdb \cdots\}$ (the event trajectory). However, in some applications the timing information is essential and must be included in the model. This leads to a "timed" discrete event system model. In general, discrete event systems can be accurately modeled as

$$G = (\mathcal{X}, \mathcal{E}, f_e, \varphi, \boldsymbol{E}_v), \tag{7.3}$$

where $\mathcal{X}$ is the set of states

$$x \in \mathcal{X} := \{x^0, x^1, x^2, \cdots\}$$

and $\mathcal{E}$ is the set of events

$$e \in \mathcal{E} := \{e^1, e^2, e^3, \cdots\}.$$

State transitions are defined by the operators

$$f_e : \mathcal{X} \to \mathcal{X}. \tag{7.4}$$

An event e may only occur if it is in the set defined by the enable function,

$$\varphi : \mathcal{X} \to \mathcal{P}(\mathcal{E}) - \emptyset, \tag{7.5}$$

where $\boldsymbol{\mathcal{P}}(\boldsymbol{\mathcal{E}})$ denotes the power set of $\boldsymbol{\mathcal{E}}$.

A state trajectory is written as any sequence

$$\{\boldsymbol{x}_k\} \in \mathcal{X}^{\mathbb{N}} \tag{7.6}$$

such that

$$\boldsymbol{x}_{k+1} = \boldsymbol{f}_e(\boldsymbol{x}_k), \tag{7.7}$$

for all $k \in \mathbb{N} := \{0, 1, 2, \cdots, N\}$. Although (7.7) corresponds to (7.1), the subscripts k of $\boldsymbol{x}_k$ and e of $\boldsymbol{f}_e$ denote a "logical" time determined by an event occurring and an event-driven function, respectively. In (7.6), $\mathcal{X}^{\mathbb{N}}$ is the set of state sequences, e.g.,

$$\mathcal{X}^{\mathbb{N}} = \{\{x^0, x^1\}, \{x^1, x^2, x^3\}, \cdots\}.$$

On the other hand, an event trajectory is any sequence

$$\{\boldsymbol{e}_k\} \in \mathcal{E}^{\mathbb{N}} \tag{7.8}$$

such that there exists a state trajectory, $\{\boldsymbol{x}_k\} \in \boldsymbol{\mathcal{X}}^{\mathbb{N}}$, where for every $k \in \mathbb{N}$, $\boldsymbol{e}_k \in \varphi(\boldsymbol{x}_k)$. Here, $\mathcal{E}^{\mathbb{N}}$ is the set of event sequences, e.g.,

$$\mathcal{E}^{\mathbb{N}} = \{\{e^1, e^1, e^2\}, \{e^2, e^3\}, \cdots\}.$$

The set of all such event trajectories is denoted by $\boldsymbol{E} \subset \mathcal{E}^{\mathbb{N}}$. In the model of discrete event systems (7.3), $\boldsymbol{E}_v \subset \boldsymbol{E}$ represents the set of "valid" event trajectories that are physically possible in G. When the initial state is $\boldsymbol{x}_0 \in \mathcal{X}$, it is written as $\boldsymbol{E}_v(\boldsymbol{x}_0)$.

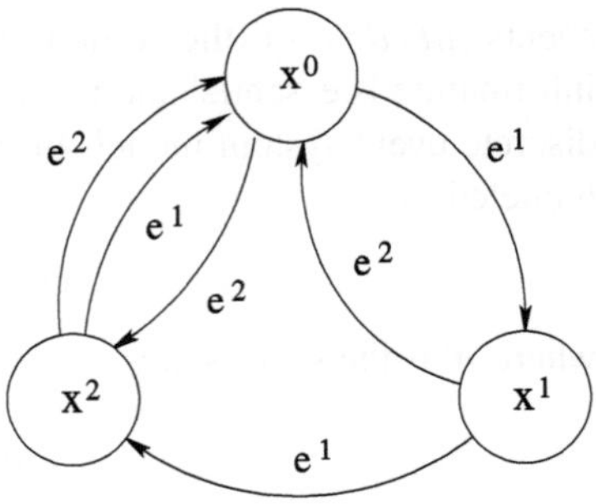

Fig. 7.2 Vending machine

Furthermore, $\boldsymbol{E}_a \subset \boldsymbol{E}_v$ denotes the set of "allowed" event trajectories [6]. Note that there can be only one state trajectory corresponding to a given event trajectory. However, an event trajectory that produces a given state trajectory is not unique in general.

In the following subsection, some examples of discrete event systems are provided.

7.3.1 Vending Machine

The first example is a simple vending machine. The set of states is defined as

$$\mathcal{X} = \{x^0, x^1, x^2\} = \{0, 1, 2\},$$

where $x = 0, 1, 2$ denote the stored amount in the machine by coins (e.g., a **1** or **2** euro coin). The set of events is given by

$$\mathcal{E} = \{e^1, e^2\}.$$

Here $e^1(e^2)$ denotes a **1** (**2**) euro coin being inserted into the coin slot. The state transition graph is as shown in Fig. 7.2. When three euros are stored in this machine, the item (e.g., ticket) is obtained as an output. The sets of event trajectories are given as

$$\begin{cases} (1)\ \{e^1, e^1, e^1\},\quad \{e^1, e^2\},\quad \{e^2, e^1\}, \\ (2)\ \{e^1, e^1, e^2\},\quad \{e^2, e^2\}. \end{cases}$$

Here, in case (2), one euro will be returned as change. Thus, the sets of state trajectories become the following three cases:

$$\{x^0, x^1, x^2, x^0\},\quad \{x^0, x^1, x^0\},\quad \{x^0, x^2, x^0\}.$$

Note that the case of $\boldsymbol{x}_0 \neq \{x^0\} = \emptyset$ should not be considered in this example.

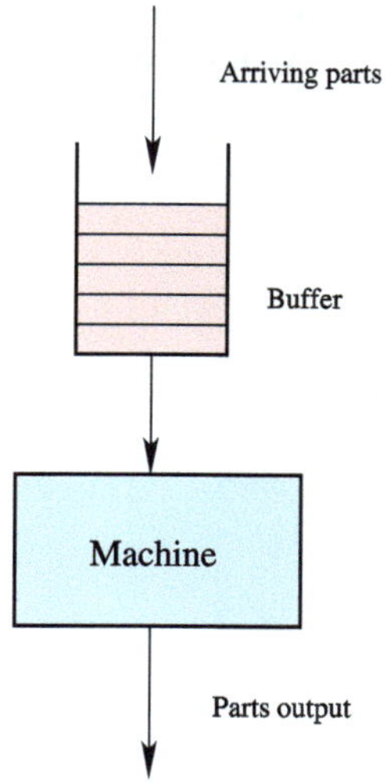

Fig. 7.3 Machine with a single buffer

7.3.2 *Buffer Machine*

The second example is that of a single buffer machine. As shown in Fig. 7.3, the machine processes only one part at a time. The set of states can be assigned as

$$\mathcal{X} = \{x^0, x^1, x^2, \cdots\} = \{0, 1, 2, 3, \cdots\}. \tag{7.9}$$

Suppose that the event set is given by $\mathcal{E} = \{e^1, e^2, e^3\}$, where

- e^1 = "a part arrives"
- e^2 = "the machine has finished processing a part"
- e^3 = "the machine has finished processing a part *and* a part arrives at the same time".

The state transition function is defined by

$$x_{k+1} = f_{e^1}(x_k) = x_k + 1$$
$$x_{k+1} = f_{e^2}(x_k) = x_k - 1$$
$$x_{k+1} = f_{e^3}(x_k) = x_k.$$

Here, the enable function is defined by

$$\varphi(x) = \{e^1, e^2, e^3\}$$

if the buffer is not empty (i.e., $x > 0$), and

$$\varphi(x) = \{e^1, e^3\}$$

if the buffer is empty (i.e., $x = 0$).

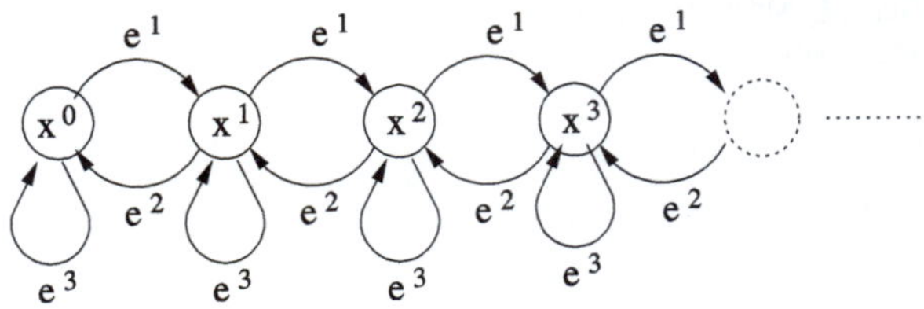

Fig. 7.4 State transition of the buffer machine

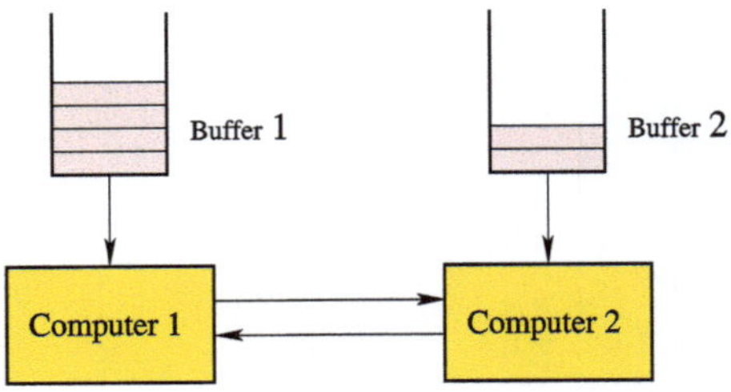

Fig. 7.5 Two-computer load balancing system

A graphical representation of the machine is shown in Fig. 7.4. The circles represent the states and the events are labeled the arrows which represent the state transition. If a set of valid event trajectories is given by

$$\{e^1, e^1, e^2, e^2, e^1, e^1, e^2, e^2, \cdots\} \subset \boldsymbol{E}_v,$$

the sets of state trajectories are obtained as

$$\{x^0, x^1, x^2, x^1, x^0, x^1, x^2, \cdots\} \text{ or } \{x^1, x^2, x^3, x^2, x^1, x^2, x^3, \cdots\} \quad \cdots.$$

If a set of event trajectories is given by

$$\{e^2, e^2, e^1, e^1, e^2, e^2, \cdots\},$$

the set of state trajectories is obtained, for instance, when the initial state $\boldsymbol{x}_0 \geq x^2 = 2$,

$$\{x^2, x^1, x^0, x^1, x^2, x^1, x^0, \cdots\}.$$

However, when $\boldsymbol{x}_0 < x^2 = 2$, the set of trajectories is not obtained because of a deadlock.

7.3.3 Load Balancing System

The next example concerns a load balancing network by two computers, each of which has the task of processing identical jobs (i.e., parallel processing). As shown in Fig. 7.5, the jobs are stored in buffers; each computer can sense the other computer's buffer level and can pass jobs to the other computer. There is a load balancing mechanism on each computer that seeks to balance (equalize) the load between the two computers.

In this example, the set of states is defined as $\boldsymbol{X} = \mathbb{N}^2$ (i.e., two-dimensional vectors of natural numbers), e.g., $\boldsymbol{x} = [x^1, x^2]^T = [1, 5]^T$. This means that Computer 1 has one job and Computer 2 has five jobs. The set of the event is given by

$$\mathcal{E} = \{e_{12}^{\ell}, e_{21}^{\ell}, e^0\}, \quad \ell \in \mathbb{N}, \tag{7.10}$$

where

- e_{12}^{ℓ} = "pass ℓ jobs from Computer 1 to Computer 2"
- e_{21}^{ℓ} = "pass ℓ jobs from Computer 2 to Computer 1"
- e^0 = "pass no jobs" (null event).

The enable function is written as

$$\varphi(\boldsymbol{x}) = \begin{cases} \{e_{12}^1, e_{12}^2, \cdots, e_{12}^{\ell}\} \text{ if } x^1 > x^2 \text{ and } \ell = \frac{1}{2}|x^1 - x^2| \\ \{e_{21}^1, e_{21}^2, \cdots, e_{21}^{\ell}\} \text{ if } x^2 > x^1 \text{ and } \ell = \frac{1}{2}|x^1 - x^2| \\ \{e^0\} \text{ otherwise.} \end{cases}$$

This means that if the difference of the number of jobs for each computer is "even," then ℓ jobs are passed from one computer to another.

The state transition function is given by

$$\begin{bmatrix} x_{k+1}^1 \\ x_{k+1}^2 \end{bmatrix} = f_{e_{12}^{\ell}}(\boldsymbol{x}_k) = \begin{bmatrix} x_k^1 - \ell \\ x_k^2 + \ell \end{bmatrix}$$

$$\begin{bmatrix} x_{k+1}^1 \\ x_{k+1}^2 \end{bmatrix} = f_{e_{21}^{\ell}}(\boldsymbol{x}_k) = \begin{bmatrix} x_k^1 + \ell \\ x_k^2 - \ell \end{bmatrix}$$

$$\begin{bmatrix} x_{k+1}^1 \\ x_{k+1}^2 \end{bmatrix} = f_{e^0}(\boldsymbol{x}_k) = \begin{bmatrix} x_k^1 \\ x_k^2 \end{bmatrix}.$$

7.4 Petri Nets

In this section, Petri net systems are introduced. A Petri net is a special case of discrete event system. It is constructed by the following sets:

- $\mathcal{P} = \{p_1, p_2, \cdots\}$; a finite set of *places* (represented with circles)
- $\mathcal{T} = \{\tau_1, \tau_2, \cdots\}$; a finite set of *transitions* (represented with line segments)
- $\mathcal{F} \subset (\mathcal{P} \times \mathcal{T}) \cup (\mathcal{T} \times \mathcal{P})$; a set of *arcs* (represented with arrows)
- $\mathcal{W} : \mathcal{F} \to \{1, 2, 3, \cdots\}$; a *weight function* (represented with numbers labeling arcs)

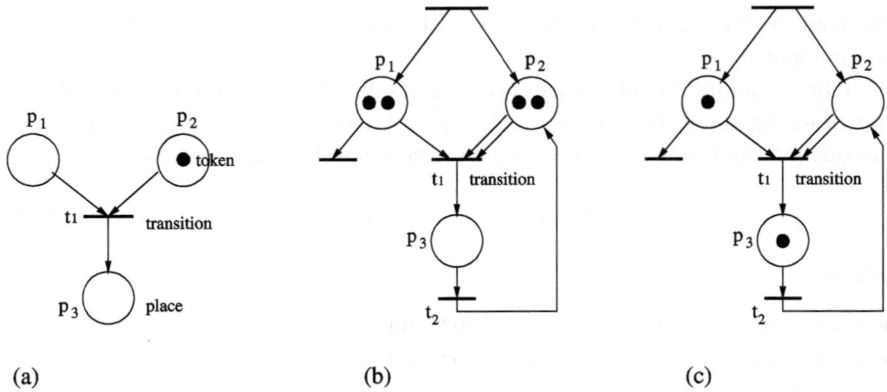

Fig. 7.6 Petri net systems

- $\mathcal{M}_k : \mathcal{P} \to \mathbb{N}$; a marking at time k (represented with dark dots, i.e., *tokens*, in places).

The marking (in other words, state) of a Petri net is written as

$$\boldsymbol{\mathcal{M}}_k = [\mathcal{M}_k(p_1), \mathcal{M}_k(p_2), \cdots]^T, \tag{7.11}$$

where $\mathcal{M}_k(p_i)$ denotes the marking (i.e., the number of tokens) at place $p_i \in \mathcal{P}$. A transition $\tau_j \in \mathcal{T}$ is said to be *enabled* if

$$\mathcal{M}_k(p_i) \geq \mathcal{W}(p_i, \tau_j), \quad i, j = 1, 2, \cdots \tag{7.12}$$

for all $p_i \in \mathcal{P}$ such that $(p_i, \tau_j) \in \mathcal{F}$. If a transition τ_j is enabled (fires) at time k, the next marking is given by

$$\mathcal{M}_{k+1}(p_i) = \mathcal{M}_k(p_i) + \mathcal{W}(\tau_j, p_i) - \mathcal{W}(p_i, \tau_j), \quad i, j = 1, 2, \cdots \tag{7.13}$$

where $(\tau_j, p_i) \in \mathcal{F}$ and $(p_i, \tau_j) \in \mathcal{F}$.

7.4.1 Simple Petri Net and Graph

Graphical representations of Petri nets are, e.g., as shown in Figs. 7.6(a), (b), and (c). Figure 7.6(a) is a basic element of the graph that is composed of three places, one token, and one transition. If the initial situation is as shown in the graph, the marking is expressed as

$$\boldsymbol{\mathcal{M}}_0 = [\mathcal{M}_0(p_1), \mathcal{M}_0(p_2), \mathcal{M}_0(p_3)]^T = [0, 1, 0]^T,$$

and the weighting functions are given as follows:

$$\mathcal{W}(p_1, \tau_1) = 1, \quad \mathcal{W}(p_2, \tau_1) = 1, \quad \mathcal{W}(\tau_1, p_3) = 1, \quad \text{other combinations} = 0.$$

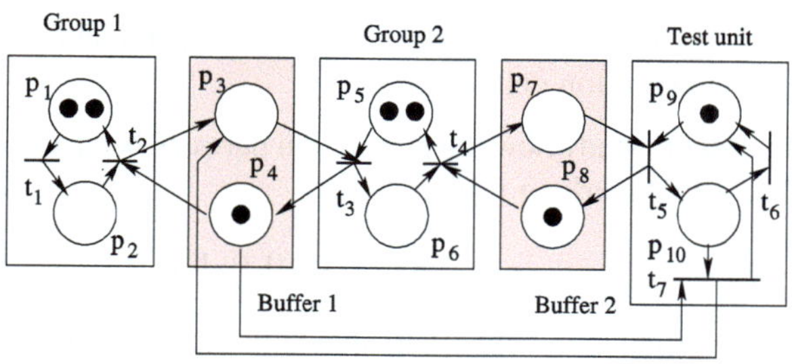

Fig. 7.7 Petri net of a production network

Since $\mathcal{M}_0(p_2) \geq \mathcal{W}(p_2, \tau_1)$, the transition τ_1 fires. Thus, the next marking will be given as

$$\boldsymbol{\mathcal{M}}_1 = [0, 0, 1]^T.$$

Next, consider an example of recurrent types of Petri nets as shown in Fig. 7.6(b), (c). In this case, the initial marking is expressed as

$$\boldsymbol{\mathcal{M}}_0 = [\mathcal{M}_0(p_1), \mathcal{M}_0(p_2), \mathcal{M}_0(p_3)]^T = [2, 2, 0]^T,$$

and the weighting functions are:

$$\mathcal{W}(p_1, \tau_1) = 1, \quad \mathcal{W}(p_2, \tau_1) = 2, \quad \mathcal{W}(\tau_1, p_3) = 1, \quad \cdots.$$

Since the following hold:

$$\mathcal{M}_0(p_1) \geq \mathcal{W}(p_1, \tau_1) \quad \text{and} \quad \mathcal{M}_0(p_2) \geq \mathcal{W}(p_2, \tau_1),$$

the transition τ_1 fires. Thus, the next marking will be given by

$$\boldsymbol{\mathcal{M}}_1 = [1, 0, 1]^T$$

as shown in Fig. 7.6(c).

Furthermore, when considering the next step (τ_2 fires), the marking becomes

$$\boldsymbol{\mathcal{M}}_2 = [1, 1, 0]^T.$$

7.4.2 Production Network

Consider a production network that consists of two groups of machines, two buffers, and an inspection test unit, as shown in Fig. 7.7.

- $\mathcal{M}(p_1)$ is the number of parts waiting to be processed by a machine in Group 1,
- $\mathcal{M}(p_2)$ is the number of parts being processed by some machine in Group 1,
- $\mathcal{M}(p_3)$ is the number of parts in Buffer 1,
- $\mathcal{M}(p_4)$ is a part counter for Buffer 1,
- and so on in regard to Group 2 and Buffer 2,

- $\mathcal{M}(p_9)$ and $\mathcal{M}(P_{10})$ are used to store parts and to limit the number of parts in the inspection unit, respectively.

In order to study how the production network behaves, we will consider an example. The initial marking is given as

$$\boldsymbol{\mathcal{M}}_0 = [2, 0, 0, 1, 2, 0, 0, 1, 1, 0]^T$$

in the production network as shown in Fig. 7.7. Since

$$\mathcal{M}_0(p_1) \geq \mathcal{W}(p_1, \tau_1),$$

τ_1 fires, and then in Group 1,

$$\mathcal{M}_1(p_1) = \mathcal{M}_0(p_1) + \mathcal{W}(\tau_1, p_1) - \mathcal{W}(p_1, \tau_1) = 2 + 0 - 1 = 1$$
$$\mathcal{M}_1(p_2) = \mathcal{M}_0(p_2) + \mathcal{W}(\tau_1, p_2) - \mathcal{W}(p_2, \tau_1) = 0 + 1 - 0 = 1.$$

Thus, the marking of the Petri net is given as

$$\boldsymbol{\mathcal{M}}_1 = [1, 1, 0, 1, 2, 0, 0, 1, 1, 0]^T.$$

Furthermore, if τ_2 fires, the next marking is given as follows:

$$\boldsymbol{\mathcal{M}}_2 = [2, 0, 1, 0, 2, 0, 0, 1, 1, 0]^T.$$

7.4.3 Network Computers

Consider a network of computers arranged in a "ring." This type of network is represented by the extended Petri net as shown in Fig. 7.8. Each place $p_i \in \boldsymbol{P}$ indicates a computer node in the network, and the state of the computer is given by $\mathcal{M}(p_i)$. Communications between the nodes are represented with transitions (e.g., transitions τ_1 and τ_2 represent different ways to communicate to node 2). The transition τ_{11} simply represents a null event, where if there are no tokens in any place it will fire.

If the marking is given as shown in Fig. 7.8, the τ_2, τ_4, τ_6, τ_8, and τ_{10} fire, in sequence. However, if τ_1 fired first, then only τ_4, τ_6, τ_8, τ_{10}, and $\tau_2, \cdots$ will fire. If the odd-numbered transitions fire for a nonzero initial marking, the network will eventually settle into a pattern where the even-numbered transitions fire in sequence an infinite number of times (the odd-numbered transitions drain the network of tokens).

7.5 Feedback/Supervisory Control and Stability Concepts

The relationship between continuous feedback control for time-driven systems and supervisory control for event-driven systems is described in this section [10]. The

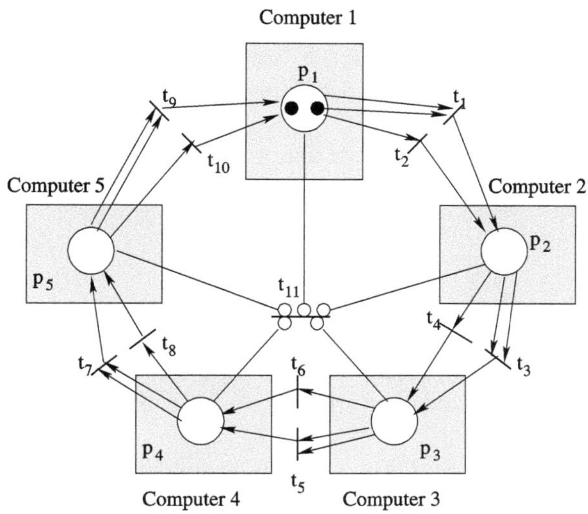

Fig. 7.8 Petri net of a network of computers

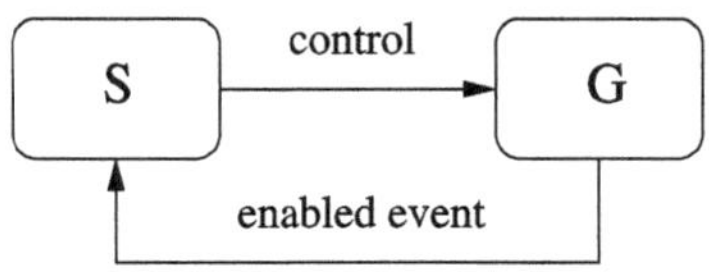

Fig. 7.9 Feedback and supervisory control

discrete event system as described so far is simply a spontaneous generator of event strings without a means of external control. In order to model the control of discrete event systems, the set of events is partitioned into *controllable* and *uncontrollable* events: $\mathcal{E} = \mathcal{E}_c \cup \mathcal{E}_u$ [8]. Control of a discrete event system G consists of switching the control input through a sequence of elements $u_1, u_2, \cdots$, in response to the observed string of previously generated events. This type of controller is called a *supervisor* S, as shown in Fig. 7.9. In introducing the concept of a supervisor, it will be necessary to distinguish between the "plant" (the object to be controlled) in control theory and the controlling agent in discrete event (or automata) theory.

Here, the difference between time-driven and event-driven systems is discussed from a viewpoint of stability [6, 7]. The motion (dynamic behavior) of a discrete event system G which begins at $\boldsymbol{x}_0$ is defined as

$$\boldsymbol{x}_k := X(\boldsymbol{x}_0, E_k, k), \tag{7.14}$$

where E_k denotes the sequence of events $e_0, e_1, e_2, \cdots, e_{k-1}$. Based on this expression for dynamic behavior, some of the stability concepts will now be described.

Stability in the Sense of Lyapunov A closed invariant set $\mathcal{X}_m \subset \mathcal{X}$ of G is called stable in the sense of Lyapunov if for any $\epsilon > 0$ it is possible to find some $\delta > 0$ such that

$$\rho(\boldsymbol{x}_0, \mathcal{X}_m) < \delta \;\Rightarrow\; \rho(\boldsymbol{x}_k, \mathcal{X}_m) < \epsilon, \quad \forall E_k = \{e_0, e_1, e_2, \cdots, e_{k-1}\}. \tag{7.15}$$

Here, $\rho(\boldsymbol{x}, \boldsymbol{y})$ denotes a metric.[2] Although the expression of definition (7.15) may be mathematically strict, it is not always easy to analyze the stability of discrete event systems. Thus, definitions of stability based on the *Lyapunov function* are presented.

The motions of G are uniformly bounded with respect to a set of allowed event trajectories, if the following condition is satisfied:

$$\psi_1(\rho(\boldsymbol{x}, \mathcal{X}_m)) \leq V(\boldsymbol{x}) \leq \psi_2(\rho(\boldsymbol{x}, \mathcal{X}_m)), \tag{7.16}$$

where $V(\boldsymbol{x})$ is a nonincreasing function for $\boldsymbol{x}_0 \in S(\mathcal{X}_m; r)$, and ψ_1, ψ_2 are strictly increasing functions that satisfy $\psi_i : [0, \infty) \to \mathbb{R}^+ (i = 1, 2)$. Here, $V(\boldsymbol{x})$ corresponds to a Lyapunov function for a continuous-time system, and $S(\mathcal{X}_m; r)$ is the r-neighborhood of an arbitrary set $\mathcal{X}_m \subset \mathcal{X}$.[3]

Asymptotic Stability If the following limit exists as $k \to \infty$:

$$\rho(\boldsymbol{x}_k, \mathcal{X}_m) \to 0, \tag{7.17}$$

then the closed invariant set $\mathcal{X}_m$ is called asymptotically stable with respect to a set of allowed event trajectories.

By using a Lyapunov function, the following condition can be given:

$$V(\boldsymbol{x}_{k+1}) - V(\boldsymbol{x}_k) \leq -\psi_3(\rho(\boldsymbol{x}_k), \mathcal{X}_m), \quad \forall \boldsymbol{x}_0 \in S(\mathcal{X}_m; r), \ \forall E_k, \tag{7.18}$$

where $\psi_3 : [0, \infty) \to \mathbb{R}^+$.

Exponential Stability Moreover, if the metric can be written as

$$\rho(\boldsymbol{x}_k, \mathcal{X}_m) \leq \zeta \mathrm{e}^{-\alpha k} \rho(\boldsymbol{x}_0, \mathcal{X}_m), \quad 0 < \zeta < \infty, \quad \alpha > 0, \tag{7.19}$$

the closed invariant set $\mathcal{X}_m$ is called exponentially stable with respect to a set of allowed event trajectories.

By using a Lyapunov function $V(\boldsymbol{x})$ and three positive constants (c_1, c_2, and c_3), the conditions can be given as follows. Corresponding to (7.16) and (7.18), we have

$$c_1(\rho(\boldsymbol{x}, \mathcal{X}_m)) \leq V(\boldsymbol{x}) \leq c_2(\rho(\boldsymbol{x}, \mathcal{X}_m)), \tag{7.20}$$

and

$$V(\boldsymbol{x}_{k+1}) - V(\boldsymbol{x}_k) \leq -c_3(\rho(\boldsymbol{x}_k), \mathcal{X}_m), \quad \forall \boldsymbol{x}0 \in S(\mathcal{X}_m; r), \ \forall E_k, \tag{7.21}$$

where $c_3 < c_2$. The relationship between the above conditions and (7.19) is shown below.

---

[2] See Appendix A and compare it with the norms in Appendix A of Chap. 2.

[3] See also Appendix A.

Since $(1/c_2)V(\boldsymbol{x}_k) \leq \rho(\boldsymbol{x}_k, \mathcal{X}_m)$,

$$V(\boldsymbol{x}_{k+1}) - V(\boldsymbol{x}_k) \leq -(c_3/c_2)V(\boldsymbol{x}_k).$$

Therefore,

$$V(\boldsymbol{x}_{k+1}) \leq \left(1 - \frac{c_3}{c_2}\right) V(\boldsymbol{x}_k).$$

The recurrent operations give the following result:

$$V(\boldsymbol{x}) \leq \left(1 - \frac{c_3}{c_2}\right)^k V(\boldsymbol{x}_0).$$

From (7.20),

$$c_1 \cdot \rho(\boldsymbol{x}_0, \mathcal{X}_m) \leq \left(1 - \frac{c_3}{c_2}\right)^k V(\boldsymbol{x}_0).$$

On the other hand,

$$V(\boldsymbol{x}_k) \leq c_2 \cdot \rho(\boldsymbol{x}_0, \mathcal{X}_m).$$

Therefore, we have the following inequality:

$$\rho(\boldsymbol{x}_k, \mathcal{X}_m) \leq \frac{c_2}{c_1}\left(1 - \frac{c_3}{c_2}\right)^k \rho(\boldsymbol{x}_0, \mathcal{X}_m). \tag{7.22}$$

From (7.22) parameters α and ζ in (7.19) can be determined for $k \to \infty$, because the following is valid:

$$\mathrm{e}^{-\alpha k} = \lim_{\nu \to \infty} \left(1 - \frac{\alpha k}{\nu}\right)^{\nu}.$$

That is, $\zeta = c_2/c_1$ and $\alpha = (c_2 \nu)/(c_3 k)$ can be assigned for any $k, \nu = 1, 2, \cdots$. Hence, the system is exponentially stable.

7.6 Multiple Metrics and Stability

The metric (or Lyapunov function) that has so far been considered in this chapter is a scalar number. However, those metrics do not always have clear physical and practical meanings. Although metrics ρ_1 and ρ_2, and norms, i.e.,

$$\rho_1(\boldsymbol{x}, \boldsymbol{y}) = \sum_{i=1}^{n} |x_i - y_i|$$

$$\rho_2(\boldsymbol{x}, \boldsymbol{y}) = \left(\sum_{i=1}^{n} |x_i - y_i|^2\right)^{1/2}$$

or

$$\|\boldsymbol{x}\|_1 = \sum_{i=1}^{n} |x_i|$$

$$\|\boldsymbol{x}\|_2 = \left(\sum_{i=1}^{n} |x_i|^2 \right)^{1/2}$$

may be meaningful, the inequality assumption becomes too severe. The stability condition that is derived based on these metrics will be a weak one. Therefore, in this last section, a vector of metrics (or norms) and a norm of time sequence (state trajectory) are considered. Their notation is given in Appendix B.

If the elapsed time t_k is explicitly considered as shown in Fig. 7.1, a state trajectory of the discrete event system in (7.1) and (7.7) can be written as:

$$\begin{aligned} &\boldsymbol{x}(t_{k+1}) = \boldsymbol{f}_{\boldsymbol{e}}(\boldsymbol{x}(t_k)) = \boldsymbol{f}(\boldsymbol{x}(t_k), \boldsymbol{e}(t_k)), \\ &\boldsymbol{f}_{\boldsymbol{e}} : \mathbb{Z}^n \to \mathbb{Z}^n, \quad \boldsymbol{f} : \mathbb{Z}^n \times \mathbb{Z}^m \to \mathbb{Z}^n, \end{aligned} \tag{7.23}$$

by corresponding $\boldsymbol{e} = \boldsymbol{u} \in \mathbb{Z}^m$. Moreover, assume that the nominal system is the following linear time-dependent discrete system with transition matrix $\boldsymbol{\Phi}$:

$$\begin{aligned} &\boldsymbol{x}(t_{k+1}) = \boldsymbol{\Phi}(t_{k+1}, t_k)\boldsymbol{x}(t_k) + \boldsymbol{f}(\boldsymbol{x}(t_k), \boldsymbol{e}(t_k)), \\ &\boldsymbol{\Phi} \in \mathbb{Z}^{n \times n}, \quad \boldsymbol{f} : \mathbb{Z}^n \times \mathbb{Z}^m \to \mathbb{Z}^n \\ &k = 0, 1, 2, \cdots, \infty. \end{aligned} \tag{7.24}$$

Here, the following vectors of ℓ_1 norms that are defined for each element of state vector $\boldsymbol{x}$ are defined:

$$\|x_i(t_k)\|_1 = \sum_{k=0}^{\infty} |x_i(t_k)|, \qquad i = 1, 2, \cdots, n \tag{7.25}$$

and

$$\|\boldsymbol{x}(t_k)\|_{\ell_1} = \begin{bmatrix} \|x_1(t_k)\|_1 \\ \|x_2(t_k)\|_1 \\ \vdots \\ \|x_n(t_k)\|_1 \end{bmatrix}. \tag{7.26}$$

From the recurrence equation (7.24),

$$\begin{aligned} &\boldsymbol{x}(t_k) = \boldsymbol{\Phi}(t_k, t_0)\boldsymbol{x}(t_0) + \sum_{l=1}^{k} \boldsymbol{\Phi}(t_k, t_l)\boldsymbol{f}(\boldsymbol{x}(t_{l-1}), \boldsymbol{e}(t_{l-1})), \\ &k = 0, 1, 2, \cdots, \infty, \end{aligned} \tag{7.27}$$

where transition matrix $\boldsymbol{\Phi}(t_k, t_l)$ is assumed to be

(1) $\boldsymbol{\Phi}(t_k, t_i)\boldsymbol{\Phi}(t_i, t_l) = \boldsymbol{\Phi}(t_k, t_l)$
(2) $\boldsymbol{\Phi}(t_k, t_k) = \boldsymbol{I}$.

With respect to $\boldsymbol{f}(\cdot, \cdot)$ in (7.27), the following matrix can be considered:

$$\boldsymbol{\Psi}(t_k)) = \begin{bmatrix} \psi_{11}(t_k) & \psi_{12}(t_k) & \dots & \psi_{1n}(t_k)) \\ \psi_{21}(t_k) & \psi_{22}(t_k) & \dots & \psi_{2n}(t_k) \\ \vdots & \vdots & \ddots & \vdots \\ \psi_{n1}(t_k) & \psi_{n2}(t_k) & \dots & \psi_{nn}(t_k) \end{bmatrix}, \tag{7.28}$$

where

$$\psi_{ij}(t_k) = \frac{f_{ij}(\boldsymbol{x}(t_k), \boldsymbol{e}(t_k))}{x_j(t_k)}. \tag{7.29}$$

Therefore, (7.27) is rewritten as

$$\boldsymbol{x}(t_k) = \boldsymbol{\Phi}(t_k, t_0)\boldsymbol{x}(t_0) + \sum_{l=1}^{k} \boldsymbol{\Phi}(t_k, t_l)\boldsymbol{\Psi}(t_{l-1})\boldsymbol{x}(t_{l-1}), \tag{7.30}$$

where the following equality is defined:

$$f_i(\boldsymbol{x}(t_k), \boldsymbol{e}(t_k)) := \sum_{j=1}^{n} f_{ij}(\boldsymbol{x}(t_k), \boldsymbol{e}(t_k)) \tag{7.31}$$

for $\boldsymbol{f}(\cdot, \cdot) = [f_1, f_2, \cdots, f_n]$ [4]. Note that a special case of (7.31) is

$$f_i(\boldsymbol{x}(t_k), \boldsymbol{e}(t_k)) = \begin{cases} f_{ij}(\boldsymbol{x}(t_k), \boldsymbol{e}(t_k)), & \text{for } j = i \\ 0, & \text{for } j \neq i \end{cases} \tag{7.32}$$

and thus (7.28) and (7.29) are written as a diagonal matrix as follows:

$$\boldsymbol{\Psi}(t_k)) = \begin{bmatrix} \psi_{11}(t_k) & 0 & \dots & 0 \\ 0 & \psi_{22}(t_k) & \dots & 0 \\ \vdots & \vdots & \ddots & \vdots \\ 0 & 0 & \dots & \psi_{nn}(t_k) \end{bmatrix}, \tag{7.33}$$

where

$$\psi_{ii}(t_k) = \frac{f_i(\boldsymbol{x}(t_k), \boldsymbol{e}(t_k))}{x_i(t_k)}. \tag{7.34}$$

If the upper bound of the absolute value of each element in (7.28) is given by

$$|\psi_{ij}(t_k)| = \frac{|f_{ij}(\boldsymbol{x}(t_k), \boldsymbol{e}(t_k))|}{|x_j(t_k)|} \leq \bar{\psi}_{ij}, \tag{7.35}$$

the following matrix with non-negative elements can be defined:

$$\bar{\boldsymbol{\Psi}} := \begin{bmatrix} \bar{\psi}_{11} & \bar{\psi}_{12} & \dots & \bar{\psi}_{1n} \\ \bar{\psi}_{21} & \bar{\psi}_{22} & \dots & \bar{\psi}_{2n} \\ \vdots & \vdots & \ddots & \vdots \\ \bar{\psi}_{n1} & \bar{\psi}_{n2} & \dots & \bar{\psi}_{nn} \end{bmatrix}. \tag{7.36}$$

Here, when (7.33) and (7.34) are applied, (7.35) and (7.36) should be rewritten as follows:

$$|\psi_{ii}(t_k)| = \frac{|f_i(\boldsymbol{x}(t_k), \boldsymbol{e}(t_k))|}{|x_i(t_k)|} \leq \bar{\psi}_{ii} \tag{7.37}$$

and

$$\bar{\boldsymbol{\Psi}} := \begin{bmatrix} \bar{\psi}_{11} & 0 & \dots & 0 \\ 0 & \bar{\psi}_{22} & \dots & 0 \\ \vdots & \vdots & \ddots & \vdots \\ 0 & 0 & \dots & \bar{\psi}_{nn} \end{bmatrix}. \tag{7.38}$$

On the other hand, by using (7.25) and (7.26) the following inequalities are obtained from (7.30):

$$\begin{aligned} \|\boldsymbol{x}(t_k)\|_{\ell_1} &\leqq \|\boldsymbol{\Phi}(t_k, t_0)\boldsymbol{x}(t_0)\|_{\ell_1} + \left\| \sum_{l=1}^{k} \boldsymbol{\Phi}(t_k, t_l)\boldsymbol{\Psi}(t_{l-1})\boldsymbol{x}(t_{l-1}) \right\|_{\ell_1} \\ &\leqq \|\boldsymbol{\Phi}(t_k, t_0)\|_{\ell_1} |\boldsymbol{x}(t_0)| + \left(\sum_{l=1}^{k} |\boldsymbol{\Phi}(t_k, t_l)| \right) \bar{\boldsymbol{\Psi}} \cdot \|\boldsymbol{x}(t_k)\|_{\ell_1}, \end{aligned} \tag{7.39}$$

where $\leqq$ denotes the inequality symbol for each element of a vector. Then,

$$\left[\boldsymbol{I} - \left(\sum_{l=1}^{k} |\boldsymbol{\Phi}(t_k, t_l)| \right) \bar{\boldsymbol{\Psi}} \right] \|\boldsymbol{x}(t_k)\|_{\ell_1} \leqq \|\boldsymbol{\Phi}(t_k, t_0)\|_{\ell_1} |\boldsymbol{x}(t_0)|. \tag{7.40}$$

Since the first term on the right side of (7.27) is a (nominal) linear discrete system, the norm $\boldsymbol{\Phi}(t_k, t_0)\boldsymbol{x}(t_0)$ can be assumed to be bounded. Thus, if the matrix

$$\boldsymbol{A} = \boldsymbol{I} - \left(\sum_{l=1}^{k} |\boldsymbol{\Phi}(t_k, t_l)| \right) \bar{\boldsymbol{\psi}} \tag{7.41}$$

is Ostrowski's M-matrix for $k \to \infty$, all the elements of $\|\boldsymbol{x}(t_k)\|_{\ell_1}$ become bounded [4, 5]. Hence, we can say that the discrete system is stable.

7.7 Exercises

(1) Draw the state transition graph for a vending machine with four states and three events. Show the event and state trajectories.

(2) Consider a machine with two buffers that can only process one part of one type at a time.

  (i) Define the set of states and events.
  (ii) Draw the state transition graph for the buffer machine.

(3) Produce a Petri net model of the single buffer machine that is modeled in Sect. 7.3.2.

(4) Describe a candidate for a Lyapunov (nonincreasing and non-negative) function in Sect. 7.6.

(5) Describe the difference between sectors (7.35) and (7.37).

(6) Show that the matrix condition (7.41) is invariant, when considering (7.39) in the ℓ_∞ space, i.e.,

$$\|x_i(t_k)\|_\infty = \sup_{k \in \mathbb{Z}} |x_i(t_k)|$$

and

$$\|\boldsymbol{x}(t_k)\|_{\ell_\infty} = \begin{bmatrix} \|x_1(t_k)\|_\infty \\ \|x_2(t_k)\|_\infty \\ \vdots \\ \|x_n(t_k)\|_\infty \end{bmatrix}.$$

(7) Show that the vector-matrix expression with non-negative elements as shown in Appendix B can also be defined for multiple metrics or Lyapunov functions.

Appendix A: Metric Space and Invariant Set

$\{\mathcal{X}; \rho\}$ is a metric space [9] if

- $\rho(\boldsymbol{x}, \boldsymbol{y}) = \rho(\boldsymbol{y}, \boldsymbol{x})$ for all $\boldsymbol{x}, \boldsymbol{y} \in \mathcal{X}$ (i.e., the metrics are commutative) and
- $\rho(\boldsymbol{x}, \boldsymbol{z}) \leq \rho(\boldsymbol{x}, \boldsymbol{y}) + \rho(\boldsymbol{y}, \boldsymbol{z})$ for all $\boldsymbol{x}, \boldsymbol{y}, \boldsymbol{z} \in \mathcal{X}$ (i.e., the triangle inequality is satisfied).

If $\boldsymbol{x} = [x_1, \cdots, x_n]^T \in \mathbb{R}^n$ and $\boldsymbol{y} = [y_1, \cdots, y_n]^T \in \mathbb{R}^n$, the examples of metrics are given as

- $\rho_1(\boldsymbol{x}, \boldsymbol{y}) = \sum_{i=1}^{n} |x_i - y_i|$
- $\rho_2(\boldsymbol{x}, \boldsymbol{y}) = \left(\sum_{i=1}^{n} |x_i - y_i|^p \right)^{1/p}$

- $\rho_\infty(\boldsymbol{x}, \boldsymbol{y}) = \sup_i |x_i - y_i|$.

The concept of a metric can be extended to the distance from point $\boldsymbol{x}$ to an arbitrary set $\mathcal{X}_s \subset \mathcal{X}$ as follows:

$$\rho(\boldsymbol{x}, \mathcal{X}_s) := \inf\{\rho(\boldsymbol{x}, \boldsymbol{x}') : \boldsymbol{x}' \in \mathcal{X}_s\}.$$

Moreover, the *r-neighborhood* of subset $\mathcal{X}_s$ is defined as

$$S(\mathcal{X}_s; r) := \{\boldsymbol{x} : 0 < \rho(\boldsymbol{x}, \mathcal{X}_s) < r\},$$

where $r > 0$.

A set is called invariant with respect to the model G, if all motions originating in the set remain in the set. Mathematically, the set $\mathcal{X}_m \subset \mathcal{X}$ is an invariant set with respect to G, if $\boldsymbol{x}_0 \in \mathcal{X}_m$ implies $\boldsymbol{x}_k := X(\boldsymbol{x}_0, E_k, k) \in \mathcal{X}_m$ for all $k \in \mathbb{N}$ and all event sequences E_k. (That is, all invariant sets are closed with respect to $\{\mathcal{X}; \rho\}$.)

Appendix B: Absolute Value of Each Element of Vector-Matrix

In order to simplify the expression, the notation of a matrix that is composed of the absolute value of each element is defined as follows:[4]

$$|\boldsymbol{\Psi}(t_k)| := \begin{bmatrix} |\psi_{11}(t_k)| & |\psi_{12}(t_k)| & \dots & |\psi_{1n}(t_k)| \\ |\psi_{21}(t_k)| & |\psi_{22}(t_k)| & \dots & |\psi_{2n}(t_k)| \\ \vdots & \vdots & \ddots & \vdots \\ |\psi_{n1}(t_k)| & |\psi_{n2}(t_k)| & \dots & |\psi_{nn}(t_k)| \end{bmatrix}.$$

For a vector of the absolute value of each element, the following notation is defined:

$$|\boldsymbol{x}(t_k)| := [|x_1(t_k)|, |x_2(t_k)|, \cdots, |x_n(t_k)|]^T.$$

Note that each non-negative element can be replaced by some metric or norm in Euclidean space. Furthermore, each non-negative element can be replaced by some norm in a function space, e.g., (7.25), generally,

$$\|x_i(t_k)\|_{\ell_p} = \left(\sum_{k=0}^{\infty} |x_i(t_k)|^p\right)^{1/p}$$

[4]In mathematics, since vectors, norms, quadratic forms, and function spaces are regarded as important, these definitions may be despised.

and

$$\|\boldsymbol{x}(t_k)\|_{\ell_p} \begin{bmatrix} \|x_1(t_k)\|_{\ell_p} \\ \|x_2(t_k)\|_{\ell_p} \\ \vdots \\ \|x_n(t_K)\|_{\ell_p} \end{bmatrix}$$

in a vector representation. Regarding matrices, for example, we can define the following norm:

$$\|\boldsymbol{\Psi}(t_k)\|_{\ell_1} = \begin{bmatrix} \sum_{k=0}^{\infty}|\psi_{11}(t_k)| & \sum_{k=0}^{\infty}|\psi_{12}(t_k)| & \dots & \sum_{k=0}^{\infty}|\psi_{1n}(t_k)| \\ \sum_{k=0}^{\infty}|\psi_{21}(t_k)| & \sum_{k=0}^{\infty}|\psi_{22}(t_k)| & \dots & \sum_{k=0}^{\infty}|\psi_{2n}(t_k)| \\ \vdots & \vdots & \ddots & \vdots \\ \sum_{k=0}^{\infty}|\psi_{n1}(t_k)| & \sum_{k=0}^{\infty}|\psi_{n2}(t_k)| & \dots & \sum_{k=0}^{\infty}|\psi_{nn}(t_k)| \end{bmatrix}.$$

These expressions can be defined for continuous systems as follows:

$$\|x_i(t_k)\|_{L_p} = \left(\int_{k=0}^{\infty}|x_i(\tau)|^p \mathrm{d}\tau\right)^{1/p}$$

and

$$\|\boldsymbol{x}(t_k)\|_{L_p} \begin{bmatrix} \|x_1(t_k)\|_{L_p} \\ \|x_2(t_k)\|_{L_p} \\ \vdots \\ \|x_n(t_K)\|_{L_p} \end{bmatrix}.$$

For a matrix expression, for example, in an L_1 space,

$$\|\boldsymbol{\Psi}(\tau)\|_{L_1} = \begin{bmatrix} \int_{k=0}^{\infty}|\psi_{11}(\tau)|\mathrm{d}\tau & \int_{k=0}^{\infty}|\psi_{12}(\tau)|\mathrm{d}\tau & \dots & \int_{k=0}^{\infty}|\psi_{1n}(\tau)|\mathrm{d}\tau \\ \int_{k=0}^{\infty}|\psi_{21}(\tau)|\mathrm{d}\tau & \int_{k=0}^{\infty}|\psi_{22}(\tau)|\mathrm{d}\tau & \dots & \int_{k=0}^{\infty}|\psi_{2n}(\tau)|\mathrm{d}\tau \\ \vdots & \vdots & \ddots & \vdots \\ \int_{k=0}^{\infty}|\psi_{n1}(\tau)|\mathrm{d}\tau & \int_{k=0}^{\infty}|\psi_{n2}(\tau)|\mathrm{d}\tau & \dots & \int_{k=0}^{\infty}|\psi_{nn}(\tau)|\mathrm{d}\tau \end{bmatrix}.$$

References

1. Cassandras CG, Lafortune S (1999) Introduction to discrete event systems. Springer, Berlin

2. Fukao T (1972) Introduction to systems theory. Shoko-do, Tokyo (in Japanese)
3. Kalman RE, Falb PL, Arbib MA (1969) Topics in mathematical system theory. McGraw-Hill, New York
4. Okuyama Y (1974) Relative stability of linear dynamic systems with bounded uncertain coefficients. Trans SICE 10:527–532 (in Japanese)
5. Ostrowski AM (1955) Note on bounds for some determinants. Duke Math J 22:95–102
6. Passino KM, Burgess KL (1998) The stability analysis of discrete event systems. Wiley, New York
7. Passino KM, Michel AN, Antsaklis PJ (1994) Lyapunov stability of a class of discrete event systems. IEEE Trans Autom Control 37:269–279
8. Ramadge PJ, Wonham WM (1989) The control of discrete event systems. Proc IEEE 77:81–98
9. Rudin W (1987) Real and complex analysis, 3rd edn. McGraw-Hill, New York
10. Uzam M (2004) Synthesis of feedback control elements for discrete event systems. Int J Adv Manuf Technol, 48–69

ERRATUM

Errata and Comments

The following 'red' letters are errata that should be corrected (or inserted).

Chapter 1: Mathematical Descriptions and Models

1. Page 5: Eq. (1.10),

$$\begin{aligned} &y(k+n)+\cdots+a_{n-1}y(k+1)+a_n y(k) \\ &\quad = b_0 u(k+n)+\cdots+b_{n-1}u(k+1)+b_n u(k) \end{aligned}$$

2. Page 6: Eq. (1.18),

$$y(k)=\begin{bmatrix} b_n-a_n b_0 & b_{n-1}-a_{n-1}b_0 & \dots & b_1-a_1 b_0 \end{bmatrix}\begin{bmatrix} x_1(k) \\ x_2(k) \\ \vdots \\ x_n(k) \end{bmatrix}+b_0 u(k).$$

3. Fig. 1.5: Time sequences of the solution for Example 1.6
4. Below Fig. 1.5: Note that the response is delayed by one step as shown in Fig. 1.5 if $y(k+1)=x_1(k)$ (i.e., $y(k)=x_1(k-1)$) is applied to the computer program for (1.55). This response corresponds to the result of Exercise (7).
5. Page 7: Eq. (1.21),

$$\boldsymbol{x}(k)=\boldsymbol{A}\boldsymbol{x}(k-1)+\boldsymbol{B}u(k-1),$$

The online version of the original chapters can be found under doi:10.1007/978-1-4471-5667-3.

Y. Okuyama, *Discrete Control Systems*,
DOI 10.1007/978-1-4471-5667-3_8, © Springer-Verlag London 2014

6. Page 13: Eq. (1.35),

$$y(k+n)+\cdots+a_{n-1}y(k+1)+a_n y(k)$$
$$=b_0u(k+n)+\cdots+b_{n-1}u(k+1)+b_n u(k).$$

7. Page 13:

$$\begin{cases} \mathcal{Z}\{y(k+n)\}=z^n\hat{y}(z)-(y(0)z^n+\cdots+y(n-1)z) \\ \quad\cdots \\ \mathcal{Z}\{a_{n-1}y(k+1)\}=a_{n-1}(z\hat{y}(z)-y(0)z) \\ \mathcal{Z}\{a_n y(k)\}=a_n\hat{y}(z). \end{cases}$$

8. Page 13: For simplicity, the initial conditions are assumed to be zero (i.e., $y(0)=y(1)=\cdots=y(n-1)=0$ and also $u(0)=u(1)=\cdots=u(n-1)=0$).

   Comments:
   There would be no contradiction, because $y(\kappa)$ and $u(\kappa)$ in (1.35) defined for $\kappa=k+n$ ($\kappa\geq n$). However, in a computer simulation, backward expressions (1.19), (1.21), and (1.40) should be used.

9. Page 13: Eq. (1.36),

$$(z^n+a_1z^{n-1}+\cdots+a_{n-1}z+a_n)\hat{y}(z)=(b_0z^n+b_1z^{n-1}+\cdots+b_{n-1}z+b_n)\hat{u}(z).$$

10. Page 14: Eq. (1.38), The z-transform with respect to $\kappa=k+2$ is given as

$$(z^2-z+0.5)\hat{y}(z)-y(0)z^2-y(1)z+y(0)z=(z+1)\hat{u}(z)-u(0)z.$$

and

$$(z^2-z+0.5)\hat{y}(z)-y(0)z^2-y(1)z+y(0)z=\frac{z(z+1)}{z-1}-u(0)z$$

11. Page 18:

$$\frac{z^2+z)}{z^3-2z^2+1.5z-0.5} \Rightarrow \frac{z^2+z}{z^3-2z^2+1.5z-0.5}$$

12. Page 18:

$$\begin{cases} \begin{bmatrix} x_1(k+1) \\ x_2(k+1) \end{bmatrix}=\begin{bmatrix} -a_1 & 1 \\ -a_2 & 0 \end{bmatrix}\begin{bmatrix} x_1(k) \\ x_2(k) \end{bmatrix}+\begin{bmatrix} b_1 \\ b_2 \end{bmatrix}u(k), \\ y(k)=x_1(k), \quad \text{where } a_1=-1,\ a_2=0.5, \text{ and } b_1=b_2=1 \end{cases}$$

   Tips:
   The z-transforms of the above equations are given as:

$$\begin{bmatrix} z-1 & -1 \\ 0.5 & z \end{bmatrix}\begin{bmatrix} \hat{x}_1(z) \\ \hat{x}_2(z) \end{bmatrix}=\begin{bmatrix} \hat{u}(z) \\ \hat{u}(z) \end{bmatrix}$$

Then,

$$\begin{bmatrix} \hat{x}_1(z) \\ \hat{x}_2(z) \end{bmatrix} = \frac{1}{z^2 - z + 0.5} \begin{bmatrix} z & 1 \\ 0.5 & z-1 \end{bmatrix} \begin{bmatrix} z/(z-1) \\ z/(z-1) \end{bmatrix}.$$

Thus,

$$\hat{y}(z) = \hat{x}_1(z) = \frac{z^2 + z}{(z^2) - z + 0.5)(z-1)} = \frac{z^2 + z}{z^3 - 2z^2 + 1.5z - 0.5}.$$

13. Page 19: The caption in Fig. 1.6,
Fig. 1.6 Block diagram for Example 1.6, where $a_1 = 1$, $a_2 = -0.5$, and $b_1 = b_2 = 1$
14. Page 23: Eq. (1.68),

$$G_1(z) := \tilde{\mathcal{Z}}\{G_1(s)\} = \frac{K_0}{1 - z^{-1}} + \frac{K_1}{1 - \mathrm{e}^{p_1 h} z^{-1}} + \cdots + \frac{K_n}{1 - \mathrm{e}^{p_n h} z^{-1}}.$$

15. Page 24: In Table 1.2, The fourth line in 'Discrete time',

$$\mathrm{e}^{p_k h} \quad \Rightarrow \quad kh\mathrm{e}^{p_k h}$$

16. Page 25: Eq. (1.76),

$$\boldsymbol{\Phi}(\tau) := \mathrm{e}^{\boldsymbol{A}\tau} = \boldsymbol{I} + \boldsymbol{A}\tau + \frac{\boldsymbol{A}^2\tau^2}{2!} + \cdots,$$

where

$$\boldsymbol{I} = \begin{bmatrix} 1 & 0 \\ 0 & 1 \end{bmatrix}$$

17. Page 26: Eqs. (1.78), (1.79), and (1.80),
My first manuscript was written as follows:

$$\begin{cases} \boldsymbol{x}(k+1) = \boldsymbol{\Phi}(h)\boldsymbol{x}(k) + \boldsymbol{\Gamma}(h)u(k), \quad \boldsymbol{\Gamma}(h) = \int_0^h \boldsymbol{\Phi}(\tau)\boldsymbol{B}\mathrm{d}\tau \\ y(k) = \boldsymbol{C}\boldsymbol{x}(k). \end{cases}$$

$$\begin{cases} z\hat{\boldsymbol{x}}(z) = \boldsymbol{\Phi}\hat{\boldsymbol{x}}(z) + \boldsymbol{\Gamma}\hat{u}(z) \\ \hat{y}(z) = \boldsymbol{C}\hat{\boldsymbol{x}}(z). \end{cases}$$

$$\hat{y}(z) = \boldsymbol{C}[\boldsymbol{I} - \boldsymbol{\Phi}z^{-1}]^{-1}\boldsymbol{\Gamma}z^{-1}\hat{u}(z).$$

These expressions might be preferable to (1.78), (1.79), and (1.80).

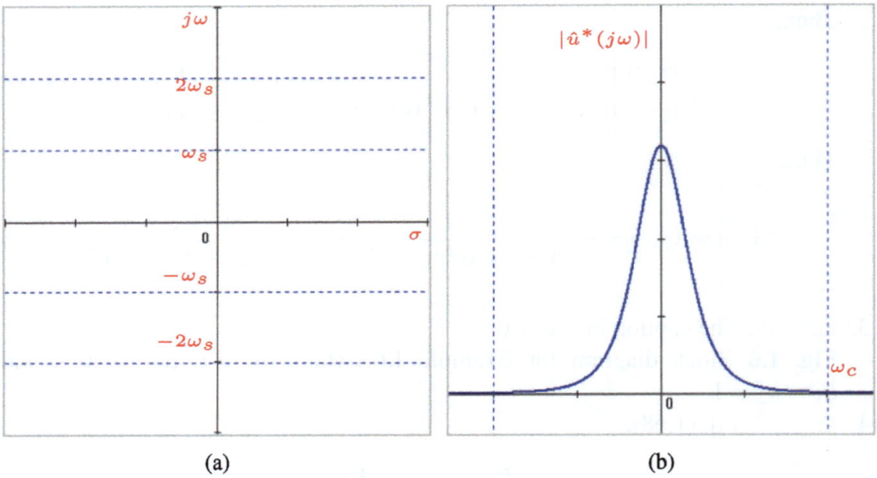

Fig. 1.23 Frequency shifting and spectrum

Chapter 2: Discretized Feedback Systems

1. Page 50: Eq. (2.7),

$$-\beta e^2 \leq g(e)e \leq \beta e^2.$$

2. Page 56: In Eq. (2.25),

$$e^{*\dagger}(z) \quad \Rightarrow \quad \hat{e}^{*\dagger}(z)$$

3. Page 69: Eq. (2.62),

$$\sum_{k,l=1,k\neq l}^{N} |x(k)|^2|y(l)|^2 - 2\sum_{k,l=1,k\neq l}^{N} |x(k)y(k)|\cdot|x(l)y(l)|$$
$$= \sum_{k,l=1,k\neq l}^{N} |x(k)y(l) - x(l)y(k)|^2,$$

4. Page 71: In the last line
   $\cdots$ and the inequality problem is proved.

Chapter 3: Robust Stability Analysis

1. Page 75: Eq. (3.8),

$$\|e(k)\|_2 \leq \|r(k)\|_2 + \sup_{|z|=1} |G(z)|(\rho\cdot\|e(k)\|_2 + \|d(k)\|_2).$$

2. Page 94: Eq. (3.46) in Theorem 3.3,

$$\eta(q_0, \omega_0) - \max_q \min_\omega \eta(q, \omega)$$

3. Page 94: The verification of robust stability using the above modified Hall diagram (off-axis M-circles) is based on the following theorem.

Chapter 4: Model Reference Feedback and PID Control

1. Page 114: In Eqs. (4.16), (4.17), (4.20), and (4.21),

$$G_I(s) \quad \Rightarrow \quad G_1(s)$$
$$G_I(z) \quad \Rightarrow \quad G_1(z).$$

2. Page 117: In Eq. (4.22),

$$\hat{f}(z) \quad \Rightarrow \quad \mathcal{F}(z).$$

3. Page 117: In the first line under Fig. 12,
$\cdots$, characteristic equation of the nominal feedback system is given as
4. Page 132:

$$\tilde{y}_3^{(3)} = \tilde{y}_3^{(2)} - \frac{a_{32}^{(2)}}{a_{22}^{(2)}} \tilde{y}_2^{(2)}, \quad \Rightarrow \quad \tilde{y}_3^{(3)} = \tilde{y}_3^{(1)} - \frac{a_{31}^{(1)}}{a_{11}^{(1)}} \tilde{y}_1^{(1)} - \frac{a_{32}^{(2)}}{a_{22}^{(2)}} \tilde{y}_2^{(2)},$$

5. Page 140: In Exercise (3), determine the characteristic equation of the nominal system, $\mathcal{F}(z) = 0$, for Example 4.1 (A) $\cdots$.
6. Page 140: (4) Show that the approximate PID control system in Fig. 4.18 is obtained from the model-reference feedback system in Fig. 4.17, when $\mathcal{D}_m(\cdot)$ and $\mathcal{D}_f(\cdot)$, are in high resolution.

Chapter 5: Multi-Loop Feedback Systems

1. Page 153:

$$\tilde{y}_n^{(n)} = \tilde{y}_n^{(n-1)} - \frac{a_{n,n-1}^{(n-1)}}{a_{n-1,n-1}^{(n-1)}} y_{n-1}^{(n-1)}$$

$$\Rightarrow \quad \tilde{y}_n^{(n)} = \tilde{y}_n^{(1)} - \frac{a_{n1}^{(1)}}{a_{11}^{(1)}} \tilde{y}_1^{(1)} - \frac{a_{n2}^{(2)}}{a_{22}^{(2)}} \tilde{y}_2^{(2)} - \cdots - \frac{a_{n,n-1}^{(n-1)}}{a_{n-1,n-1}^{(n-1)}} \tilde{y}_{n-1}^{(n-1)}$$

2. Page 153: $\cdots$, where $y_j^{(j)} \quad \Rightarrow \quad \cdots$, where $0 < \tilde{y}_j^{(j)} \le y_j^{(j)}$

3. Page 153: $\cdots$ all principal minors of matrix $\mathcal{A}$ $\Rightarrow$ $\cdots$ all principal minors of matrix (5.15)
4. Page 163:

$$\tilde{y}_n^{(n)} = \tilde{y}_n^{(n-1)} - \frac{a_{n,n-1}^{(n-1)}}{a_{n-1,n-1}^{(n-1)}} y_{n-1}^{(n-1)}$$

$$\Rightarrow \quad \tilde{y}_n^{(n)} = \tilde{y}_n^{(1)} - \frac{a_{n1}^{(1)}}{a_{11}^{(1)}} \tilde{y}_1^{(1)} - \frac{a_{n2}^{(2)}}{a_{22}^{(2)}} \tilde{y}_2^{(2)} - \cdots - \frac{a_{n,n-1}^{(n-1)}}{a_{n-1,n-1}^{(n-1)}} \tilde{y}_{n-1}^{(n-1)}$$

5. Page 163: $\cdots$, where $y_j^{(j)}$ $\Rightarrow$ $\cdots$, where $0 < \tilde{y}_j^{(j)} \leq y_j^{(j)}$
6. Page 163: $\cdots$ all principal minors of matrix $\mathcal{A}$ $\Rightarrow$ $\cdots$ all principal minors of matrix (5.48)
7. Page 177:

$$y_j \geq 0, \;\; J = 1, 2, \cdots, n \text{ and } a_{ij} \leq 0, \;\; i \neq j$$

$$\Rightarrow \quad \tilde{y}_j \geq 0, \;\; j = 1, 2, \cdots, n \text{ and } a_{ij} \leq 0, \;\; i \neq j$$

Chapter 6: Interval Polynomials and Robust Performance

1. Page 186: Equation (6.17) should be written as follows:

$$\begin{aligned}
&\tilde{F}(z) = D_{c1}(z)D_{c2}(z)D_{11}(z)D_{22}(z)D_{12}(z)D_{21}(z) \\
&+[K_1^-, K_1^+]N_{c1}(z)N_{11}(z)D_{c2}(z)D_{22}(z)D_{12}(z)D_{21}(z) \\
&+[K_2^-, K_2^+]N_{c2}(z)N_{22}(z)D_{c1}(z)D_{11}(z)D_{12}(z)D_{21}(z) \\
&+[K^-, K^+]N_{c1}(z)N_{c2}(z)N_{11}(z)N_{22}(z)D_{12}(z)D_{21}(z) \\
&-[K^-, K^+]N_{c1}(z)N_{c2}(z)N_{12}(z)N_{21}(z)D_{11}(z)D_{22}(z) = 0.
\end{aligned}$$

2. Page 191: Fig. 6.3 $\cdots$ for discrete control system $\Rightarrow$ Fig. 6.3 $\cdots$ for discrete control systems
3. Page 192:

$$\phi = \tan^{-1}\left(\frac{-\gamma + 2(1+\gamma^2)\theta - \gamma(1+\gamma^2)\theta^2}{1-(1+\gamma^2)\theta^2}\right)$$

The proof of Lemma 6.1 is given as follows:

$$x^2 + (y-\gamma)^2 = \frac{[1-(1+\gamma^2)\theta^2]^2 + [-\gamma + 2\theta(1+\gamma^2) - \gamma(1+\gamma^2)\theta^2]^2}{[1-2\gamma\theta + (1+\gamma^2)\theta^2]^2}$$

$$= \frac{(1+\gamma^2)[1+(1+\gamma^2)\theta^4 - 2\theta^2 + 4(1+\gamma^2)\theta^2 + \gamma^2(1+\gamma^2)\theta^4 - 4\gamma\theta - 4\gamma(1+\gamma^2)\theta^3 + 2\gamma^2\theta^2]}{[1-2\gamma\theta + (1+\gamma^2)\theta^2]^2}$$

$$= \frac{(1+\gamma^2)[1+4\gamma^2\theta^2+(1+\gamma^2)^2\theta^4-4\gamma\theta-4\gamma(1+\gamma^2)\theta^3+2(1+\gamma^2)\theta^2]}{[1-2\gamma\theta+(1+\gamma^2)\theta^2]^2} = 1+\gamma^2.$$

Thus, Lemma 6.1 has been proved. □

4. Page 204:

$$\tilde{F}(s) = [a_0^-, a_0^+]s^3 + [a_1^-, a_1^+]s^2 + [a_2^-, a_2^+]s + [a_3^-, a_3^+]$$

Chapter 7: Relation to Discrete Event Systems

1. Page 228: **Fig. 7.6** Petri net systems ⇒ **Fig. 7.6** Petri net systems (In the following figures, transitions τ_i are written in t_i)
2. Page 232: In (7.21),

$$\forall x0 \in S(\mathcal{X}_m; r) \quad \Rightarrow \quad \forall x_0 \in S(\mathcal{X}_m; r)$$

3. Page 234: Their notation is ⋯ ⇒ The notations are ⋯
4. Page 234:

$$\|x_i(t_k)\|_{\ell_1} = \sum_{k=0}^{\infty} |x_i(t_k)|$$

and

$$\|\boldsymbol{x}(t_k)\|_{\ell_1} = \begin{bmatrix} \|x_1(t_k)\|_{\ell_1} \\ \|x_2(t_k)\|_{\ell_1} \\ \vdots \\ \|x_n(t_k)\|_{\ell_1} \end{bmatrix}.$$

5. Page 235: In (7.28) and (7.33),

$$\boldsymbol{\Psi}(t_k)) \quad \Rightarrow \quad \boldsymbol{\Psi}(t_k)$$

6. Page 236:

$$\boldsymbol{I} - \left(\sum_{l=1}^{k} |\boldsymbol{\Phi}(t_k, t_l)| \right) \bar{\boldsymbol{\psi}} \quad \Rightarrow \quad \boldsymbol{I} - \left(\sum_{l=1}^{k} |\boldsymbol{\Phi}(t_k, t_l)| \right) \bar{\boldsymbol{\Psi}}$$

7. Page 237: ⋯ and three events. ⇒ ⋯ and two events.
8. Page 239:

$$\|\boldsymbol{x}(t_k)\|_{\ell_p} = \begin{bmatrix} \|x_1(t_k)\|_{\ell_p} \\ \|x_2(t_k)\|_{\ell_p} \\ \vdots \\ \|x_n(t_k)\|_{\ell_p} \end{bmatrix}$$

9. Page 239: The equations for continuous-time systems should be corrected as follows:

$$\|x_i(\tau)\|_{L_p} = \left(\int_0^\infty |x_i(\tau)|^p \mathrm{d}\tau\right)^{1/p}.$$

and

$$\|\boldsymbol{x}(\tau)\|_{L_p} = \begin{bmatrix} \|x_1(\tau)\|_{L_p} \\ \|x_2(\tau)\|_{L_p} \\ \vdots \\ \|x_n(\tau)\|_{L_p} \end{bmatrix},$$

furthermore,

$$\|\boldsymbol{\Psi}(\tau)\|_{L_1} = \begin{bmatrix} \int_0^\infty |\psi_{11}(\tau)|\mathrm{d}\tau & \int_0^\infty |\psi_{12}(\tau)|\mathrm{d}\tau & \dots & \int_0^\infty |\psi_{1n}(\tau)|\mathrm{d}\tau \\ \int_0^\infty |\psi_{21}(\tau)|\mathrm{d}\tau & \int_0^\infty |\psi_{22}(\tau)|\mathrm{d}\tau & \dots & \int_0^\infty |\psi_{2n}(\tau)|\mathrm{d}\tau \\ \vdots & \vdots & \ddots & \vdots \\ \int_0^\infty |\psi_{n1}(\tau)|\mathrm{d}\tau & \int_0^\infty |\psi_{n2}(\tau)|\mathrm{d}\tau & \dots & \int_0^\infty |\psi_{nn}(\tau)|\mathrm{d}\tau \end{bmatrix}.$$

Index

1. Page 242: Four discrete-type equation ⇒ Forward discrete-time equation

Index

Y. Okuyama, *Discrete Control Systems*,
DOI 10.1007/978-1-4471-5667-3, © Springer-Verlag London 2014

MIX
Papier aus verantwortungsvollen Quellen
Paper from responsible sources
FSC® C105338

If you have any concerns about our products,
you can contact us on
ProductSafety@springernature.com

In case Publisher is established outside the EU,
the EU authorized representative is:
Springer Nature Customer Service Center GmbH
Europaplatz 3, 69115 Heidelberg, Germany

Printed by Libri Plureos GmbH
in Hamburg, Germany